Solar Energy in Buildings

T0348614

Solar Energy in Buildings
Thermal Balance for Efficient Heating and Cooling

Dorota Chwieduk D.Sc., Ph.D. M.Sc.,
Mech. Eng. Energy and Buildings

AMSTERDAM • BOSTON • HEIDELBERG • LONDON
NEW YORK • OXFORD • PARIS • SAN DIEGO
SAN FRANCISCO • SINGAPORE • SYDNEY • TOKYO

Academic Press is an imprint of Elsevier

Academic Press is an imprint of Elsevier
525 B Street, Suite 1800, San Diego, CA 92101—4495, USA
225 Wyman Street, Waltham, MA 02451, USA
32 Jamestown Road, London, NW1 7BY, UK
The Boulevard, Langford Lane, Kidlington, Oxford, OX5 1GB, UK
Radarweg 29, PO Box 211, 1000 AE Amsterdam, The Netherlands

Notice

Knowledge and best practice in this field are constantly changing. As new research and experience
broaden our understanding, changes in research methods, professional practices, or medical treatment
may become necessary. Practitioners and researchers must always rely on their own experience and
knowledge in evaluating and using any information, methods, compounds, or experiments described
herein. In using such information or methods they should be mindful of their own safety and the safety of
others, including parties for whom they have a professional responsibility. To the fullest extent of the law,
neither the Publisher nor the authors, contributors, or editors, assume any liability for any injury and/or
damage to persons or property as a matter of products liability, negligence or otherwise, or from any use
or operation of any methods, products, instructions, or ideas contained in the material herein.

Library of Congress Control Number
Chwieduk, Dorota, author.
  Solar energy in buildings : thermal balance for efficient heating and cooling / Dorota Chwieduk. —
First edition.
    pages cm
  Includes index.
  ISBN 978-0-12-410514-0
1. Building-integrated photovoltaic systems. 2. Solar thermal energy. 3. Solar buildings. 4. Energy
conservation. I. Title.
  TK1087.C46 2014
  690'.83704724–dc23
                                    2014012869

British Library Cataloguing-in-Publication Data
A catalogue record for this book is available from the British Library

ISBN: 978-0-12-410514-0

For information on all Academic Press publications
visit our website at store.elsevier.com

This book has been manufactured using Print On Demand technology. Each copy is produced to order
and is limited to black ink. The online version of this book will show color figures where appropriate.

Working together
to grow libraries in
developing countries

www.elsevier.com • www.bookaid.org

Contents

Acknowledgments

The book has been written in English with the help of translator, Adam Rajewski.

The main aim of this publication is to present fundamentals of solar energy for buildings and show what modern solar energy technologies may offer to the building sector applications. The book demonstrates that a modern building is not only a construction structure, but also an integrated and complex energy system.

Solar energy reaches the Earth by various ways, depending on geographical location and local terrain features. Each building is affected by solar radiation. Therefore it is important to plan and manage this interaction. The book presents certain recommendations for shaping building envelopes in the context of solar radiation availability, and shows how passive and active technologies can be integrated with a building structure. A method for determining influence of solar radiation on the building heat balance as well as quantitative energy input in selected cases is shown as well. Moreover possibilities for utilizing solar radiation in the construction technology by implementation of active systems involving photothermal and photovoltaic conversion are analyzed.

It may be expected that incorporation of active and passive solar energy systems will enable rapid decrease of energy consumption in buildings. Adopting improved technologies for collecting solar energy and converting it into useful heat and electrical power, as well as coupling those achievements to other renewable energy technologies and efficiency solutions, will soon result with practical realization of nearly zero-energy buildings, then followed by structures independent from energy supplies or even energy-plus ones.

Thanks to well-designed envelopes and structures, correct choice of materials and suitable siting (including orientation in reference to cardinal directions as well as neighboring structures) combined with a thought-through concept for utilizing specific zones and rooms, the buildings will soon considerably decrease their energy needs. Moreover, thanks to implementing highly efficient and reliable equipment and systems, the final energy demand will also drop considerably. Then if the final energy demand can be covered by renewable energy—solar energy in particular—the fossil fuel primary energy consumption will be significantly reduced or even totally nullified. A building structure of the future will be one, which generates energy for its own needs, and sells a surplus to an external grid. Nowadays a building is not just a structure anymore, it is also an energy system which links anticipated behavior of the structure and utilized materials to the energy systems and related equipment. Energy management in a building no longer boils down to supplying power from external networks. Now it is also about local energy generation following actual demand, where key role is played by solar systems which are the focus of this publication.

The Directive *2010/31/EC* of the European Parliament and of the Council *on the energy performance of buildings (a recast* of the earlier Directive *2002/ 91/EC)* obliges member states to carry out analyses related to determining energy characteristics of buildings. The Directive creates a general framework for a methodology of calculation of the integrated energy performance of buildings and specifies certification of compliance with minimum requirements on the energy performance of new and existing buildings. Article 1 of Section 1 stipulates that the Directive promotes the improvement of the energy performance of buildings, taking into account outdoor climatic and local conditions, as well as indoor climate requirements and cost-effectiveness. Another notable stipulation of the Directive is its article on "nearly zero-energy buildings" (Article 9). According to this article, from the year 2021 all new buildings will have to be nearly zero-energy buildings, and in case of public utility buildings the rule will come into force already in 2019. EU Member States are obliged to draw up and implement national plans for introducing nearly zero-energy buildings.

Therefore adopting rules of nearly zero-energy construction technology, including solar energy utilization in building structures, is no longer just a concept for implementing general energy conservation principles and minimizing environmental footprint, but also a legal necessity. This publication is intended to—at least partially—assist in implementation of new EU requirements.

The main aim of this book is to systematically present knowledge on phenomena related to collecting, converting, storing, and utilizing solar energy radiation occurring in a building, i.e., in elements of its envelope and interior, as well as within integral systems. Issues, which rarely come into focus of publications on both construction technology and solar energy systems, are covered in detail.

Special attention was given to the issue of solar radiation availability at the building envelope in reference to individual components, especially transparent envelope elements—windows, and then also opaque elements. Detailed discussion based on simulation analyses performed by the author are presented and graphically illustrated. Developed results may become an important and interesting source of information for specialists designing buildings and solar energy systems, as well as those willing to participate in conceptual planning of their own building structures.

Much attention is given to the issue of energy balance of a building and its rooms, with focus on solar energy interactions. Energy balances of sample rooms in different locations within a building are analyzed basing on own simulation calculations. In particular their heating and cooling needs are compared. Obtained results are extensively illustrated with figures and followed by qualitative and quantitative conclusions concerning influence of solar radiation on room energy balances. It is shown, how significant role in such a balance may be played by solar energy, stressing the importance of its planned utilization, including interaction blockades whenever necessary. Presented data may be a source of information for all building users, they may also be used as a base for detailed architectural and structural engineering works. It may also serve as a foundation for further analysis on shaping building envelopes, planning their material structure, as well as internal and external layouts, especially for optimization studies.

Except for author's own theoretical and computational studies on availability, collection, conversion, and utilization of solar energy in active and passive ways, the book also contains a review of fundamentals of solar energy systems for buildings and presents some application examples. Those examples are real-life structures utilizing solar energy as well as selected information on state-of-the-art solar energy systems and their further development prospects.

Chapter 1 gives the terminology utilized in reference to the solar energy use. The chapter also covers physical fundamentals of selected phenomena related to solar radiation, its travel to the Earth's surface including transmission through the atmosphere and the greenhouse effect. The final section presents classification of key technologies that utilize in direct or indirect way solar energy.

Chapter 2 covers the issue of solar radiation availability on the Earth, nature of the solar radiation and its properties, as well as conditions enabling practical utilization of solar energy. It describes fundamentals of the spherical geometry for Sun—Earth relations and methods for measuring solar radiation. Key astronomical and geographical parameters used to describe the Earth position and movement in reference to the Sun are described together with methods for determining irradiation of arbitrarily defined reception surfaces. Various solar radiation models, including isotropic and anisotropic diffuse models, are presented. The chapter also discusses questions of shading and presents sun charts, which describe position of the Sun over the horizon and may be used as a practical aid when determining solar radiation attenuation due to shading.

Chapter 3 discusses ability to shape building envelope in reference to incident solar radiation. This chapter compares results of irradiation simulations for variously oriented surfaces in a selected location (higher latitudes) when using isotropic and anisotropic models. Recommendations for designing building envelopes in the context of solar energy availability are given.

Chapter 4 discusses fundamentals of the phenomena occurring during collecting and converting solar energy into useful heat, as well as during storage and utilization of that heat. Fundamentals of optics related to the solar radiation going through transparent media, including radiation transmission, absorption, emission, reflection, and refraction are discussed. Phenomena occurring in solar energy receivers are analyzed. Energy balance for a generic solar energy receiver is presented. Thermal resistance method, which may be used for modeling phenomena occurring in solar energy receivers, is described. Also optical and thermal phenomena occurring in transparent glass panes (glazing) and in a composite system of a transparent body and absorbing surface are described in detail. Impact of the absorbing surface type on solar energy gained is analyzed, including discussion of selective surfaces. Heat exchange between the solar energy receiver and its surroundings, including interaction between surrounding and covered or uncovered receiver, is discussed. Solar energy collection and conversion into heat in a receiver in form of a solar collector implemented in both active and passive solar systems is described. Solutions for storage of gained heat are analyzed. Also thermo-diffusion phenomena occurring in thermosiphon (natural gravity) installations are presented.

Chapter 5 concerns passive solar energy utilization in a building. Solutions for reducing energy demand of a building are presented. Passive solar systems classification is given and selected solar technologies for building applications are described together with experimental research of phenomena occurring in passive solar systems. Particular attention is given to the selected glazing technologies, transparent insulations, and natural daylighting systems— applications that are still not commonly used, but at the same time enable considerable decrease of building's electricity demand. Also utilization of phase-change materials (PCMs) integrated with the building structure in various manners and used for storage of heat, including heat gained from solar radiation, is investigated.

Chapter 6 provides various forms of a heat balance equation for a room within a building. General form of a balance equation is presented together with simplifications for steady and quasi-steady states. Steady state heat transfer through building envelope is described with particular attention given to its thermal resistance and building's heat capacity. Phenomena occurring inside envelope structure and its surrounding are described using the thermal resistance method. This is followed by a description of unsteady processes occurring in the envelope and its surroundings, taking into account solar radiation influence, both for transparent and opaque elements. Energy transfer through glazing, heat exchange with external and internal surrounding, and heat exchange in the gas cavity between glass panes are discussed in detail. Solar energy and heat flow through the central part and edges of glazing as well as its frame are described in a spatially diverse way. Solar energy passing via building's envelope is investigated. More complex forms of the energy equation for unsteady states are given together with solutions and exemplary results of simulation calculations for opaque and transparent elements, with special attention given to solar radiation influence on the thermal balance. Solutions for exemplary energy equations are presented. Heating and cooling needs for room examples are analyzed and the results are graphically illustrated. Energy consumption parameters describing annual heat demand for space heating and cooling in exemplary rooms differently situated within a building are compared. It is highlighted that the cooling needs during summer season is considerably higher at certain room locations, which results primarily from the role of windows in the building thermal balance.

Chapter 7 describes active solar heating systems utilized in buildings. State-of-the-art solutions are presented together with development directions. Basic configurations of active solar heating systems are presented together with their functional description and operation in various energy supply and demand conditions. Structure and operation of main components of active solar heating systems, namely collectors and storage systems, are discussed too. Roles of key solar collector components are presented and so are different collector types, primarily flat and vacuum collectors, together with scopes of application. Thermal characteristics for selected solar collectors are presented. The focus is on liquid-based systems, which are primary application arrangements for active solar heating in construction industry. Modern integrated heating systems are discussed, where a solar subsystem is just one component of a more complex solution for heating (space heating and domestic hot water) and possibly also cooling (air-conditioning). Various methods of cooperation with other conventional and renewable energy systems are described.

Chapter 7 also covers photovoltaics (PV) and its applications in buildings. It describes physical fundamentals of the internal photovoltaic effect. Key PV technologies and application examples in buildings are presented. Also, solar cooling technologies and their possible applications are presented, giving an idea of solar combi-systems very perspective for future energy efficient solutions. An important observation made in this chapter, and at the same time an important conclusion concerning solar energy applications in a building, is that implementation of solar systems in a building should be taken into account already at the stage of architectural and construction concept, allowing to integrate components of solar systems with the building's envelope and its internal structure, creating a coherent complex system. This is significant for both technical and aesthetic purposes.

The author would like to show what the modern solar engineering, which integrates solutions developed by energy engineering, architecture, and civil engineering, is really about. Solar engineering understood in such a way covers the very concept of building's shape and siting, its structural design and materials selection for both transparent and opaque elements. It may be expected that transparent envelope elements will prevail in the envelopes of future buildings, and it is those very elements that are particularly sensitive to the solar radiation influence. Solar engineering is also about utilizing natural lighting and planning ambient surrounding to influence availability of solar radiation, including light. Finally solar engineering includes integration of building's internal systems with its envelope, combined or alternating utilization of different energy sources, collecting energy available in the environment, storing it and utilizing in an efficient manner, as well as waste energy recovery. Therefore solar engineering in a building is a complex problem of its structure with regard to energy conservation and environment protection.

This publication is intended for architects, engineers, and other experts in areas like construction, energy, civil, and environmental engineering, as well as students pursuing degrees in science and engineering, and all other individuals interested in solar energy systems in general and possibilities of utilizing them in buildings in particular. This book may become a source of information for those who touch its subject for the first time, as well as readers who already have some knowledge on solar energy and its applications in buildings. It contains information on both scientific fundamentals and practical applications.

Chapter 1

Solar Radiation—Fundamentals

1.1 TERMINOLOGY

Solar energy terminology is systematized in the standard ISO 9488 "Solar energy—Vocabulary" [1]. This section describes certain selected terms related to solar energy, which will be used throughout this publication. Vocabulary used in this book is consistent with the above-mentioned standard.

Solar radiation is characterized by various parameters. Certain vocabulary incompatibilities may be observed in publications on solar energy and civil engineering. Vocabulary related to the description of solar energy quantities was proposed in the 1980s by Duffie, Beckman, and Klein in one of the best known publications on solar energy fundamentals [2]. In numerous later publications, vocabulary varied, depending on the research center involved. In 1999, standard EN ISO 9488:1999 [1] concerning solar energy vocabulary was published in order for it to be standardized. Vocabulary given below is defined according to this standard.

Solar radiation is radiation emitted by the Sun, whereas solar energy is the energy emitted by the Sun in the form of electromagnetic waves. According to the EN ISO 9488 standard, solar energy may also be understood as any energy available thanks to collecting and converting solar radiation.

Global solar radiation is the solar radiation incident on the horizontal Earth's surface. According to the EN ISO 9488 standard, the global radiation term may only be applied to a horizontal surface. In case of any tilted surface, the term hemispherical radiation should be used instead. In fact, the global radiation is a case of hemispherical radiation too, but only that which reaches the horizontal Earth's surface. Hemispherical solar radiation incident on a horizontal surface consists of beam radiation (also known as direct radiation) and radiation scattered by the atmosphere. Beam radiation is incident at small solid angle starting from the Sun's disk.

Hemispherical radiation is radiation received from the entire hemisphere above (i.e., solid angle 2π sr). Hemispherical solar radiation incident on a tilted surface consists of direct and scattered radiation, as well as radiation reflected by the ground. Scattered and reflected components form diffuse radiation that reaches any tilted surface. Diffuse radiation may be determined as hemispheric radiation minus beam radiation.

Solar Energy in Buildings. http://dx.doi.org/10.1016/B978-0-12-410514-0.00001-3

Radiant energy is the amount of energy carried by radiation. The radiation itself is energy emission or transfer in the form of electromagnetic waves. Radiant energy flux (radiant flux, flux of radiation) is the total power emitted, transferred, or received in the form of radiation. Then, irradiance is a power density of radiation incident on a surface (i.e., the amount of incident energy in the time unit-per-unit area of the given surface).

Spectral solar irradiance is irradiance for a specific wavelength (i.e., irradiance of monochromatic solar radiation). Solar spectrum is presented as a spectral solar irradiance function of radiation wavelength.

Radiant exitance is the radiant flux leaving a unit area of a surface in the form of emission, reflection, or transmission. Then, radiant self-exitance is the radiant flux density of the surface's own emission, meaning that it refers to the flux leaving the surface only by means of emission.

A parameter that is very extensively used to describe radiation quantity is irradiation (radiation exposure). It is defined as the incident energy per unit area of a surface, and found by integration of irradiance over a certain time interval, often in a single hour or day. Meteorology also uses descriptions like hourly or daily sum of radiation per unit area of a surface. Hourly sum is found by integration of irradiance over a time interval of 1 h. Daily sum, in turn, is obtained by integration of irradiance over a time interval limited by the times of sunrise and sunset. Another term used in meteorology is diurnal sum, which refers to the radiation balance during a 24-h period, thus also including longwave Earth radiation. The term "insolation" currently is not recommended for use, although it may be often encountered in literature on solar energy. Insolation should be considered a colloquial term, which generally describes conditions of solar radiation availability.

Another term often used in solar energy science and applications is "sunshine duration" (solar hours). It is a sum of hours during which the Sun's disk is directly visible over a specified time interval.

Table 1.1 compiles key characteristic parameters describing solar radiation as per the EN ISO 9488 standard. The names and symbols as shown in the table will be utilized throughout this book.

Other terms used when discussing solar energy, not listed in Table 1.1, are explained below.

Total radiation is defined as a sum of shortwave solar radiation with wavelengths between 0.28 and 3 μm, and thermal longwave radiation with wavelengths above 3 μm. Longwave radiation is usually emitted by sources of terrestrial temperature, like clouds, atmosphere, ground, and other terrestrial objects. Atmospheric radiation is longwave radiation emitted by the atmosphere and dispersing in it.

Discussions of solar energy problems also take into account radiation that reaches the outer boundary of the Earth's atmosphere, so-called extraterrestrial radiation. Then, in the atmosphere itself, atmospheric attenuation of solar

TABLE 1.1 Key Characteristic Parameters Describing Solar Radiation

Key Parameters of Solar Radiation	Symbol	Unit
Solar radiation Solar energy		
Global solar radiation • Beam (direct) solar radiation • Diffuse solar radiation	E E_b E_d	(J)

Hemispherical solar radiation
• Beam solar radiation
• Diffuse solar radiation
  • Scattered
  • Reflected
	E
	E_b
	
	E_d
	E_r
		(J)

Radiant energy	E	(J)
Radiant energy flux Radiant power Flux of radiation	ϕ	(W)
Irradiance • Solar irradiance • Hemispheric solar irradiance • Global solar irradiance • Beam solar irradiance • Diffuse solar irradiance	 G G_b G_d	(W/m^2)
Spectral solar irradiance	E_λ	(W/m^2/μm)
Radiant exitance		(W/m^2)
Radiant self-exitance Specific emission	M	(W/m^2)
Irradiation (radiance exposure) • For a day • For an hour or determined period of time	 H I	(J/m^2)
Solar hours ≡ sunshine duration	U_s	(h)

radiation is observed. This attenuation is defined as a loss of flux of beam radiation during transfer through the atmosphere, attributable to absorption and scattering by the atmosphere's components.

Another significant parameter not shown in Table 1.1 is the sky temperature. It is the temperature of longwave radiation of sky (considered to be a black body) incident on the horizontal surface of the Earth.

Further sections of this chapter provide definitions and descriptions (both physical and mathematical) of more parameters used in solar engineering, which are not listed above.

1.2 SOLAR ENERGY

Thermonuclear reactions occurring in the Sun (thermonuclear fusion of deuterium and tritium) are the source of solar energy. Because of those reactions, the temperature of the Sun's core is at the level of 10^7 K. Density of the Sun's core is approximately 100 times higher than that of water. The Sun consists approximately of 80% hydrogen and 20% helium; other elements constitute only 0.1%. Thermonuclear fusion reactions occurring in the Sun have been, and still remain, a subject of multiple research studies. It is believed that a similar process could be replicated in a controlled manner on the Earth, thus helping to solve our energy demand problems. During the fusion process, four hydrogen nuclei (protons ^{1}p) form single helium nucleus (alpha particle $^4\alpha$). An alpha particle consists of two neutrons ^{1}n and two positively charged protons ^{1}p. Moreover, each such reaction releases two positrons e^+ and two neutrinos ν_e as well as energy. Thermonuclear fusion reaction in its entirety is described by the equation [3,4]:

$$4^1_1p \rightarrow {}^4_2\alpha + 2e^+ + 2\nu_e + \Delta E$$

Comparison of nuclear particles before and after the reaction yields a difference—loss of mass occurring during the reaction. If we neglect the mass of a neutrino, and assume that masses of positron and electron are equal, we conclude that:

$$\Delta m = 4m(^1p) - m(^4\alpha) - 2m(e^+)$$

Loss of mass is converted into energy $\Delta E = \Delta m_s c^2$. Every second the Sun loses approximately 4.3 million tonnes of mass ($\Delta m_s = 4.3 \times 10^9$ kg/s). Taking into account the speed of light wave c, the solar radiant power Φ_s of the energy released by the Sun into space is around 3.845×10^{26} W. Total surface of the Sun A_s is equal to 6.0874×10^{12} km^2. Around 0.2 km^2 of the Sun's surface emits approximately 400 EJ/year, which corresponds with current total primary energy consumption on the Earth.

Self-exitance of the Sun's radiation (Sun's specific emissivity) corresponds to:

$$M_{eS} = \frac{\Phi_s}{A_s} = 63.11 \ (\text{MW}/\text{m}^2)$$

Spectral distribution of internal radiation is very irregular. Internal radiation is absorbed by outer, passive layers of the Sun. Those outer (external) layers heat up to the temperature around 5800 K known as the Sun's effective temperature, and become a source of radiation with continuous spectrum. This spectrum corresponds to the radiation emitted by a black body at a temperature of 5800 K. Effective temperature of the Sun may be found in a simple way. According to the Stefan–Boltzmann law, the density of the radiant flux emitted by the Sun (self-exitance) is:

$$M_{eS}(T) = \sigma T^4$$

Taking into account the Boltzmann constant ($\sigma = 56.7 \times 10^{-9}$ W/(m^{2}K^4)), the effective temperature of the Sun is then:

$$T_s = \sqrt[4]{\frac{M_s}{\sigma}} = 5777 \text{ K}$$

Specific emissivity of the Sun is variable in time, thus so is the effective temperature of the Sun. Often it is assumed to be roughly equal to 5800 K.

Figure 1.1 shows spectral distribution of solar radiation not disturbed by the influence of the Earth's atmosphere (data to draw the figure were taken from the Website: http://rredc.nrel.gov/solar/spectra/am1.5). It is similar to distribution of a black body with a temperature of 5800 K. Area under the curve reflects total flux density of solar radiation reaching the outer boundary of the atmosphere, known as the solar constant, whose value is assumed at $G_{sc} = 1367$ W/m^2. This value first appeared in the literature in the late 1970s as a result of applying improved measurement methods and taking measurements from spacecraft platforms. The World Radiation Center, and in consequence all world standards, assume this value as a valid one, although until quite recently $G_{sc} = 1353$ W/m^2 was used.

The point of the Earth's orbit, where our planet is farthest from the Sun, is called aphelion. At the aphelion, the distance between the Earth and Sun is approximately 152×10^9 m. The point of the Earth's orbit, where the planet is closest to the Sun, is called perihelion and then the distance is 147×10^9 m.

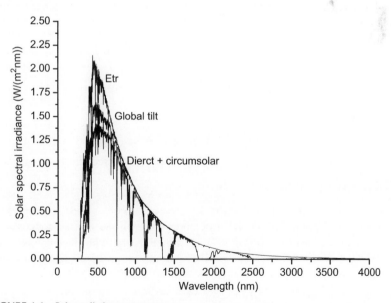

FIGURE 1.1   Solar radiation spectrum.

FIGURE 1.2   Near-elliptical Earth's orbit in reference to the Sun.

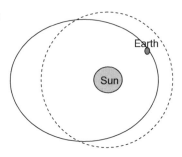

Due to the size difference between the Earth and the Sun and the distance separating them, the power of solar radiation reaching Earth's atmosphere boundaries is 1.75×10^{17} W. Energy reaching the outer boundary of the atmosphere in a unit of time per unit area of surface perpendicular to the incident radiation, at average Sun-to-Earth distance of (1495×10^6 km), is 1367 W/m^2. It is the solar constant mentioned earlier.

Extraterrestrial radiation may differ by $\pm 2\%$ from the solar constant because of fluctuations of the Sun's emissions, and approximately 3% because of changes of mutual position of the Sun and Earth (i.e., distance separating them), which is related to the Sun's apparent movement along the ecliptic, as illustrated—in a purely conceptual manner—in Figure 1.2.

Variability of solar radiation flux density reaching the upper layers of the Earth's atmosphere attributable to the variable distance between the Earth and the Sun over a year is expressed (acc. to Duffie, Beckman [2]) by the following relation:

$$G_{on} = G_{sc}\left(1 + 0.033 \cos\frac{360\,n}{365}\right) \tag{1.1}$$

G_{on}, irradiance at the surface normal to the radiation direction.

This value is fundamental for discussions of solar radiation availability on the Earth.

1.3 RADIATION TRANSMISSION THROUGH THE ATMOSPHERE

Before reaching the Earth's surface, solar radiation is subjected to various interactions and attenuation. At first, upon reaching the Earth's atmosphere, the radiation is partially reflected by its outer layers, with the reflected part returning to space. Then, when passing through the atmosphere, radiation is further attenuated, as the atmosphere partially absorbs and scatters it.

The Earth's atmosphere [4] is a gaseous layer surrounding the planet's surface. It consists of a homogenous (approximately) mixture of various gases and certain amount of suspended matter, so-called atmospheric aerosol. Aerosol particle radii are between 10^{-7} and 10^{-3} cm, and approximately 50% of aerosol particles with radii between 1.5×10^{-6} and 0.5×10^{-4} cm contain radioactive substances. Composition of the atmosphere is roughly constant up to the altitude of approximately 80 km, except for the aerosol, which primarily concentrates near the Earth's surface, water vapor contained mainly up to 10 km up, and ozone, which is mainly present at 20–40 km of altitude. Above the altitude of 80 km, the atmosphere's composition is much more diverse, primarily because of photochemical phenomena driven by solar and space radiation. Solar radiation inflow to the Earth is neither steady nor continuous, and therefore it becomes a source of mechanical energy of moving masses of air (which, among other things, is the driving force for winds). Optical and electrical phenomena also occur in the atmosphere.

By average, some 30% of solar radiation reaching the Earth's atmosphere is reflected back into space. The average density of solar radiation flux, equal to the solar constant ($G_{sc} = 1367$ W/m^2), when adjusted for reflectivity of the Earth's atmosphere known as albedo, decreases to roughly 1000 W/m^2.

Further processes to which solar radiation is subjected are related to absorption. Those are very significant, as they alter the spectrum of radiation that reaches the Earth's surface in reference to the spectrum of extraterrestrial solar radiation. Spectral curve is lowered (because of attenuation) and becomes more "peaky" (because of changing monochromatic flux for different wavelengths). As the absorption processes occur in various areas of the spectrum (wavelength intervals), the full spectrum may be divided into several key bands.

The ultraviolet radiation band corresponds to wavelengths λ lower than 0.3 μm. Radiation of this band is fully absorbed by ions and particles of oxygen, ozone, and nitrogen. It is assumed that this kind of radiation does not reach the sea level (unless the ozone layer is damaged). Near ultraviolet radiation, between 0.3 and 0.38 μm, reaches the Earth's surface in small amounts (allowing people to tan).

Visible light corresponds with wavelengths higher than 0.38 μm and lower than 0.78 μm. Pure atmosphere is perfectly transparent for this band of solar radiation, which may therefore reach the Earth's surface not attenuated. However, if aerosols and gaseous contaminants are present in the air, they can intensively absorb solar radiation in this range, considerably attenuating it.

Wavelength of 0.78 μm marks the start of the infrared band. Solar radiation fits into the near infrared band up to 3 μm, which is also known as shortwave infrared. In the spectrum of radiation reaching the outer layers of the atmosphere, this kind of radiation has similar share as the visible light. However, by average, 20% of near infrared is absorbed in the atmosphere, mainly by water vapor and carbon dioxide particles. This is so-called selective absorption of infrared solar radiation. CO_2 particles absorb wavelengths between 2 and 2.8 μm. Water vapor particles absorb radiation with the following wavelengths: 0.72–0.93 μm interval, and then specific values of 1.1; 1.4; 1.8 μm; and finally from 2.3 to 2.5 μm [4]. If we assume the CO_2 concentration to be relatively constant, then its volumetric share is, by average, 0.03%. Because of increasing environment pollution, CO_2 concentration in some areas of the world is higher than the assumed average. Water vapor content can—in a fully natural way—even reach 4%. Variations of water vapor content are related to the variability of air humidity and cloud cover. It should be added that the share of infrared radiation's energy carried at wavelengths above 2.5 μm is very small making it negligible.

Besides the selective absorption, also nonselective absorption of solar radiation by atmospheric aerosols present at different altitudes occurs in the atmosphere. It affects the entire solar radiation spectrum (i.e., all wavelengths). Attenuation of solar radiation because of absorption by atmospheric aerosols depends on the type of aerosols, particle dimensions, and their volumetric concentration.

Except for being absorbed, solar radiation passing through the atmosphere is also scattered. This phenomenon affects visible and infrared spectra. It occurs when photons of the solar radiation prove unable to break bonds of atmospheric particles and, as an effect, are scattered in all directions. Scattering is a process dependent on the wavelength in question. It is interaction between the radiation and matter, which results with a change of wave propagation direction, without affecting its energy or wavelength. Radiation is scattered by gaseous particles of the atmosphere's components, mainly those of oxygen O_2, nitrogen N_2, and water vapor H_2O, as well as by atmospheric aerosols.

As mentioned earlier, scattering alters radiation's propagation direction. Scattering intensity depends on the atmosphere's composition, its humidity, local cloud cover, dustiness, but also distance of solar radiation's travel through the atmosphere, which is expressed by the so-called air mass. This distance depends on the zenith angle of incident radiation (therefore also on the Sun's altitude, which is a difference between the right angle and zenith angle) and on the local altitude of the investigated spot on the Earth's surface.

Thus, the distance traveled by solar radiation through the atmosphere depends on the Sun's position on the hemisphere. This fact is taken into account by introducing the value of AM_X, already mentioned air mass. Air mass is defined as a ratio of mass of the atmosphere, which needs to be crossed by the beam radiation, so it can reach the investigated point on the Earth's surface, to the mass of the atmosphere that the radiation would have to cross if the Sun was at its zenith. For angles of incidence on horizontal surface (zenith angles) between 0° and 70°, it is assumed [2,5] that the air mass is the inverse of cosine of the zenith angle between the normal to the Earth's surface and the direction of beam, i.e.,

$$AM_X = \frac{1}{\cos(\theta_z)} \tag{1.2}$$

as illustrated by Figure 1.3.

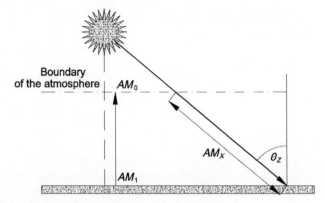

FIGURE 1.3   Simplified determination of air mass value AM_X.

For zenith angles between $0°$ and $70°$, the following approximation is used [2]:

$$AM_X = p/(p_o \cos \theta_z)$$

Symbols p represent the atmospheric pressure whereas $p_o = 1013$ hPa is the standard pressure and the p symbol is the actual pressure value (hPa). If we assume that the actual atmospheric pressure p is equal to the standard atmospheric pressure (i.e., $p = p_o$) a simplified relation described by Eqn (1.2) may be derived.

At the sea level and when the Sun is at its zenith, the air mass is equal to 1 ($AM_X = 1$), as in such a case the zenith angle θ_z is 0. Air mass of 1, AM_1, corresponds with the power density of solar radiation of 1000 W/m^2. Air mass of 2, AM_2, describing a case when the Sun's altitude (above the horizon) is $30°$ (zenith angle θ_z of $60°$) translates into power density of solar radiation of 740 W/m^2. The air mass of 0, AM_0, describes a location beyond the Earth's atmosphere where the extraterrestrial solar irradiance is 1367 W/m^2—equal to the solar constant.

The presented discussion applies to zenith angles lower than $70°$. In case of angles over $70°$, the approximate relation (1.2) may no longer be used, as the Earth's curvature and light refraction become significant factors that need to be taken into account. Dickinson and Cheremisinoff proposed the following approximate relation for use throughout the entire range of zenith angles:

$$AM_X = (1 - 0.1 \, H) \left[\sin \alpha + 0.15 \, (\alpha + 3.9)^{-1.253} \right]^{-1}$$

where the symbols stand for: α, altitude the Sun (where $\alpha = 90 - \theta_z$) ($°$); and H, altitude (km).

Kasten [6], in turn, proposed to use the following relation:

$$AM_X = \left[\cos (\alpha) + \frac{0.15}{(93.885 - (\alpha))^{1.253}} \right]^{-1} \frac{p}{p_o}$$

Just like before, symbols p denote relevant atmospheric pressure.

Appropriate solar radiation spectrum graphs are created for specific air mass values. As it was mentioned, the air mass of 0 corresponds with the spectrum reaching the outer boundaries of the atmosphere. Spectrum of radiation reaching the Earth's surface is, however, altered by absorption bands of ozone, water vapor, and CO_2. The higher the air mass gets, the lower the spectral curve is (atmospheric attenuation increases).

Phenomenon of solar radiation scattering in the atmosphere depends on the amount of particles encountered by radiation along its path, and dimensions of those particles in reference to the wavelength in question. Air particles have very small dimensions (e.g., oxygen and nitrogen) in relation to the radiation wavelengths, and, according to the Rayleigh theory [4,7], scattering on air particles is significant only for short wavelengths (i.e., those below 0.6 μm). For this band, the scattering coefficient is a function of the negative fourth power of the wavelength (λ^{-4}). Therefore, scattering intensifies with decreasing wavelengths. In the visible light spectrum of the solar radiation, violet and blue bands have the lowest wavelengths and at the same time the highest monochromatic irradiance, and therefore its scattering is also most intense resulting with the apparent blue color of the sky.

In case of wavelengths over 0.6 μm, scattering on air particles is negligibly small. In perfectly pure atmosphere, when the Sun is at the zenith, Rayleigh scattering reduces irradiance by some 9.4%. According to the Rayleigh scattering theory, angular distribution of scattered radiation is symmetrical about the direction of incident radiation and displays rotational symmetry about the axis normal to the direction of incident radiation.

When passing through the Earth's atmosphere, the solar radiation is also scattered on large particles present because of water and dust presence (i.e., on aerosols). Scattering on those particles affects the entire spectrum of solar radiation. Dust and water tend to form large particles in the atmosphere because of the concentration of water particles in certain locations, as well as water condensation on dust particles. Description of the scattering process on water and dust particles is complex, as the nature of those particles and their concentration in the air strongly depends on time and location.

Description of this scattering process uses the Ångström model (turbidity equation), in which the atmosphere transmittance is described as follows [2]:

$$\tau_{\alpha, \lambda} = \exp\left(-\beta \lambda^{-\alpha} m\right)$$

The equation above uses two coefficients. One of them is the atmosphere purity coefficient β, which may vary from 0 to 0.4 (i.e., from very clean to very turbid atmosphere). The other coefficient α is a single lumped wavelength exponent and is related to the distribution of aerosol dimensions in the atmosphere (in most cases assumed as equal to 1.3).

Coefficients α and β vary with changing weather conditions. In case of clear sky, scattering intensity depends on the Sun's position on the sky described by the zenith angle (or altitude angle) and, even more importantly, on the aerosol content in the atmosphere.

Aerosol scattering was investigated, among others, by Mie [4,7]. According to the Mie theory, scattering is investigated for particle sizes larger than 0.1 wavelengths, with the scattering coefficient being inversely proportional to the radiation wavelength. Thus, it is assumed that, in the real atmosphere, the scattering coefficient is proportional to $1/\lambda^n$, with n varying between 1 and 4. The cleaner the air, the closer n gets to the value of 4 (increasing dustiness lowers the n value).

Irradiance of radiation scattered in the atmosphere (on gas particles and aerosols) depends on the angle between the beams of incident and scattered radiation. This irradiance is described by the scattering function known as scattering indicatrix. The shape of the indicatrix characterizes the magnitude of its dissymmetry Dis (i.e., ratio of the radiation scattered at the angle φ and $\pi - \varphi$) to the incident radiation, which is expressed as:

$$Dis\,(\varphi) = I(\varphi)/I(\pi - \varphi)$$

Shape on an indicatrix depends on the type of medium. If the dimensions of particles causing scattering are much smaller than wavelengths, then—according to the Rayleigh theory—$Dis\,(\varphi) = 1$ and angular distribution is symmetrical about the radiation direction and displays rotational symmetry about the axis normal to the direction of incident radiation. As a result, irradiance of radiation scattered in both directions (forward and backward) is identical. This, in turn, causes radiation attenuation, as part of scattered radiation goes back outside the atmosphere, into space. If the particles are larger and have spherical shape, scattering is determined according to the Mie theory, dissymmetry grows, indicatrix lengthens forward, and irradiance of forward scatter increases.

In case of clear sky, scattering intensity depends on the Sun's apparent position on the sky (i.e., zenith angle or altitude), which, in turn, is related to the geographic location on the Earth and therefore also on the air mass, as well as on the aerosol content in the atmosphere.

Problems of solar radiation attenuation in the Earth's atmosphere was comprehensively investigated by Monteith. In his discussions on changes of radiation flux in the atmosphere, he assumed that those changes may be expressed as a function of previously discussed air mass, and also as a function of atmospheric extinction.

Extinction e_λ [4] is a value used to express changes of irradiance of monochromatic radiation passing through any medium, which is described by a logarithmic relation:

$$e_\lambda = \ln\,(\phi_o/\phi)_\lambda$$

where ϕ_o and ϕ denote incident and leaving radiation fluxes, respectively. Using the relation shown above, Monteith assumed that the flux of monochromatic radiation that passes through the Earth's atmosphere fulfills the relation:

$$E_\lambda = E_{o\lambda} \exp(-e_\lambda AM_x)$$

Budyłowski, in turn, proposed [8] to rewrite this relation for ideal direct panchromatic radiation, taking into account effective distance traveled by the radiation instead of referring to the geometrical path. He introduced a total solar radiation transmittance, which describes attenuation of solar radiation in the atmosphere and is a function of all phenomena of absorption and scattering occurring in the atmosphere. Total atmospheric transmittance is described by the relation:

$$\tau = \exp(a_o P AM_x^k) \qquad (1.3)$$

where the symbols stand for: a_o, extinction of an ideal atmosphere; M_x, air mass; P, atmospheric clearness coefficient.

Atmospheric extinction for ideal atmosphere is a function of atmospheric pressure; literature provides relations allowing to calculate its value [4]. Exponent k is obtained as a linear function of the coefficient P. The value of cleanliness coefficient P for a clear atmosphere corresponds to a hypothetical number of layers of ideal atmosphere whose extinction is equal to that of a real atmosphere. This value is found taking into account processes of scattering and absorption in the atmosphere [8]. Those processes result from the natural state of the atmosphere, which, on one hand, is subjected to cyclic changes, and, on the other, can be altered temporarily. Those temporary changes result from natural phenomena like volcanic eruptions, forest fires, and high concentrations of desert dust during sandstorms, but also from interactions caused by human activity.

Using Eqn (1.3), it is possible to express the beam radiation flux after passing through the Earth's atmosphere during a clear sky as:

$$G_b = G_o \exp(a_o P AM_x^k) \qquad (1.4)$$

Irradiance G_o present in the Eqn (1.4) is a function of the angle of incidence θ for the surface under consideration and of extraterrestrial irradiance G_{on}, as shown in the relation:

$$G_o = G_{on} \cos \theta$$

If cloud cover is present, beam radiation is further attenuated. Influence of clouds is taken into account by introducing arbitrary coefficient f_{cloud}, with value changing from 1 (total overcast) to 0 (clear sky). Irradiance at any surface on the Earth may now be expressed as:

$$G_b = G_o \exp(a_o P AM_x^k)(1 - f_{cloud}) \qquad (1.5)$$

Reduction of the atmosphere's transparency, related to emission of polluting gases and fluids, is the highest in those areas where certain natural phenomena—like already mentioned forest fires, desert storms, or volcanic eruptions—tend to occur, as well as in the industrial and urban areas. The highest transparency is observed in the countryside and in the mountainous regions. The transparency is also much higher in the winter than in the summer.

Scatter depends on cloudiness, type of clouds, and the Sun's altitude. In case of a total overcast with low- and medium-family clouds, practically only diffuse radiation reaches the Earth's surface. It is assumed that its distribution is isotropic. In case of a total overcast with high-family clouds, however, some small amount of beam radiation reaches the Earth's surface as well [8].

During sunrise and sunset, when the Sun is just above the horizon, incident radiation only consists of diffuse radiation. As the Sun's altitude increases, so does the share of beam radiation. At altitudes exceeding 40°, the share of beam radiation in total radiation reaches its maximum. However, even at very clean and dry air, the share of diffuse radiation is 20—25%. At higher latitudes, the Sun's altitudes over 40° are only observed in the summer season, at times close to solar noon; at those times, the share of the beam radiation may be the highest.

After passing through the atmosphere, attenuated solar radiation reaches an arbitrarily inclined surface as hemispherical radiation composed of beam radiation, radiation diffused by the atmosphere, and radiation reflected from the Earth's surface. The highest reflectivity is observed on ice and snow covers, and then somewhat lower on bodies of water. Except from the reflection, also radiation absorption occurs on the Earth's surface. Because of absorption, the Earth's temperature rises and it becomes a source of longwave thermal radiation. Absorption occurring in the atmosphere has already been discussed. Solar radiation attenuated while passing the atmosphere still carries considerable energy potential. Any surface on the globe, therefore, will be characterized by certain conditions for collecting and utilizing solar energy, depending on its location, shape, material it is made of, and a number of other factors.

1.4 LONGWAVE RADIATION: GREENHOUSE EFFECT

As mentioned in the previous section, the Earth's surface and atmosphere, as well as objects located on the Earth, absorb part of the solar radiation, depending on their absorptivity (for solar radiation). As a result of absorption, the temperature of those media increases, and so does the heat transfer by longwave thermal radiation with other nearby objects. Atmosphere particles also absorb part of the Earth's radiation and heat from mixing particles of air and water vapor during the process of condensation. Atmosphere becomes a secondary source of radiation with wavelengths 3.0–50 μm. Atmosphere's longwave radiation depends on the vertical distribution of particle temperatures, atmosphere's composition (absorption and emission-related properties of gases), as well as wavelength and direction of radiation propagation.

Atmosphere's absorptive spectrum is not continuous. For the wavelengths 8–14 μm, the atmosphere is practically transparent. This range is known as the so-called atmospheric window. At perfectly clear sky, thermal radiation of such wavelengths emitted by the Earth's surface passes through the atmosphere without any attenuation. For other wavelengths, however, the atmosphere is practically impermeable. At increasing cloud cover density, caused by increased water vapor content in the atmosphere, its ability to absorb radiation increases. Thus, the clouds formation "closes the atmospheric window." As a result, the atmosphere becomes opaque for the whole spectrum of Earth's longwave radiation, and, therefore, heat losses of the Earth by radiation decrease and the planet's temperature grows.

Using these principles, along with certain simplifications in physical and mathematical description of the phenomena, the whole concept of the greenhouse effect may be expressed in a relatively simple way [9]. If we assume the Earth's radius to be R (not including atmosphere's thickness) and extraterrestrial irradiance to be G_o (i.e., equal to the solar constant), then the energy reaching the Earth can be expressed in a simplified way as:

$$\Phi_c = \pi R^2 (1 - \rho_0) G_o$$

If the absorptivity (and therefore also emissivity) of the Earth for solar radiation was equal to one, then the entire energy reaching the planet's surface would be absorbed and then radiated back. Moreover, if we neglect the influence of geothermal energy and tidal energy on the Earth's energy balance, the balance may be expressed by the following simplified equation:

$$\pi R^2 (1 - \rho_0) G_o = 4\pi R^2 \sigma T_e^4$$

One of the elements in the equation is the average Earth's temperature T_e. Using known values like: G_o, solar constant; R, Earth's radius; ρ_0, albedo, in the equation enables us to calculate the approximate average of the Earth's temperature T_e. Its value is approximately 250 K (i.e., −23 °C). Spectral distribution of the Earth's longwave radiation should therefore match the spectrum of a black body with calculated temperature of 250 K. Such a curve peaks at the wavelength of 10 μm, and the entire spectrum is contained within longwaves, way beyond the solar radiation spectrum, as shown in Figure 1.4. In fact, the average Earth's temperature is approximately 14 °C, so it is 37° more than the calculated value. It is so because the calculated value of 250 K corresponds with the outer layers of the atmosphere, not the surface temperature (because of assumed model simplifications). In fact, the Earth's atmosphere reduces energy losses, creating specific "infrared" thermal insulation layer. This phenomenon is known as the greenhouse effect.

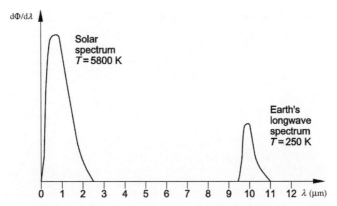

FIGURE 1.4   Spectral distribution of solar (shortwave) and Earth's (longwave) radiation.

The greenhouse effect occurs also in any room with a window or glazed facade. It is put into practical application in greenhouses. Glazing (glasses) are mostly transparent for solar radiation (reflectivity and absorptivity of several percent). This allows slightly attenuated solar radiation to reach the room's interior. There it is absorbed by internal walls and other room elements. As a result, the temperature of those elements—and therefore also the temperature inside the room—grows. Glass envelope is practically opaque for longwaves of the thermal radiation. Thanks to photothermal conversion, gained solar energy stays trapped in the room. Therefore, with some simplification, it may be stated that glazing fulfills a similar role as the Earth's atmosphere.

1.5 THE SKY TEMPERATURE

When analyzing heat transfers by radiation into any object on the Earth's surface, it is assumed that this object exchanges heat with closer and farther surroundings. The farther surrounding in this context is the sky (hemisphere). It is assumed that the sky behaves like a black spherical surface with the globe in its center. Sky temperature T_{sky} is lower than the temperature T_a of the air surrounding the objects on the Earth's surface (see the discussion on the average temperature of the Earth and outer atmospheric layers in Section 1.4). When analyzing the processes related to interactions between the Earth and sky (atmosphere) in a simplified way [2], it is assumed that the sky temperature T_{sky} is lower than the ambient air temperature T_a by 6°, which may be expressed as:

$$T_{sky} = T_a - 6° \tag{1.6}$$

In more detailed discussions [10,11], the sky is approximated as a black body (with a temperature $T_{sky}(t)$ (K)), which emits thermal radiation with intensity L_{sky} (W/m^2), equal to:

$$L_{sky}(t) = \sigma \left(T_{sky}(t)^4 \right) \tag{1.7}$$

Flux of the sky radiation depends on the cloud cover. Flux density of sky radiation L_{clear} (W/m^2) for a clear sky may be expressed as a function of the ambient temperature $T_a(t)$ (K) using the Swinbanck formula [12] given as:

$$L_{clear}(t) = 5.31 \times 10^{-13}\left(T_a(t)^6\right) \tag{1.8}$$

The Swinbanck model assumes that the temperature of clouds is five degrees lower than the ambient temperature and their emissivity ε_c for longwave radiation is $\varepsilon_c = 0.96$. Flux density of the sky radiation for cloudy sky L_{cloud} (W/m^2) is determined using the formula:

$$L_{cloud}(t) = \varepsilon_c \cdot \sigma(T_a(t) - 5)^4 \tag{1.9}$$

The actual sky condition lies between a clear sky and a totally overcast one. Sky radiation intensity for the sky close to real conditions is approximated [11] with the relation:

$$L_{sky}(t) = (1 - c(t)) \cdot L_{clear}(t) + c(t) \cdot L_{cloud}(t) \tag{1.10}$$

Coefficient c reflects the condition of the atmosphere, primarily cloud cover, and is assumed to be a function of coefficient K_d that describes the share of diffuse radiation in global radiation. Value K_d is determined using approximate equations or measurements, as a function of the sky clearness index K_T. Its average values for different latitudes may be found in the literature [2]. In the Swinbanck formula, if the sky is clear, the share of diffuse radiation is at the level of 16.5% ($K_d = 0.165$) and the coefficient c as a function of K_d is found as:

$$c(t) = (K_d(t) - 0.165)/0.835 \tag{1.11}$$

After transformations using Eqns (1.7) to (1.10), the sky temperature may be found with Eqn (1.6) as:

$$T_{sky}(t)$$
$$= \sqrt[4]{\frac{\left(1 - \frac{K_d(t) - 0.165}{0.835}\right)\left(5.31 \times 10^{-13} \cdot T_a(t)^6\right) + \frac{K_d(t) - 0.165}{0.835}\left(\varepsilon_c \sigma(T_a(t) - 5)^4\right)}{\sigma}}$$

$$\tag{1.12}$$

Sometimes a simplified Swinbanck formula (T_{sky} in a function of T_a (K)) is expressed as:

$$T_{sky}(t) = 0.0552 \cdot (T_a(t))^{1.5}$$

Detailed analyses of solar energy utilization in the building sector should use the Eqn (1.12) to determine the sky temperature, as for certain elements of a building envelope at certain weather conditions, heat exchange with the sky (hemisphere) significantly influences their thermal conditions.

1.6 CLASSIFICATION OF KEY SOLAR ENERGY TECHNOLOGIES

Energy of solar radiation naturally determines climatic conditions, lighting, and biological life on the Earth. This energy is utilized by all organisms living on the Earth. Human civilization, however, has started to utilize this energy in a planned manner to fulfill specific useful requirements and is doing that with increasing intensity and effectiveness. Modern technologies combined with innovative engineering allow us to utilize solar energy with increasing energy efficiency and economic feasibility. Some solutions, which until quite recently had seemed impossible, are becoming more and more common. Others, which used to stir a lot of hopes, still have not gone beyond the research and development stage.

Discussions of solar energy utilization involve various classification criteria. In a most general manner, two different forms of solar energy utilization may be defined:

- direct use of solar radiation,
- indirect use of solar energy by utilizing forms of energy driven by solar radiation (e.g., wind energy, hydro energy, and biomass energy).

Solar energy may be used directly using:

- thermal solar power plants,
- solar chimney power plants,
- solar furnace,
- active heating systems with solar collectors,
- passive solar systems integrated into solar architectural systems,
- photovoltaic systems for electrical power generation.

Thermal solar power plants generate electricity. The working principle of such plants is similar to that of fossil fuel power stations, where primary energy is used to generate high-temperature steam. This steam is subsequently used to drive turbines coupled to electric generators. In the case of solar thermal power plants, the working fluid of high temperature may be generated in systems equipped with parabolic concentrating collectors and in thermal solar power towers. Concentrated solar power plants and solar power towers are high-capacity systems and are not directly used in the building sector, although the power they generate may obviously be used to cover the energy demands of a building.

Another solution enabling direct use of solar energy not related to local energy generation in a building is a dish Stirling system [3,13].

This section will shortly describe key solutions for indirect use of solar energy [3,8,13—15].

One of the most popular solutions used by human civilization for a long time is hydropower. Solar energy affects the water circulation on the Earth.

Radiation causes vaporization of surface waters, including water of rivers, lakes, seas, and oceans, as well as water contained in soil. Approximately 22% of solar energy reaching the Earth's surface is responsible for driving the water circulation. Some 20% of vaporized water returns to the land masses in the form of precipitation and then may vaporize again. The rest falls back to the oceans, seas, rivers, and groundwater. Using hydropower is about utilizing kinetic energy of flowing surface waters.

Another known renewable energy (derived from solar energy), also used for many centuries, is wind power. Solar radiation causes the heating effect of masses of atmospheric air unevenly, thus creating pressure differences. Those, in turn, result with circulating air movements. Approximately 1−2% of solar energy reaching the Earth's surface is converted into the wind kinetic energy.

Another commonly used derivative of solar energy is the energy of biomass. Solar radiation is the driver of the photosynthesis, a key process in the life of plants. Biomass used for energy generation is an organic substance derived from plants or animals. Its forms include wood and wood processing waste, straw, energy crops, as well as wastes from agriculture or animal husbandry. Biomass is also created by a process of "social metabolism." In this case, we refer to the waste biomass generated by human activity. It may have a form of wastewater sludge (similar to peat) and municipal waste (containing scrap paper). Biomass is usually utilized after preliminary treatment or conversion into more convenient forms. Sometimes it is converted to the form of biogas generated in a process of digestion of organic matter. The source of that matter may include manure, wastewater sludge, land filled waste (so-called landfill gas). There are also other useful products of biomass processing, like pyrolysis oil with properties similar to mineral fuel oils, as well as rapeseed oil, wood gas, methanol, or ethanol. These may be used as additives to fuel oils consumed by internal combustion engines, both stationary and traction.

Recently, utilization of another form of solar energy derivative has been gaining popularity: using energy contained in environment (i.e., ambient air, water, or ground) via heat pumps. The potential of using energy from renewable sources, like air, ground, surface waters, and groundwater, is evaluated by analyzing functionality and performance of a heat pump. Heat pumps utilizing renewable energy, mainly ground heat pumps, are becoming increasingly popular in the building sector.

Other derivatives of solar energy are energy of waves driven by wind and energy of marine currents caused by water temperature gradients. These, of course, cannot be applied directly in the building sector.

Finally, it should be added that there are also other kinds of renewable energy, which are neither a direct nor indirect effect of solar radiation. Those include geothermal energy—internal energy of the planet Earth—and tidal energy. The latter is a result of gravity forces and Earth's movement or mutual interactions between the Earth, Moon, and—to a much lower degree—also the Sun.

In case of direct use of solar radiation, three basic methods may be used to convert it into useful forms of energy. Those are photothermal conversion, photovoltaic conversion, and photochemical conversion. Photothermal conversion is conversion of solar energy into heat. Photovoltaic conversion is conversion of solar energy into electricity. Photochemical conversion is conversion of solar energy into chemical energy or other energy forms related to chemical processes (e.g., biochemical energy). It should be noted that different kinds of conversion may occur simultaneously. Usually, practical applications involve the first two processes. This publication discusses those technologies for solar energy conversion that are and may be used in the building sector.

REFERENCES

[1]   EN ISO 9488:1999. Solar energy. Vocabulary. European Standard, CEN, Brussels.
[2]   Duffie JA, Beckman WA. Solar engineering of thermal processes. New York: John Wiley & Sons, Inc.; 1991.
[3]   Quaschning V. Understanding renewable energy systems. London (UK): EARTHSCAN; 2006.
[4]   Encyklopedia Fizyki (Encyclopedia in Physics), t. 1—3, vol. 1—3. Warszawa: PWN; 1974.
[5]   Sherry JE, Justus CG. A simply hourly clear — sky solar radiation model based on meteorological parameters. Sol Energy 1983;30:425.
[6]   Kasten F. A new table and approximate formula for the relative optical air mass. Arch Meteorol Geophys 1966;B14:206.
[7]   Feynman RP, Leighton RB, Sands M, Gottlieb MA. The Feynmann lectures on physics, vol. 1, part 2. Pearson/Addison-Wesley; 1963.
[8]   Budyłowski J. Model matematyczny promieniowania słonecznego dla potrzeb heliotechniki [Mathematical model of solar radiation for solar engineering application]. Archiwum Termodynamiki (Arch Thermodyn) 1984;5(1).
[9]   Twidell J, Weir T. Renewable energy resources. London: E&FN SPON; 1996. University Press Cambridge.
[10] Kalogirou S, volume editor. Solar thermal systems: components and applications. In: Comprehensive renewable Energy, vol. 3. Elsevier; 2012.
[11] Clean energy project analysis: RETScreen® engineering & cases textbook, solar water heating project analysis, Minister of Natural Resources Canada, 2001—2004.
[12] Swinbank WC. J Royal Metrol Soc 1963;89:339—48.
[13] Nelson V. Introduction to renewable energy. CRC Press, Taylor and Francis Group; 2011.
[14] Sayigh A, editor. Comprehensive renewable energy. Elsevier; 2012.
[15] Energy Visions 2050. VTT EDITA. 2009.

Chapter 2

Availability of Solar Radiation on the Earth

2.1 FUNDAMENTALS OF SPHERICAL GEOMETRY OF THE EARTH IN RELATION TO THE SUN

Solar radiation that reaches the boundaries of the Earth's atmosphere is beam radiation, also known as direct radiation. This radiation travels directly from the Sun to any point on the Earth along a straight line connecting the Sun with the selected point on the Earth. Upon passing through the Earth's atmosphere, some radiation gets scattered. As a result, any point on the Earth's surface is reached by both beam radiation and radiation scattered by the atmosphere. Beam radiation is strictly directional, while the scattered radiation reaches the Earth's surface from all directions. Any tilted surface (in reference to horizontal one) on the Earth is reached not only by beam and scattered radiation, but also radiation reflected by the Earth's surface and various objects located nearby. A surface on the Earth may have different conditions for reception and capturing solar energy, depending on its geographic location, nearby surroundings, inclination, material, and other factors.

Location of any surface in reference to the Sun is determined with fundamental concepts of spherical geometry related to the Earth's movement in relation to the Sun [1–4]. The Earth spins around it own axis, which is perpendicular to the equatorial plane. One full revolution takes 24 h. Earth's spinning movement is used as a basis for time measurements. Time measurements used on the Earth are related to the local meridian of the observation point. Along the entire meridian the local time is identical, as the Sun only passes the plane of the single meridian once per day (24 h). Commonly used time measurements apply a relation between local time and geographic longitude. One degree of longitude corresponds with the time difference of 4 min, so 15° corresponds with 1 h. In practical applications, time zones are used. Twenty-four such longitudinal zones have been designated, each encompassing 15° of longitude (360°/24 h = 15°/h), with time in each next zone differing by 1 h from the previous one (although there are some exceptions; e.g., Australia).

Solar Energy in Buildings. http://dx.doi.org/10.1016/B978-0-12-410514-0.00002-5

One of the key parameters used to describe the Earth's movement in reference to the Sun and their mutual configuration is the hour angle of the Sun ω. It may be determined for any point on the Earth as a function of solar time. The hour angle ω is the angular displacement of the Sun east or west of the local meridian caused by the Earth's spinning movement. One hour time difference corresponds to the 15° change of the hourly angle. The hourly angle is equal to zero for the 12 o'clock (noon) and each hour of deviation is corresponding with the angle change of $-15°$ (before noon) or $+15°$ (after noon). The hourly angle is defined by the equation:

$$\omega = 15(t_{sol} - 12). \tag{2.1}$$

The solar time t_{sol} is determined by the apparent angular movement of the Sun in the sky, with the solar noon used as a reference point (12 o'clock in Eqn (2.1.)) and is expressed in hours. Meteorological stations carry out their solar radiation measurements in reference to the solar time. Solar noon occurs at the moment when the Sun is in the point where its apparent path intersects the local meridian plane. Time in the local time zone (standard time) is the main component of the solar time, but is not equal to it. The relation between the solar time t_{sol} and the standard time t_{std} (expressed in minutes) is given by the equation:

$$t_{sol} = t_{std} + 4(L_{std} - L_{loc}) + E. \tag{2.2}$$

As the Eqn (2.2) shows, solar time may be described as a standard time adjusted by the E value, so-called the equation of time, and geographic location component, which takes into account the difference between the observation point and the point for which the standard time is defined. The geographic location component is fixed for certain observation point. It is expressed as a difference (in degrees) between the local meridian of the observer L_{loc} and the meridian L_{std} for which the standard time is defined. In case of using summer and winter time, this component has two different values for relevant seasons. The E value is variable over time, although it is assumed to be constant for every day of the year. The equation of time E takes into account the variability of the Earth's motion around the Sun, which affects the moment when the Sun crosses the local meridian plane of the observation point. When orbiting the Sun, the Earth simultaneously spins around its own axis and travels along the elliptical orbit (which is the reason why the distance between the Earth and the Sun during the year varies). The speed of the orbiting movement is decreasing slightly when the Earth is farther from the Sun, and then increases when the distance is shorter. Value of the E parameter may be determined with the Spencer equation [1] in the following manner:

$$E = 229.2(0.000075 + 0.001868 \cos B - 0.032077 \sin B - 0.014615 \cos 2B$$
$$- 0.04089 \sin 2B)$$

The B value is calculated as:

$$B = (n-1)\frac{360}{365}$$

where n is the number of days during a year.

While the Earth orbits the Sun, its axis is constantly tilted at $66°33'$ to the orbit plane (Figure 2.1). The path of the apparent Sun's movement on the celestial sphere, as seen by an observer on the Earth during 1 year, is called the ecliptic. While the Earth moves, insolation conditions on the Earth vary in daily and annual cycles. Because of the tilt of the Earth's axis, the equatorial plane is different from the orbit plane. The Sun, while "traveling" along the ecliptic (in fact it's the Earth's movement, not the Sun's) finds itself in the equatorial plane only twice: on March 21 and September 23, spring equinox and fall equinox, respectively. After those days, the Sun "traveling" along tilted ecliptic distances itself from the equatorial plane, while approaching the plane of one of the tropics—on the northern or southern hemisphere. Upon reaching the plane of a tropic, the Sun starts to "travel back" to the equator. The fact that the location where the Sun culminates at the zenith, causes variability of the Sun's culmination altitude on other latitudes. This, in turn, results with variable length of days and nights. Intersection between the ecliptic and the equatorial plane marks the boundary of a zone where the Sun may possibly culminate at the zenith. Therefore, the tilt of the Earth's axis is the cause of existence of seasons and related variability of day length and the Sun's maximum altitude.

Seasonal changes in the mutual location of the Earth and the Sun are described by the parameter known as the Sun's declination δ. The declination is defined as the angle between the line connecting the Earth and the Sun, and

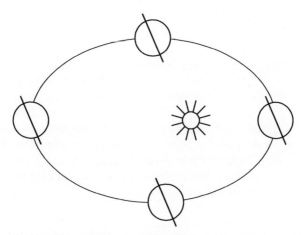

FIGURE 2.1   Earth's orbiting movement around the Sun.

the equatorial plane at the solar noon. The declination value is positive for the northern hemisphere in summer. Depending on the declination, the flux of radiation reaching the irradiated surface varies. The value of the Sun's declination varies within the following limits:

$$-23.45° \text{ (December 22)} \leq \delta \leq +23.45° \text{ (June 22)}.$$

The Sun's declination is zero on the equinox days. Often, its value is calculated with an approximate Cooper formula [3]:

$$\delta = 23.45 \sin\left(360\frac{284 + n}{365}\right) \tag{2.3}$$

where n is the number of the day during a year (for January 1 $n = 1$), just like in the Spencer equation for E.

Solar energy reaches points of the Earth's surface only during daylight hours. Radiant energy flux decreases as the Sun lowers over the horizon. Flux is at its maximum when the Sun is at the zenith. If the solar altitude angle is 30°, the amount of the energy reaching the Earth halves. This means that in case of perfectly clear weather, irradiance cannot exceed 500 W/m². For the countries located at higher latitudes, the Sun does not exceed the altitude angle of 30° during 5 months of fall and winter.

Determining the position of the Sun, in reference to an observer located at the horizontal Earth's surface, requires more parameters. One significant value is the solar zenith angle, θ_z. It is an angle between the normal to the surface (horizon's plane) and the direction of beam radiation (i.e., it is the incident angle of beam radiation on the horizontal surface). The difference between this angle and 90° is called the solar altitude angle α_s or solar elevation angle. It is the angle between the horizontal plane and the straight line toward the Sun (which coincides with the beam direction). Another important parameter is the solar azimuth angle γ_s (solar azimuth). This is the angle between the plane passing through the Sun and observer's zenith, and the local meridian plane, measured on the Earth's horizontal surface. This angle expresses deviation of the horizontal projection of the beam direction from the south; if the projection is due south, the azimuth value is zero, the eastern deviation (from the south) is considered negative, and the western is positive. Therefore, the azimuth of −90° corresponds to the east, +90° to the west, and ±180° to the north.

Angular parameters determining the Sun's position in the sky in reference to an observer on the Earth's horizontal surface are presented in Figure 2.2.

The observer's position on the discussed surface of the Earth is described by geographic coordinates. Value significant for determining solar radiation availability is the latitude φ (i.e., angular coordinate of the location in question in reference to the equator; positive values in the north).

Usually, surfaces of a building envelope, which are the main focus of this publication, are variously situated (i.e., tilted against the horizontal surface and

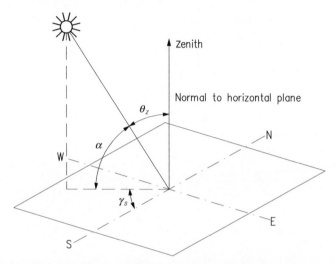

FIGURE 2.2   Key angles describing position of the Sun in reference to an observer.

variously oriented against geographic directions). Tilt is described by the tilt angle β against the horizontal plane, which is the angle between the discussed surface (plane) and the horizontal plane. Tilt angle may theoretically vary in the range of $0 \leq \beta \leq 180°$. Geographic orientation related to the position of the surface in reference to the Sun is described by the azimuth angle γ of that surface. It is an angle between the plane normal to the discussed surface and the plane of the local meridian. The azimuth may vary in range of $-180° \leq +180°$ (negative values to the east, positive to the west, $0°$ to the south). Inclination and orientation of a surface within a building envelope described by the tilt angle and azimuth angle is constant over time. Value variable over time is the angle of incidence, the θ angle, on the discussed surface of the envelope. This is the angle between the direction of incident beam radiation and the direction normal to the surface.

Figure 2.3 presents the geometric relations between the incident solar radiation and a surface receiving radiation, which may be any surface of a building envelope. In the figure, the discussed surface is tilted at the angle β to the horizontal Earth's surface and oriented toward the equator with the deviation of the azimuth angle γ. Normal to the horizontal Earth's surface and normal to the discussed tilted surface are also drawn. Figure 2.3 also presents the beam radiation incident angle θ on the surface and the zenith angle θ_z equal to the angle of beam radiation incidence on the horizontal plane. Position of the Sun in reference to the observer is described by the solar azimuth angle γ_s and the solar altitude angle α_s.

In winter, differences between the daily solar irradiation reaching surfaces located at various latitudes are considerable and, in extreme cases, can even

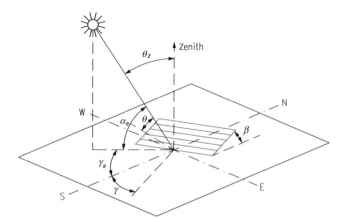

FIGURE 2.3 Key relations between the Sun's position and a tilted irradiated surface.

reach $25\,\mathrm{MJ/(m^2 day)}$. Differences in irradiation of the horizontal surface during winter and summer at the countries located at higher latitudes (φ of some 50° and more) are also quite large. This primarily results from the variable time of daily solar operation, variable zenith angle, and variable attenuation of the solar radiation in the Earth's atmosphere. For instance, the length of a solar day T_d for the latitude of 48° in summer is twice longer than in winter (16 h against 8 h, respectively). Moreover, the incident angle of beam radiation on a horizontal surface (i.e., the zenith angle) is very big in winter (the Sun's altitude is low). This angle is bigger than 60° and usually at least twice bigger than in summer (the value most desirable for capturing solar energy is 0° incident angle; i.e., radiation along the normal direction). In winter, the Sun is "located" over the southern hemisphere, so the distance through the atmosphere that the radiation needs to cross, which corresponds with the air mass coefficient, is much bigger than during summer. This results with stronger radiation attenuation in the Earth's atmosphere.

The incident angle θ of beam radiation on a freely selected tilted surface of any orientation is expressed as a function of key angular parameters of the Sun's location in reference to the investigated surface, and, according to Benford and Back [5], can be expressed as:

$$\cos(\theta) = \sin(\delta)[\sin(\varphi)\cos(\beta) - \cos(\varphi)\sin(\beta)\cos(\gamma)]$$
$$+ \cos(\delta)[\cos(\varphi)\cos(\beta)\cos(\omega) + \sin(\varphi)\sin(\beta)\cos(\gamma)\cos(\omega)$$
$$+ \sin(\beta)\sin(\gamma)\sin(\omega)] \tag{2.4}$$

$$\cos(\theta) = \cos(\theta_z)\cos(\beta) + \sin(\theta_z)\sin(\beta)\cos(\gamma_s - \gamma) \tag{2.5}$$

In certain specific surface location conditions, Eqn (2.4) may be simplified. In case of a surface facing south or north, the azimuth angle is 0° or 180°, respectively. This means that the last part of Eqn (2.4) may be discarded. For a surface facing south (main orientation of a facade of a building aimed at capturing solar energy in case of low-energy and passive solar buildings) the azimuth angle $\gamma = 0°$ and then the incident angle of the solar radiation reaching that surface may be determined using equation:

$$
\begin{aligned}
\cos(\theta) &= \sin(\delta)[\sin(\varphi)\cos(\beta) - \cos(\varphi)\sin(\beta)] \\
&\quad + \cos(\delta)[\cos(\varphi)\cos(\beta)\cos(\omega) + \sin(\varphi)\sin(\beta)\cos(\omega)] \\
&= \sin(\delta)[\sin(\varphi - \beta)] + \cos(\delta)\cos(\omega)\cos(\varphi - \beta)
\end{aligned} \tag{2.6}
$$

In case of a surface facing north $\gamma = 180°$, then Eqn (2.4) takes a form of:

$$
\begin{aligned}
\cos(\theta) &= \sin(\delta)[\sin(\varphi)\cos(\beta) + \cos(\varphi)\sin(\beta)] \\
&\quad + \cos(\delta)[\cos(\varphi)\cos(\beta)\cos(\omega) - \sin(\varphi)\sin(\beta)\cos(\omega)] \\
&= \sin(\delta)[\sin(\varphi + \beta)] + \cos(\delta)\cos(\omega)\cos(\varphi + \beta)
\end{aligned} \tag{2.7}
$$

In case of a vertical surface ($\beta = 90°$) with any orientation (outer walls of building envelopes), Eqn (2.4) takes a simplified form of:

$$
\begin{aligned}
\cos(\theta) &= \sin(\delta)\cos(\varphi)\cos(\gamma) + \cos(\delta)\sin(\varphi)\cos(\gamma)\cos(\omega) \\
&\quad + \cos(\delta)\sin(\gamma)\sin(\omega)
\end{aligned} \tag{2.8}
$$

Moreover, if a vertical surface of a building envelope is oriented directly toward one of the four cardinal directions, then Eqn (2.4) is further simplified. Thus, in case of a south-facing vertical surface ($\beta = 90°$, $\gamma = 0°$) the angle of incidence may be obtained with equation:

$$
\cos(\theta) = -\sin(\delta)\cos(\phi) + \cos(\delta)\sin(\varphi)\cos(\omega) \tag{2.9}
$$

In case of a vertical surface facing north ($\beta = 90°$, $\gamma = 180°$), east ($\beta = 90°$, $\gamma = -90°$) or west ($\beta = 90°$, $\gamma = 90°$) relation for the angle of incidence may be expressed as, respectively:

$$
\cos(\theta) = \sin(\delta)\cos(\varphi) - \cos(\delta)\sin(\phi)\cos(\omega) \tag{2.10}
$$

$$
\cos(\theta) = -\cos(\delta)\sin(\omega) \tag{2.11}
$$

$$
\cos(\theta) = \cos(\delta)\sin(\omega) \tag{2.12}
$$

A special case is a horizontal surface, where $\beta = 0$. In this case, the angle of incidence of beam radiation θ is equal to the zenith angle θ_z. During the day, when the Sun is above the horizon, this angle must be between 0 and 90°. Using Eqn (2.4), it is possible to determine the angle of incidence (cosine of this angle), which is the solar zenith angle corresponding with:

$$
\cos(\theta_z) = \sin(\delta)\sin(\varphi) + \cos(\delta)\cos(\varphi)\cos(\omega) \tag{2.13}
$$

The sunrise hour angle ω_{rise} for the horizontal surface may be calculated with Eqn (2.13). During a sunrise and sunset, the Sun's zenith angle θ_z for incident beam radiation is a right angle (90°, which means that the solar altitude angle is $0°$ and the Sun is on a horizon line). At the zenith angle of $90°$, the expression on the left side of the equation is equal to zero. The considered equation may be transformed into a form that will allow determining the hour angle of the sunrise ω_{rise} or one for the sunset ω_{set}. This angle (its cosine) after transformation may be expressed as:

$$\cos \omega_{rise} = - tg(\varphi)tg(\delta) \tag{2.14}$$

It may be noted that the hour angle ω_{rise} or ω_{set} for certain (fixed) locations described by certain latitude φ varies as a function of declination, which means changes at the following days of a year.

Just like the sunrise (or sunset) hour angle, also the azimuth angle at sunrise γ_{rise} and at sunset γ_{set} changes over the year. The solar azimuth angle may theoretically vary from $-180°$ to $+180°$. For latitudes between the Tropic of Cancer and the Polar Circle (latitudes from 23.45 to 66.45° N) for the solar day duration (sunrise to sunset) lower than 12 h, the solar azimuth angle is contained in the range of $-90° \leq \gamma_s \leq +90°$. When a solar day is exactly 12 h long, the azimuth angle of a sunrise γ_{rise} and sunset γ_{set} is $-90°$ and $+90°$, respectively. In case of a solar day longer than 12 h, the Sun rises and sets north of the east-west line (for northern latitudes), for example:

$$\gamma_{s\ rise} < -90° \text{ and } \gamma_{s\ set} > +90°,$$

which, therefore, leads to the appropriate variation of the solar azimuth angle during the day:

$$\gamma_{s\ rise} \leq \gamma_s \leq \gamma_{s\ set}$$

This means that the Sun may have a different position in the sky at the same time (e.g., at the same hour) in different days of the year. In order to determine that position, by finding the solar azimuth angle γ_s, the Braun and Mitchel formula [6] may be used. This formula takes into account division of the celestial space into quarters and describes the solar azimuth angle γ_s as a function of the so-called pseudo solar azimuth angle γ_s' (for the first and fourth quadrant) as follows:

$$\gamma_s = C_1 C_2 \gamma_s' + C_3 \left(\frac{1 - C_1 C_2}{2} \right) 180 \tag{2.15}$$

where:

$$\sin \gamma_s' = \frac{\sin \omega \cos \delta}{\sin \theta_z} \quad \text{or}: \quad \tan \gamma_s' = \frac{\sin \omega}{\sin \varphi \cos \omega - \cos \varphi \tan \delta} \tag{2.16}$$

Coefficients C_1, C_2, and C_3 are constant and equal to $+1$ or -1. The coefficient C_1 is equal to either $+1$ or -1, when the hour angle ω for the specified time (or, more precisely, its absolute value) is respectively higher or lower than the hour angle ω_s for which the Sun is located directly at the east or west (i.e., when $\gamma_s = \pm 90°$), which may be expressed as:

$$C_1 = \begin{cases} +1 & if \quad |\omega| < \omega_s \\ -1 & if \quad |\omega| \geq \omega_s \end{cases} \tag{2.17}$$

For example, in high-latitude countries in winter, this coefficient will be always equal to $+1$, while between the equinox days (from March 21 until September 23) in the morning and evening it will be -1, while during the rest of the day it will be $+1$.

The C_2 coefficient is, in fact, constant during longer periods of time (i.e., during the seasons). It is equal to either $+1$ or -1, depending on declination and latitude, according to the following equation:

$$C_2 = \begin{cases} +1 & if \quad \varphi(\varphi - \delta) \geq 0 \\ -1 & if \quad \varphi(\varphi - \delta) < 0 \end{cases} \tag{2.18}$$

It may be noted that, in the countries located north of the Tropic of Cancer, this coefficient will always be equal to $+1$.

Finally, the C_3 value is changing during a day: it assumes a value of -1 before noon and $+1$ at noon and afterward, which is described in relation to the hour angle as:

$$C_3 = \begin{cases} +1 & if \quad \omega \geq 0 \\ -1 & if \quad \omega < 0 \end{cases} \tag{2.19}$$

Determining the Sun's apparent position in the sky is important when analyzing the influence of solar radiation on a building. At higher latitudes in summer, the hour of "sunrise" and "sunset" on a vertical building facade is not the same as the actual time of sunrise and sunset. A southern facade will "see" the Sun for a period much shorter than the duration of daylight.

If the surface reached by the solar radiation is located at the latitude φ, tilted at the angle β, and facing south, then the incident angle of beam radiation may be described with geometric relations the same as for a horizontal surface located at latitude equal to $(\varphi - \beta)$. This may be concluded from the triangle congruence which may be seen in Figure 2.4. Thus, Eqn (2.13) may be transformed into:

$$\cos(\theta) = \sin(\delta)\sin(\varphi - \beta) + \cos(\delta)\cos(\varphi - \beta)\cos(\omega) \tag{2.20}$$

This form of the equation allows us to determine the angle of incidence that is used for creating the Sun path diagrams [1,7].

In order to obtain the best possible insolation conditions, it is necessary to try to locate the investigated surface in such a way that the beam incident direction is perpendicular to that surface, which means that the beam radiation incident angle θ is equal to zero. In practice, this means that the surface used to capture solar radiation should be moving and following the apparent Sun's movement. Such

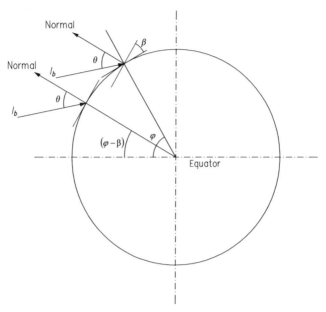

FIGURE 2.4 Geometric relations between the latitude φ, beam radiation incident angle θ, and surface tilt β in relation to the horizontal plane.

solutions utilize continuous or gradual (e.g., every hour) adjustment of position of the surface used to capture solar energy toward the incident radiation. This concept is difficult to implement in case of buildings (envelope surfaces naturally have different orientation in reference to cardinal directions and ground surface). However, in the case of some buildings, the envelope elements do really track the Sun's movement. An example of such a building is shown in Figure 2.5. The figure shows a residential house located in Freiburg and called Heliotrope (building in the background, left side). It has a cylindrical shape and revolves around own axis. Active receivers of solar energy are installed on the roof and these track the position of the Sun in the sky.

Equation (2.20) allows us to determine the incident angle of beam radiation θ on a surface tilted at the angle β in reference to the horizontal plane, oriented toward the south, and located at the northern latitude φ. Using this equation, it is, for example, possible to determine the solar radiation incident angles for a solar noon in extreme insolation conditions (i.e., during summer and winter). For the northern hemisphere, the best and the worst solar irradiance conditions correspond with the declination of $+23.45°$ in summer and $-23.45°$ in winter. For instance, for a latitude of $\varphi = 50°$, the incident angle on a horizontal surface at solar noon is $26.55°$ in summer and $73.45°$ in winter, which clearly proves unfavorable insolation of horizontal surfaces (e.g., roofs) in both seasons. Then, for a vertical surface, the incident angle is 63.45 and $16.55°$ for summer and winter, respectively, which means that, thanks to a tilt, southern

FIGURE 2.5   The Heliotrope, the Sun tracking residential house in Freiburg, Germany.

vertical surfaces (e.g., building facades) have naturally smaller access to the solar radiation in summer and larger in winter. Because of the availability of solar energy, the surface tilt is particularly important in high-latitude countries.

A significant parameter used in solar radiation availability studies is the solar day duration T_d, which is the time difference between the sunrise and sunset. This parameter is calculated from the hour angle of sunrise or sunset described by Eqn (2.14). Taking into account the fact that sunrise and sunset times are symmetrical in relation to the astronomic noon, the solar day duration T_d may be expressed as:

$$T_d = \frac{2}{15}\arccos[-\tan(\varphi)\tan(\delta)] \tag{2.21}$$

It may be noted that the duration of the daytime (with daylight) is a function of declination and latitude. For example, for latitudes beyond the Polar Circle ($\varphi > 66.5°$) Eqn (2.21) yields distinctive results. For those areas, $\tan\varphi$ is higher than one, which, in practice, means that T_d in summer is 24 h, and in winter 0 h (polar night).

Irradiance for any surface on the Earth may be expressed in relation to the radiation reaching the outer boundaries of the atmosphere. As discussed in Chapter 1, irradiance at the outer atmosphere boundaries changes together with the distance between the Earth and the Sun. Variability of this irradiance over a year on a plane normal to the beam direction G_{on} is described by Eqn (1.1). Then, in case of an arbitrarily situated plane beyond the atmosphere, irradiance also depends on the radiation incident angle, for example:

$$G_t = G_{on}\cos\theta = G_{sc}\left(1 + 0.033\cos\frac{360n}{365}\right)\cos\theta \tag{2.22}$$

A special case of irradiance on a plane parallel to the Earth's horizontal surface may be considered. The radiation incident angle on such a surface is equal to the zenith angle θ_z defined by Eqn (2.13). As a result, at a specific time the irradiance on the investigated plane beyond the Earth's atmosphere is:

$$G_o(t) = G_{sc}(t) \cdot \left(1 + 0.033 \cos \frac{360n(t)}{365}\right)(\sin(\delta(t)) \cdot \sin(\varphi)$$

$$+ \cos(\delta(t)) \cdot \cos(\varphi) \cdot \cos(\omega(t))) \tag{2.23}$$

Integration of this equation (irradiance) over a certain period of time allows calculating irradiation for that period. For any horizontal (parallel to the Earth's surface) plane beyond the atmosphere, daily irradiation can be expressed as:

$$H_o = \int G_{on} \cos(\theta_z) dt \tag{2.24}$$

Integration is carried out for a time interval starting at the time of sunrise t_{rise} and ending at sunset time t_{set}. Using Eqn (2.14), describing the sunset hour angle ω_{set} the integral from Eqn (2.24) may be transformed into:

$$H_o = G_{on} \int_{t_{rise}}^{t_{zset}} \cos(\theta_z) d\varpi \frac{dt}{d\varpi} = G_{on} \frac{24}{2\pi} 2 \int_0^{\omega_{set}} (\sin(\delta)\sin(\phi)$$

$$+ \cos(\delta)\cos(\varphi)\cos(\varpi)) d\varpi \tag{2.25}$$

Eventually, the daily irradiation of a horizontal plane beyond the Earth's atmosphere in a location related to a horizontal surface on the Earth surface described by the latitude φ on the day n may be expressed as:

$$H_o = G_{sc} \left(1 + 0.033 \cos \frac{360n}{365}\right) \frac{24}{\pi} [\sin(\delta)\sin(\phi)\varpi_{set}$$

$$+ \cos(\delta)\cos(\varphi)\sin(\varpi_{set})] \tag{2.26}$$

The discussion presented above refers to the radiation reaching the outer boundaries of the atmosphere, which is purely beam radiation. In fact, radiation reaching the Earth's surface is attenuated by interactions with the atmosphere and consists of both beam and scattered components. If the investigated surface on Earth was only reached by beam radiation, then irradiance of that surface could be expressed as:

$$G_b = G_b^* \cos \theta \tag{2.27}$$

The b index refers to the beam radiation, while the asterisk * denotes a plane perpendicular to the beam direction. If the investigated surface is horizontal, then the incident angle θ is equal to the zenith angle θ_z. In fact, however, Earth is reached by both beam and scattered radiation. Daily hemispheric irradiation of the discussed surface H_c is therefore a sum of beam and diffuse irradiation, for example:

$$H_s = \int \left(G_b^* \cos\theta + G_d\right)dt \qquad (2.28)$$

The value of irradiation is, in most cases, specified for a single hour or single day. Meteorology also uses the terms of hourly and daily solar radiation sums. Daily irradiation is obtained through integration of irradiance over a period of time starting at sunrise and ending at sunset. Irradiation may be determined using the measured data. Key data obtained from meteorological stations are hourly irradiation for beam and diffuse radiation.

2.2 SOLAR RADIATION MEASUREMENT

Solar radiation reaching a horizontal surface of the Earth carries certain energy content. Its value is measured by meteorological stations with appropriate instruments [1,3,8,9]. Radiation measurements are carried out with radiometers. Depending on the instrument's design it may measure irradiance or irradiation. There are various types of radiometers, those that are used to measure solar radiation and radiation created by the interaction between the solar radiation and atmosphere and the Earth are briefly described below.

The most popular instrument is pyranometer. It is a radiometer used to measure the irradiance of a horizontal surface. It may measure solar radiation reaching the measured surface (horizontal or tilted). With a special ring to block the beam radiation, it is possible to use the device to measure diffuse radiation only. The ring must be parallel to the equatorial plane. Its position has to be adjusted at regular intervals in order to compensate seasonal changes of the Sun's declination.

Another type of instrument is a spectral pyranometer. It is used for measuring the spectral distribution of total solar radiation (beam and diffuse together). There is also solarimeter, which is a special kind of a pyranometer based on a set of the Moll-Gorczynski thermocouple system.

Pyrheliometers are radiometers using a collimator to measure beam irradiation of a surface perpendicular to the solar beam direction. Its spectral reaction is approximately constant for wavelengths from 0.3 to 3 µm and the aperture angle should be smaller than 6°. Measurement of beam irradiance is related to the so-called field-of-view angle, which is equal to a round angle of a cone with a vortex at the center of the instrument's surface and base circumference at the edge of the aperture.

Beam radiation may also be measured by an active cavity radiometer, which provides an absolute reading [9]. In this device, solar radiation reaching the control surface is absorbed in it, which causes the temperature to rise. Rise of temperature of the control surface is measured in reference to the rise of temperature of identical absorbing surface, which is shaded and heated electrically. The instrument is designed in such a way that absorptivity of its surface is equal to 0.999. The value of solar irradiance G_b^* is then calculated directly from the following relation $\alpha A G_b^* = P_{electr.}$

A pyrradiometer is an instrument used to measure the radiation energy balance (i.e., sum of solar (shortwave)) radiation and the Earth's and atmosphere's radiation (longwave) reaching a horizontal surface from a solid angle of $2 \cdot \pi \cdot sr$. The longwave radiation alone (emitted by the Earth and atmosphere) on a horizontal surface may be measured by a pyrgeometer. The spectral band of this instrument corresponds with the spectrum of longwave atmospheric radiation. Spectral sensitivity of a pyrgeometer strongly depends on the material of the dome(s), which protect the receiving surface.

Another type of instrument used in the meteorology and significant for solar energy utilization is heliograph, a device that records the duration of the period, when the radiation is strong enough to cast clearly visible shadows. The World Meteorological Organization (WMO) recommends to use a threshold value for beam solar radiation of 120 ± 24 W/m^2 [10].

Radiation measurements are recorded with SI units according to the international pyrheliometric scale of the so-called World Radiometric Reference (WRR). The WRR defines a measurement unit of the global irradiation with an error lower than $\pm 0.3\%$. The WRR standard is maintained thanks to operation of a group of most stable and accurate pyrheliometers of different design, which form the World Standard Group (WSG; seven of these devices are in use). WRR value is determined during international calibration as an arithmetical mean of indications of at least three instruments of the WSG. The WRR has been adopted by the WMO and remains in force since July 1, 1980.

Solar radiation, just as all weather-related parameters, is a subject of certain regular deterministic changes. Deterministic character is due to the climate, which, in turn, is driven by geographic location, the Earth's orbiting motion around the Sun, and spinning around its own axis. At the same time, solar radiation is of stochastic character, which is also typical for climatic parameters. Experimental research of systems and devices utilizing solar energy may be performed in natural conditions in annual and multiyear cycles. However, it is difficult to obtain certain repeatability of weather conditions, therefore, equipment designed as solar radiation receivers are often tested in the same standard irradiation conditions created by solar radiation simulators (indoor). An example of such a simulator is shown in Figure 2.6.

Except for irradiation data, the meteorological stations also measure duration of sunshine (i.e., hours of bright sunshine during a certain period of time). Meteorological stations provide measurements of sunshine hours per day. In most cases, the Campell-Stokes method is used for counting the hours. With sufficiently bright sun, a hole is burned in a special card of the measurement instrument. The length of the burned hole during 1 day is then used to determine sunshine hours. This measurement is fairly simple.

FIGURE 2.6   Solar radiation simulator for testing solar radiation receivers (Warwick University, School of Engineering, UK).

Research and measurements utilize Sun-tracking systems. Such a mechanism consists of a rotating base driven by a motor or adjusted manually, which allows keeping the instrument turned toward the Sun. Alignment of the incident solar beam and direction normal to the receiving surface must be maintained.

Also equator-tracking mechanisms are used. Those are equipped with an axis parallel to the Earth's axis, where the variable parameters are: hour angle and declination. Another tracking mechanism is a device following solar altitude and azimuth, a so-called Azimuth-Altitude Dual Axis Tracker. This kind of device tracks the sun using its altitude and azimuth angle as motion coordinates.

In order to determine conditions for long-term operation of solar systems in variable solar irradiation and energy collection conditions, as well as influence of solar radiation on a building as a whole, and on its individual structural components, climate models simulating variability of actual weather conditions are developed and utilized.

2.3 SOLAR RADIATION DATA AND MODELS

As mentioned in the previous section, solar irradiance is a subject to certain regular deterministic variations and, at the same time, is considerably stochastic. Because of that stochastic character of variations, analysis of solar irradiation conditions of surfaces, including surfaces of building envelopes, require data about the radiation preprocessed in a relatively coordinated way by using proper description methodology.

Solar radiation, its measurements, and climate models are covered by many publications, among others [11−26]. Because of varying availability of climate data depending on their amount, accuracy, and form, various climate models are developed. Many climate models are based on so-called models of a typical year, month, or day, without addressing weather anomalies. Such models utilize weather data averaged over many years. The longer the averaging period, the higher the accuracy and versatility of the model. Comprehensive review of developed and utilized solar radiation models has been done by Kambezidis [3], who reviewed models of radiation incident on horizontal and tilted surfaces and evaluated them according to the errors, deviations, coefficient of convergence, etc. In this section, only some models are listed, those from which certain elements for analyzing solar radiation availability will be further used.

Klein [1] when creating a model of "typical" weather for analyzing insolation conditions used 1 day of every consecutive month of a year with an average declination value for the given month. Those days are: 17.01; 16.02; 16.03; 15.04; 15.05; 11.06; 17.07; 16.08; 15.09; 15.10; 14.11; and 10.12. Models of typical weather used for example for performing simulation calculations of processes occurring at various solar receivers and systems have also been created by other researchers: Anderson [12], Lund [16], Miguel and Bilbao [20]. A stochastic weather model for solar energy systems has been developed by Sfeir [25]. There has also been some research on using mathematical statistics methodology for developing climate models involving a very high number of meteorological variables and utilization of probabilistic methods. Currently, a continuously updated model of typical meteorological year [14] is often used, both for analyzing systems' receivers and buildings [4]. Another model also often in use is the test reference year [27].

If the values for individual measurement components of radiation reaching a horizontal surface on the Earth are not available, it is possible to use correlation equations enabling to estimate those components, which are based on the clearness index of the atmosphere. Those estimative methods describe relations between the diffuse radiation and global radiation incident on a horizontal surface and the hourly clearness index K_T [1]. If global hourly irradiation values are not available, daily values may be used. Correlations in the form of $H_d/H = f(K_T)$ have been proposed by many researches like Liu and Jordan [28], Collares-Pereira and Rabl [29], and Erbs [30]. Erbs has noted that during the winter season the correlated function $H_d/H = f(K_T)$ deviates slightly from its shape during the rest of the year. In winter, the share of diffuse radiation at higher values of clearness K_T is lower than during other seasons. According to the Collares-Pereira's and Rabl's formula, at the clearest atmosphere and air mass $AM_X = 1$, the clearness index K_T is approximately equal to 0.8. This means that for perfect sky clearness conditions share of the beam radiation is 80% and of the diffuse radiation is 20%. At very cloudy skies, when the clearness index value is low ($K_T \leq 0.17$), the total radiation incident on the horizontal surface on Earth is practically only composed of diffuse radiation (99%).

Application of specific models depends primarily on the type of available measurement data describing irradiation and required accuracy of performed analyses and calculations.

It should be mentioned that, currently, there is a strong interest in forecasting hourly solar radiation sums in a very short run (i.e., on a base of the hourly sums that have just passed to predict the following hourly sums on the same day). This is caused by rapid development of photovoltaics (PV) technology and integration of PV panels with a grid. Energy utilities are interested in receiving information on electricity output that is just about to be produced (i.e., in a few hours of the same day).

2.4 DETERMINING SOLAR IRRADIATION OF ARBITRARILY SITUATED SURFACES

2.4.1 Radiation Components

Hemispheric solar radiation reaches any surface from the entire celestial hemisphere (i.e., from a solid angle of $2 \cdot \pi \cdot sr$). Hemispheric radiation received by a horizontal surface is known as global solar radiation (EN ISO 9488 [8]). This radiation consists of beam radiation (i.e., radiation that arrives from a small solid angle protruding from the Sun's disk), and from the radiation scattered by the Earth's atmosphere, which reaches the surface from the entire celestial hemisphere. Then the hemispheric solar radiation reaching the tilted surface consists of beam radiation, radiation scattered by the atmosphere, and also radiation reflected from the surrounding surfaces and objects.

Determining the direction of incidence of scattered radiation is a complex task. It may be assumed that the scattered radiation consists of three components.

- Isotropic diffuse (scattered by atmosphere) radiation arrives from all the directions of the hemisphere at a uniform rate. It means that solar diffuse irradiance does not depend on the direction it comes from.
- Circumsolar diffuse radiation is a result from scattering solar radiation in the atmosphere (forward scattered) close to the beam radiation direction. It is assumed that it is concentrated within a theoretical cone with the vortex at the Earth's surface under consideration, whose height runs along the direction of the beam direction propagation.
- Horizon brightening is concentrated close to the horizon. It can be observed particularly intensively during a clear sky.

As it has been already mentioned, irradiation is defined as energy of solar radiation incident on the elementary surface obtained by integrating irradiance over a certain period of time (e.g., an hour or a day). The mathematical description of the hourly irradiation I_s, taking into account individual components of the solar radiation incident on an arbitrary tilted surface, takes a general form of:

$$I_s = I_{t,b} + I_{t,d,iso} + I_{t,d,cs} + I_{t,d,hz} + I_{t,r} \qquad (2.29)$$

Individual symbols refer to a tilted surface (lower index t) and denote:

$I_{t,b}$ = hourly beam irradiation;
$I_{t,d,iso}$ = hourly diffuse isotropic irradiation;
$I_{t,d,cs}$ = hourly circumsolar diffuse irradiation;
$I_{t,d,hz}$ = hourly horizon brightening diffuse irradiation;
$I_{t,r}$ = hourly reflected irradiation.

Then in reference to a single day, daily irradiation $H_{t,s}$ of the discussed tilted surface is described by the equation:

$$H_{t,s} = \int\limits_{rise}^{set} \left(I_{t,b}(t) + I_{t,d,iso}(t) + I_{t,d,cs}(t) + I_{t,d,hz}(t) + I_{t,r}(t) \right) dt \qquad (2.30)$$

There are a number of various mathematical models of varying complexity that allow determining the solar irradiation of a tilted surface. Those models differ primarily in assumptions concerning the nature of the diffuse radiation.

2.4.2 Isotropic Radiation Model

The oldest models of solar radiation incidents on arbitrary surfaces [1] assumed that beam radiation is a dominant component in the global solar radiation, whereas the diffuse radiation concentrates near the direction of the incidence of beam radiation. Calculations did not differentiate individual components of radiation and the hemispheric radiation was determined from the global radiation incident on a horizontal plane adjusted with a coefficient referring only to beam radiation.

In 1942, Hottel and Woertz [31] proposed a model of the hemispheric radiation incident on a tilted surface assuming that the diffuse radiation is isotropic (in 1955, Hottel and Whiller modified it further [32]). This model did not take into account any correction coefficient for diffuse radiation. For a tilted surface, it only took into account the diffuse radiation incident on horizontal planes. Radiation corrections were only introduced for beam radiation. The first method for calculating the hemispheric radiation incident on a tilted surface, which did take into account correction of all radiation components, was proposed by Liu and Jordan in 1963 [33]. They assumed an isotropic model for diffuse radiation and did not take into account the circumsolar and horizon brightening diffuse solar radiation.

According to the Liu-Jordan theory, the hourly solar irradiation I_s incident on a surface tilted at an angle β (in reference to horizontal plane) consists of three components: beam radiation, diffuse radiation, and radiation reflected by surroundings, as illustrated in Figure 2.7, and described by the equation:

$$I_c(t) = I_b(t)R_b(t) + I_d(t)R_d + (I_b(t) + I_d(t))\rho_o R_r \qquad (2.31)$$

where, except for the symbols defined before:

$\rho_o =$ surface (ground) reflectance;
$R_b =$ correction factor for beam radiation;
$R_d =$ correction factor for diffuse radiation;
$R_r =$ correction factor for reflected radiation.

Correction factors for diffuse radiation and reflected radiation are functions of only the tilt angle of the irradiated surface β (in reference to the horizontal plane). Factor values may be found with the following equations:

$$R_d = \frac{1 + \cos(\beta)}{2} \qquad (2.32)$$

$$R_r = \frac{1 - \cos(s)}{2} \qquad (2.33)$$

Correction factors for diffuse radiation R_d and reflected radiation R_r may be derived intuitively. No reflected radiation reaches a horizontal surface, therefore, for such a surface, the R_r value should be zero. For the same surface the diffuse correction factor R_d should be equal to one, as a horizontal surface is

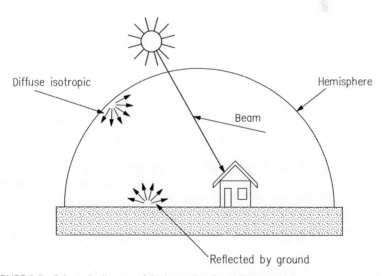

FIGURE 2.7   Schematic diagram of the isotropic solar radiation model.

"surrounded" by the sky (it only "sees" sky), from which diffuse radiation arrives at uniform rates. Then a vertical surface "sees" half of the sky and half of the ground, which means that both factors should be equal to each other with a value of 0.5. These dependencies are described by Eqns (2.31) and (2.32), which show diffuse and reflected radiation correction factors as functions of cosine of the surface tilt angle.

Correction factor for beam radiation R_b represents dependency between the irradiance of beam radiation incident on a tilted surface $G_{t,b}$ and the G_b value for a horizontal plane. The most popular method for determining the value of the correction factor for beam radiation is the Hottel-Woertz formula [33], according to which:

$$R_b(t) = \frac{G_{t,b}(t)}{G_b(t)} = \frac{\cos(\theta(t))}{\cos(\theta_z(t))} \tag{2.34}$$

The correction factor for beam radiation R_b is expressed as a ratio of cosines of two angles, the incident angle θ of beam solar radiation, measured between the beam direction and the normal to the investigated surface (in a general case, determined with Eqn (2.4) or (2.5), and in special cases with Eqns (2.6)–(2.12)), and the zenith angle θ_z, which is the incident angle of beam radiation on a horizontal surface, measured between the beam direction and the normal to a horizontal surface (determined with Eqn (2.13)).

Equation (2.34) shows that the best irradiation conditions exist if the radiation incident angle is zero, which means that beam radiation reaches the surface along the normal to the surface. Taking into account relations describing the zenith angle and incident angle of solar radiation for a tilted surface, Eqns (2.13) and (2.4), respectively, it is possible to obtain a general formula for the correction factor R_b for beam radiation incident on a surface tilted at an angle β to the horizontal plane and with orientation γ:

$$
\begin{aligned}
R_b(t) = & \frac{\sin(\delta(t))[\sin(\varphi)\cos(\beta) - \cos(\varphi)\sin(\beta)\cos(\gamma)]}{\sin(\delta(t))\sin(\varphi) + \cos(\delta)\cos(\varphi)\cos(\omega(t))} \\
& + \frac{\cos(\delta(t))[\cos(\varphi)\cos(\beta)\cos(\omega(t)) + \sin(\varphi)\sin(\beta)\cos(\gamma)\cos(\omega(t))}{\sin(\delta(t))\sin(\varphi) + \cos(\delta(t))\cos(\varphi)\cos(\omega(t))} \\
& \qquad\qquad \frac{+\sin(\beta)\sin(\gamma)\sin(\omega(t))]}{}
\end{aligned}
$$

$$\tag{2.35}$$

According to this equation, the correction factor for beam radiation is a function of the location of the investigated surface described by parameters constant over time: latitude φ, tilt angle β, and azimuth angle γ, as well as of parameters variable over time that describe the mutual position of the investigated surface and the Sun: declination δ and hour angle ω, i.e. for example:

$$R_b(t) = f(\varphi, \beta, \gamma, \delta(t), \omega(t))$$

In a special case of a surface facing south (typical orientation of solar radiation receivers in solar energy systems and main facade of passive building envelopes; i.e., for the azimuth angle $\gamma = 0$), the factor R_b for the northern hemisphere will take a form most popularly encountered in publications on solar energy systems:

$$R_b = \frac{\cos(\varphi - s)\cos(\delta)\cos(\omega) + \sin(\varphi - s)\sin(\omega)}{\cos(\varphi)\cos(\delta)\cos(\omega) + \sin(\varphi)\sin(\delta)} \qquad (2.36)$$

Except for the calculation method for determining the beam radiation correction factor as already described, Hottel and Woertz also developed a graphic method based on variability of cosines of the zenith angle and incident angle as functions of the latitude and a difference between the latitude and tilt angle of the investigated surface (presented in [1]).

Meteorological data that are typically available for the investigated area contain daily or monthly irradiation averaged over a few or even more than several years. However, equation for the hemispheric solar irradiance, according to the Liu-Jordan theory, is valid for instantaneous values. If only the averaged values for daily irradiation are available, it is necessary to integrate the formulas describing the beam radiation correction factor R_b over the entire duration of solar operation for which available data are specified. Angle variables in the equation (i.e., latitude φ and tilt β) are always constant, whereas the declination δ is constant for each single day of a year. Only the solar hour angle is a value variable over time (changing every hour). Therefore, the correction factor R_b for a single day may be expressed as:

$$R_b = \int_{t_{rise}}^{t_{set}} \frac{\cos(\varphi - \beta)\cos(\delta)\cos(\omega) + \sin(\varphi - \beta)\sin(\delta)}{\cos(\varphi)\cos(\delta)\cos(\omega) + \sin(\varphi)\sin(\delta)} dt$$

$$= \int_{t_{rise}}^{t_{set}} \frac{\cos(\varphi - \beta)\cos(\delta)\cos(\omega) + \sin(\varphi - \beta)\sin(\delta)}{\cos(\varphi)\cos(\delta)\cos(\omega) + \sin(\varphi)\sin(\delta)} \frac{dt}{d\omega} d\omega$$

$$= \frac{24}{2\pi} 2 \int_{0}^{\omega_{set}} \frac{\cos(\varphi - \beta)\cos(\delta)\cos(\omega) + \sin(\varphi - \beta)\sin(\delta)}{\cos(\varphi)\cos(\delta)\cos(\omega) + \sin(\varphi)\sin(\delta)} d\omega \qquad (2.37)$$

If we introduce a simplifying assumption that the irradiance on a plane perpendicular to the beam direction is constant during a day (in reality, high deviations are in the morning and evening), the numerator and denominator of the expression may be integrated independently, which yields an average daily correction factor for beam radiation:

$$\overline{R}_b = \frac{\sin|\omega_{set}|^* \cos(\varphi - \beta)\cos(\delta) + |\omega_{set}|^* \sin(\varphi - \beta)\sin(\delta)}{\sin|\omega_{set}|\cos(\varphi)\cos(\delta) + |\omega_{set}|\sin(\varphi)\sin(\delta)} \qquad (2.38)$$

where the new symbols stand for:

Ω_{set} = sunset hour angle for a horizontal surface;
$|\omega_{set}|^*$ = sunset hour angle for a tilted or horizontal surface—whichever is smaller; in a general case, those angles (and therefore sunset hours) may be different.

The sunset hour angle for the horizontal plane $|\omega_{set}|^*$ may be expressed as:

$$|\omega_{set}|^* = \min\{\arccos[-\tan(\varphi)\tan(\delta)], \ \arccos[-\tan(\varphi - \beta)\tan(\delta)]\}$$

Correction factors for diffuse radiation R_d and for reflected radiation R_r are constant over time and only depend on the tilt angle β of the investigated surface. Thus, eventually, the formula for the hemispheric solar irradiation to be used with average daily solar radiation sums will take a form of:

$$\sum_{day} G_s \Delta t = \overline{R_b} \sum_{day} G_b \Delta t + R_d \sum_{day} G_d \Delta t + R_r \rho_r \sum_{day} [G_d + G_b] \Delta t \qquad (2.39)$$

where:

Δt is the length of used time interval;
day is the investigated day of a year.

This equation may be expressed in a shorter form as:

$$H_s = R_b H_b + R_d H_d + R_r \rho_r (H_b + H_d) \qquad (2.40)$$

Introduced symbols H, just as before, denote daily irradiation, and according to the indexes b and d refer to beam and diffuse radiation, respectively. Daily irradiation may be calculated as integrals of irradiance between sunrise and sunset.

$$H = \int_{t_{rise}}^{t_{set}} G \, dt$$

Averaged for every month, daily irradiation $\overline{H_s}$ for the hemispheric radiation incident on a tilted surface may be determined with equation:

$$\overline{H_s} = \overline{H_b R_b} + \overline{H_d}\left(\frac{1 + \cos(\beta)}{2}\right) + \left(\overline{H_b} + \overline{H_d}\right)\rho_g\left(\frac{1 - \cos(\beta)}{2}\right) \qquad (2.41)$$

As it has been mentioned, the sunrise and sunset time for a tilted surface (determined with the Eqn (2.14)) may be completely different than the astronomic sunrise time for the relevant latitude and season. In summer, the azimuth of the astronomic sunrise point is lower than $\pi/2$ (measuring from the northerly direction) and, therefore, solar radiation would reach a tilted surface oriented to the south with delay (only after the Sun in the sky crosses the E—W line). However, during winter, the solar exposition of such a surface starts at the same time as the actual sunrise (theoretical sunrise time even occurs earlier). Sunrise and sunset time may be determined from the known duration of a solar day T_d, as described by Eqn (2.21) in the following manner:

$$t_{rise} = 12 - \frac{T_d}{2}$$

$$t_{set} = 12 + \frac{T_d}{2}$$

(2.42)

The isotropic model of diffuse radiation discussed in this section is the most frequently used calculation model. It is thought to be applicable for climates characterized by frequently occurring cloud covers (high share of diffuse radiation in global radiation), yielding relatively high convergence with measured data.

2.4.3 Anisotropic Radiation Model

It is thought [1,34] that application of the isotropic diffuse radiation model in case of a clear sky leads to an underestimation of the hemispheric radiation incident on an arbitrary tilted surface, and the obtained radiation information is not full. In addition, the isotropic model does not take cloudiness level into account. As already mentioned, in the isotropic model, irradiance of diffuse solar radiation incident on a surface is independent from the direction it comes from. In fact, however, the diffuse radiation has a directional character. For that reason, research on anisotropic radiation models has been continuing for many years. One of such models is the HDKR model named so after the names of its developers: Hay, Davies, Klucher, and Reindl.

Hay and Davies [35] have proposed a calculation model for solar radiation incident on a tilted surface, taking into account circumsolar diffuse radiation. They assumed that the diffuse radiation consists of isotropic diffuse radiation and circumsolar diffuse radiation, and the direction of propagation of circumsolar diffuse radiation is identical as that of the beam radiation. Somewhat later, Reindl (with a research team) enhanced the Hay and Davies model with horizon diffuse radiation according to the earlier Klucher theory [17]. This combination created the HDKR model [24].

Share of the circumsolar diffuse radiation in the total diffuse radiation was taken into account by introducing the so-called anisotropy index A_i. This index depends on transmittance (transparency) of the atmosphere for beam radiation and can be determined as:

$$A_i(t) = \frac{I_b(t)}{I_o(t)}$$

(2.43)

The index o in this equation refers to the radiation on the outer boundary of the atmosphere (i.e., the extraterrestrial radiation). With Eqn (1.1), the daily extraterrestrial irradiation I_o may be expressed as:

$$I_o(t) = I_{sc}(t) \cdot \left(1 + 0.033 \cos \frac{360n(t)}{365}\right) \cdot (\sin(\delta(t))\sin(\varphi)$$
$$+ \cos(\delta(t))\cos(\varphi)\cos(\omega(t))) \tag{2.44}$$

Irradiation I_{sc} may be obtained by integration of the extraterrestrial irradiance over the period of time equal to 1 h. In case of a clear sky, values of the A_i index are high. This means that the diffuse radiation is dominated by its circumsolar component. At high cloudiness, if there is no beam radiation, then A_i is zero, and there is only isotropic diffuse radiation (as the share of isotropic diffuse radiation in the total diffuse radiation is $1 - A_i$). Thus, according to Hay and Davis, the diffuse hourly irradiation incident on a tilted surface in the form of isotropic and circumsolar radiation is equal to:

$$I_{t,d} = \left\{ \left[\frac{1 + \cos(\beta)}{2}\right] - A_i \left[\frac{1 + \cos(\beta)}{2}\right] + A_i R_b \right\} I_d$$
$$= \left\{ (1 - A_i) \left[\frac{1 + \cos(\beta)}{2}\right] + A_i R_b \right\} I_d \tag{2.45}$$

Then the hemispheric irradiation of a tilted surface, using Eqn (2.45), may be expressed as:

$$I_s(t) = (I_b(t) + I_d(t)A_i(t))R_b(t) + I_d(t)(1 - A_i(t))\left[\frac{1 + \cos(\beta)}{2}\right]$$
$$+ I_o(t)\rho_g \left(\frac{1 - \cos(\beta)}{2}\right) \tag{2.46}$$

Reindl, when improving the radiation model, introduced a correction factor for horizon diffuse radiation proposed by Temps and Coulson [36], defined by equation:

$$R_{d,hz} = 1 + \sin^3\left(\frac{\beta}{2}\right) \tag{2.47}$$

According to the characteristic of the $\sin^3 x$ function for small and medium tilt angles, influence of the horizon component is low, but its influence considerably grows for high tilts and reaches maximum for vertical surfaces. Equation (2.47) has been subsequently modified by Klucher [17] by introduction of the factor f, which addresses cloudiness level (as a square root of the beam radiation share in the global radiation). The f factor is equal to:

$$f = \sqrt{\frac{I_b}{I}} \tag{2.48}$$

This eventually led to the correction factor for the horizon brightening diffuse radiation:

$$R_{d,hz} = 1 + f \cdot \sin^3\left(\frac{\beta}{2}\right) = 1 + \sqrt{\frac{I_b}{I_b + I_d}}\sin^3\left(\frac{\beta}{2}\right) \tag{2.49}$$

Introduction of the impact of horizon diffuse radiation (by the correction factor as in Eqn (2.49)) at different sky cloudiness (addressed by the factor f) yields the final form of the formula describing hourly hemispheric irradiation of a tilted surface according to the HDKR model. Taking into account the variability of individual parameters in time, this form is:

$$I_s(t) = (I_b(t) + I_d(t)A_i(t))R_b(t) + I_d(t)(1 - A_i(t))\left[\frac{1 + \cos(\beta)}{2}\right]$$
$$\times \left[1 + f(t)\sin^3\left(\frac{\beta}{2}\right)\right] + I(t)\rho_g\left(\frac{1 - \cos(\beta)}{2}\right) \tag{2.50}$$

Figure 2.8 schematically presents three components of the anisotropic diffuse radiation, the beam radiation, and radiation reflected from the ground, incident on a tilted surface of a building envelope.

Another known anisotropic model of the diffuse radiation is the Perez model [21,22,37,38]. This model uses the correction factor F_1 for the circumsolar diffuse radiation, and factor F_2 for the horizon brightness diffuse radiation. The F_1 factor is corrected with two proportionality factors: a connected with the incident angle of the cone of circumsolar radiation on a surface under consideration and b with the zenith angle of this radiation (incident

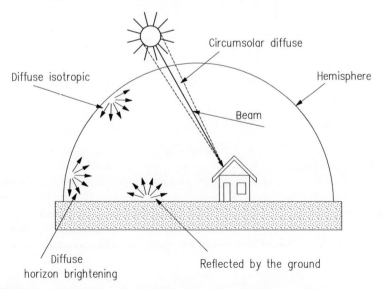

FIGURE 2.8   Schematic presentation of the anisotropic solar radiation model.

angle on a horizontal surface). According to the Perez model, the diffuse irradiation of a tilted surface is thus described by the equation:

$$I_{p,d}(t) = I_d(t)\left\{(1 - F_{1i}(t))\left[\frac{1 + \cos(\beta)}{2}\right] + F_1(t)\frac{a}{b} + F_2(t)\sin(\beta)\right\} \quad (2.51)$$

When analyzing solar thermal collectors, it may be assumed [1] that the a/b is equal to the beam radiation correction factor R_b. Factors F_1 and F_2 describe conditions of the sky. These conditions are determined by three key parameters: zenith angle θ_z, sky clearness index, and the sky brightness index. The zenith angle is determined using Eqn (2.13). The sky clearness index is expressed as a function of the zenith angle, hourly diffuse irradiation, and normal incident beam radiation. The sky brightness index is a function of the air mass, hourly diffuse irradiation, and normal incident extraterrestrial irradiation. The method for calculating the clearness and brightness indexes are provided, for example, by Duffie and Beckman [1].

The HDKR anisotropic model is recommended [39] for analyzing solar energy availability in high-latitude countries, with the dominant share of diffuse radiation in the global solar radiation.

The solar energy analyses commonly use the Liu-Jordan isotropic model. It is thought to yield acceptable results, as underestimation of diffuse radiation in solar system analyzes only acts "in favor" of expected thermal performance of solar systems. The isotropic model may be applied when analyzing small and medium scale solar thermal systems. However, comparison of calculation results of irradiation obtained with the Liu-Jordan and HDKR models clearly show that the application of the isotropic solar radiation model is not sufficient to properly estimate solar energy availability, especially when analyzing and dimensioning big scale solar thermal systems, solar passive systems, and influence of solar energy on the thermal energy balance of a building [40].

2.5 IMPACT OF SURROUNDINGS ON SOLAR RADIATION AVAILABILITY

2.5.1 Shading

Solar radiation availability for investigated surface irradiated with solar radiation depends on the surroundings and location of the surface within its surroundings. Usually, there are many nearby elements of the natural environment, like trees or hills, and also manmade structures, like buildings, walls, lampposts, etc. Those objects may restrict access of solar radiation to the investigated surface at certain times of a day or a year by shading the surface. Shading may be planned when it results from the intentional design of the surrounding elements. In case of a building envelope, shading may be caused by architectural features like overhangs or wingwalls, and other special elements in direct neighborhood (e.g., pergolas).

Therefore, shading [1,41−51], is quite a common phenomenon that may restrict access of solar radiation to the investigated surface in a planned or accidental way. Shading may affect the entire surface or part of it. If the shading obstacle is opaque, it may totally block beam radiation and also part of diffuse radiation (although the diffuse radiation also comes from the remaining part of the hemisphere, not shaded by the obstacle). If the obstacle is partially transparent, then shading is partial (depends on the obstacle's transmittance) for both beam and diffuse radiation.

The surroundings of a building determines solar energy availability and other conditions in which this building will function. When it is necessary, the surrounding and its elements may be appropriately modified to ensure required solar radiation availability and impact.

Shapes of obstacles, which shade building envelope, may be irregular. Some of them may also change in time (e.g., trees moved by the wind). At the same time, some of the obstacles may be totally opaque for solar radiation (e.g., neighboring buildings), whereas others are characterized by certain transmittance (e.g., foliage). Literature [42,43,52−54] provides data on transmittance of different tree species and plants, depending on their dimensions (described for example by height or crown span) or on the place on a tree in question (along the crown axis, at the edge of the crown). Examples of transmittance of broadleaf trees in the winter season (i.e., without leaves) are shown in Table 2.1.

Thought-through arrangement of space around the building allows to consciously limit solar radiation availability by planting proper trees at proper places. Coniferous trees create the same shading throughout the year, whereas

TABLE 2.1 Exemplary Values of Transmittance for Different Species of Broadleaf Trees (without Leaves)

Species	Tree Height (m)	Crown Span (m)	Transmittance, along Axis	Transmittance, Crown's Edge
Norway maple[a]	6.5	6.4	0.34	0.73−0.82
Sugar maple[a]	15	14.0	0.36	0.51−0.68
London pine[a]	10.8	8.5	0.35	0.72−0.73
Young ash tree[b]	5.0	6.5	0.65	0.80−0.88
Black oak[b]	10.0	7.0	0.42	0.59−0.86
Black oak[b]	12.0	7.0	0.45	0.59−0.85
Pecan[b]	14.0	6.5	0.47	0.63−0.77

[a]acc. Montgomery, Keown, and Heisler [53].
[b]acc. Holzberlein [52].

broadleaf trees enable increasing solar radiation availability at the building envelope during fall and winter. Arrangement of the green around a building plays a significant role when planning its surrounding, especially in case of standalone buildings, when the closest neighborhood may be shaped at will of users. The design intended to ensure the best possible exposure throughout the year will be obviously different than in the one intended to create only seasonal (summer) shading [38,50].

The issue of transmittance of different tree species has been researched by Holzberlein [52], Montgomery [53], and Mothloch [54]. Results of their investigation prove that the trees that do not have leaves in winter transmit by average (for the whole tree) 50–70% of the radiation that reaches them. Balcomb [43] has created a model for evaluating shading caused by a row of trees located in front of a shaded object. Burns [43] found out that in countries with more difficult climate conditions buildings equipped with passive systems should not be shaded from the south with any trees, as they restrict solar radiation access so seriously (30–50%). In result, at low irradiance in winter solar passive systems do not operate. At the same time, planting trees from eastern and western sides creates favorable results by ensuring proper shading in summer, while not restricting access to the solar radiation in winter. Shading provided by trees depends not only on the species and age (and as a consequence size), but also on the planting scheme (density) and of course location on the globe, described by the latitude. Not only geographic location, but also climate conditions cause the same tree species to grow differently in different places.

Some elements of a building envelope, in a planned manner or otherwise, cause shading of other elements, especially windows and other transparent facade components. Elements specially designed to provide shading include overhangs and wingwalls, floors, sidewalls of balconies, and loggias. The literature proposes different methodologies for determining shading caused by building envelope elements [1,43,49,55], and also by different construction structures in the vicinity and in the surroundings [48,50,56,57].

One of the methods for determining shading caused by overhangs or other horizontal protrusions has been proposed by Jones [1,43]. This method may be applied if the length of a horizontal protrusion is much bigger than its depth. In such a case, the protrusion is treated as infinitely long, which allows the shading effects of overhang's sides (ends) to be neglected. Jones analyzed shading by introducing two apparent shading planes (plane 1 and plane 2), as shown schematically in Figure 2.9, and referring his discussion to beam radiation.

Beam radiation incident angle θ on the shading plane 1, whose position is described by the azimuth angle γ and inclination $\beta = 90 + \psi$, may be determined with Eqn (2.4) or (2.5). The shading angle ψ for the shading plane 1 may be calculated as $\psi = \text{arc tan } [P/(G + H)]$, whereas the shading angle ψ for the apparent plane 2 is $\psi = \text{arc tan } (P/G)$.

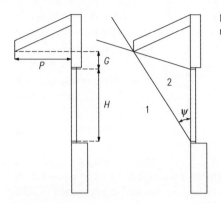

FIGURE 2.9   Shading element—overhang and two apparent shading planes 1 and 2. acc. [1]

Solar radiation reaches the front of a window with a height H and the azimuth angle γ at the time when the Sun is in a position described by the solar azimuth angle γ_s, is at the appropriate altitude α_s. Whether shading will occur or not is determined with a geometric relation describing the angular position of the Sun (γ_s, α_s) on the apparent sky in reference to the position of the investigated surface. The so-called profile angle α_p of the beam radiation incident on the surface under consideration is used to describe this geometric relationship. It changes with time and is described by the following equation:

$$\tan\big(\alpha_p(t)\big) = \frac{\tan(\alpha_s(t))}{\cos(\gamma_s(t) - \gamma)} \tag{2.52}$$

According to Duffie and Beckman [1], the profile angle α_p can be defined as a projection of the solar altitude angle α_s on a plane perpendicular to the investigated irradiated surface. It can be also defined as the angle by which an originally horizontal plane should be rotated around the exposed surface's axis (to create angle DEF in Figure 2.10) in order to contain the Sun's view (beam radiation; as it is done by the surface AEDB (Figure 2.10) that "contains" the

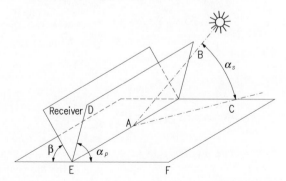

FIGURE 2.10   Graphic interpretation of the profile angle $\alpha_p(\angle \mathrm{DEF})$ in reference to the solar altitude angle $\alpha_s(\angle \mathrm{BAC})$ for the exposed surface tilted at the angle β. acc. [1].

Sun). In a special case, the profile angle is equal to the solar altitude angle, if the Sun is in a plane perpendicular to the exposed surface (for example at solar noon, for $\gamma = 0$ or $\gamma = 180°$).

It may be stated that shading is created if the profile angle α_p is larger than $(90° - \psi)$. In the discussed case, it means that the beam radiation cannot reach the window surface. Dependence of the profile angle α_p from the shading angle ψ allows determining shading of any surface at any time created by an overhang or another horizontal protrusion of a building if they are assumed to be infinitely long (which in practice means that they extend much beyond the window surface).

Utzinger and Klein [55,57] have proposed a method to determine shading, if an overhang has finite length and its outer surface has end edges parallel to the wall surface. It is assumed that an overhang is represented by one surface (its thickness is neglected) perpendicular to the wall. Overhang geometry is described by several dimensions, including projection P, gap G between the upper edge of the window and the contact between the wall and the overhang, and overhang extension lengths beyond the window areas to the left and right, E_L and E_R, respectively. Geometry of a window (or another investigated surface) that may be shaded by the overhang is described by its height H and width W. All those dimensions are shown in Figure 2.11. The scientists proposed considering height H as the characteristic dimension and present all other dimensions of the window and overhang in dimensionless format obtained through dividing their values by H. This yields relative dimensions p, g, e_L, e_R, and w.

Utzinger and Klein [55] have introduced a coefficient f_i describing the ratio of beam radiation on the surface of the window (or other investigated surface) partially shaded with an overhang to the beam radiation incident on the entire surface without shading. This coefficient is equal to the ratio of the area of the

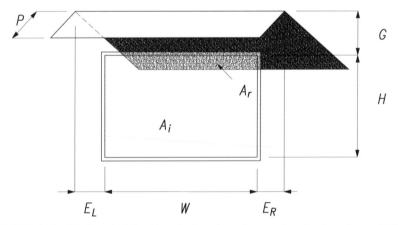

FIGURE 2.11 Window shaded with a horizontal overhang and main dimensions of this system. acc. [1].

window not shaded at the specific moment (time) to the entire window area ($f_i = A_n/A$). Its value changes over time, as the shaded area changes because of apparent movement of the Sun.

Radiation incident on a partially shaded surface is a sum of beam, diffuse, and reflected radiation. Assuming the isotropic diffuse radiation model, and taking into account shading of the window (or another investigated surface) with an overhang (as has been already described) with the coefficient f_i, Eqn (2.31) is transformed into:

$$I_s = I_b R_b f_i + I_d F_{d-s} + (I_b + I_d)\rho_o F_{r-g} \qquad (2.53)$$

In Eqn (2.53), the beam radiation component is multiplied by the discussed coefficient f_i that addresses the effect of shading. The diffuse radiation component is multiplied by the sky diffuse radiation view factor F_{d-s}. In case of a nonshaded surface, it is equal to the correction factor for diffuse radiation R_d (as described by Eqn (2.32)). Utzinger and Klein [55] provided a table of F_{d-s} values for a shaded window surface for different geometries of the overhang and window (described by dimensionless coefficients p, e, and g). For example, if the overhang is placed directly above the window ($e = 0$, $g = 0$), the window is square ($w = l$) and the overhang's protrusions are small ($p = 0.1$ – protrusion equal to 10% window height), the view factor for diffuse radiation $R_{d-s} = 0.46$. If the overhang protrusions are long (e.g., equal to the window's height, $p = 1$), then $F_{d-s} = 0.3$, and if protrusions are twice longer, then $F_{d-s} = 0.27$.

In the described method, determining the view factor for radiation reflected from the ground F_{r-g} neglects radiation reflected from the inner side of the overhang. It is assumed that F_{r-g} is equal to the correction factor for the reflected radiation R_r (calculated from Eqn (2.33)). In case of a window in a vertical building wall ($\beta = 90°$), the correction factor is equal to 0.5.

The coefficient f_i for the beam component of solar radiation changes during a day and during a year. In simplified calculations (for example during conceptual studies on an architectural-civil engineering concept), monthly average values of that coefficient are used. They are calculated as:

$$\bar{f}_i = \frac{\int G_b R_b f_i dt}{\int G_b R_b dt} \qquad (2.54)$$

Using the averaged value of the coefficient f_i and the average daily irradiation, Eqn (2.53) may be formulated for the investigated shaded vertical surface as:

$$\overline{H}_s = \overline{H}\left[\left(1 - \frac{\overline{H}_d}{\overline{H}}\right)R_b\bar{f}_i + \frac{\overline{H}_d}{\overline{H}}F_{d-s} + \frac{\rho_g}{2}\right] \qquad (2.55)$$

Equation (2.55) may be compared with Eqn (2.42), which describes the daily hemispheric irradiation on a tilted nonshaded surface averaged over a month. The so-called K-T method for determining averaged irradiation incident on tilted surfaces may be used. Duffie and Beckman [1] propose to use this method when analyzing radiation availability on shaded surfaces with average monthly irradiation (for global and diffuse radiation).

Utzinger and Klein [55] have published sets of curves representing variation of the average f_i values as functions of relative dimensions of the overhang and exposed surface for various latitudes and for all months of a year. Curve distributions are valid for the surface azimuth angles from $+15°$ to $-15°$. For angles from $+15°$ to $+30°$ and from $-30°$ to $-15°$, certain inaccuracies are observed. If the azimuth angles are beyond the range from $-30°$ to $+30°$, then the radiation availability obtained using these curves is considerably underestimated for summer and overestimated for winter. Utzinger and Klein [55] observed that in case of azimuth angles of an exposed surface larger than $+15°$ and smaller than $-15°$, the side protrusions play a larger role in shading than overhangs themselves.

Shading may be determined using so-called Sun charts. Such charts present the Sun's position (described by the solar altitude angle and the solar azimuth angle) for a specific location (latitude) for every hour of a representative day of every month of a year. Those charts are very useful for designing the building envelope and solar energy systems appropriately to changing the availability of solar radiation, and direct and diffuse circumsolar over entire year. The charts may also be used to design shading by certain envelope elements (overhangs, roofs) or by environmental features (trees, pergolas).

2.5.2 Sun Charts

Diagrams showing the ecliptic of the Sun through the sky have been introduced by Olgyay and Olgyay [49]. Their popularization has been boosted by Mazria [7], who created a methodology allowing the use of them for determination of the beam radiation shading for an arbitrary horizontal and tilted surface turned toward the south. A Sun chart is created using spherical geometry relations describing the Sun's position in the sky in reference to the observer. At the same time, it is possible to couple those relations with specific solar irradiance, which allows the determination of solar radiation availability for investigated envelope surfaces. Through appropriate transformation of the three-dimensional spherical model into a two-dimensional coordinate, it is possible to determine the shape and dimensions of the shaded areas.

It should be noted that shading evaluation is used both when determining restrictions in availability of solar radiation (beam and partially diffuse) for active and passive solar systems, and when determining daylighting (irradiation within the visible spectrum). Visible light is only a part of the solar radiation spectrum. Geometric relations, describing shading of hemispheric solar radiation and its visible part, are identical. Differences concern transmittance of different transparent materials for different parts of the spectrum (wave length), such as spectral properties of materials. If a room overheating is observed during summer, it may be reasonable to use shading for the near-infrared solar radiation (longer than 0.78 μm), while retaining high trans-mittance of the shading element in the visible spectrum (shorter than 0.78 μm).

As already mentioned, position of the Sun in the sky in reference to the observation point on the Earth is described by the solar altitude angle α and its azimuth angle γ_s. If the zenith angle θ_z in Eqn (2.13) is replaced with the relevant solar altitude angle α_s ($\theta_z = 90° - \alpha_s$) and appropriate trigonometric reduction formula is applied, a relation between the solar altitude angle α_s and other key angles of the spherical geometry of the Earth and the Sun may be obtained as:

$$\sin(\alpha_s) = \sin(\delta)\sin(\phi) + \cos(\delta)\cos(\phi)\cos(\omega) \qquad (2.56)$$

Therefore, for an arbitrary time t:

$$\alpha_s(t) = \arcsin(\sin(\delta(t))\sin(\phi) + \cos(\delta(t))\cos(\phi)\cos(\omega(t))) \qquad (2.56a)$$

At solar noon, the hour angle $\omega = 0$ and Eqn (2.56) gets simplified into the form:

$$\alpha_s(t) = \arcsin(\sin(\delta)\sin(\phi) + \cos(\delta(t))\cos(\phi)) \qquad (2.57)$$

Solar altitude angle reaches its maximum at the astronomic noon. In case of a horizontal Earth's surface for a specific latitude (φ) for any day for which declination (δ) is specified (describing angular position of the Sun at noon, as shown by Eqn (2.3)), the solar altitude angle α_s for the northern hemisphere may be expressed as:

$$\alpha_{noon} = 90° - \theta_{z,noon} = 90° - |\varphi - \delta| \qquad (2.58)$$

The solar azimuth angle may be found with Eqns (2.15)–(2.19). Appro-priate transformations yield:

$$\gamma_s = \arccos[(\sin(\alpha)\sin(\phi) - \sin(\delta))/(\cos(\alpha_s)\cos(\phi))] \qquad (2.59)$$

A Sun chart is created for a specific latitude φ and specific day of a year (described by declination δ) as a function of the solar altitude angle α_s (ordinate) and azimuth angle γ_s (abscissa). Usually, the diagram shows curves representing the sun's path on typical days of consecutive months within a year, totally 12 curves. An example of a Sun chart is shown in Figure 2.12. Azimuth angle equal to 0, corresponding with the southerly direction, is marked on the relevant axis. The topmost curve describes the month when the Sun is "traveling" on the sky along the "highest" path, which in the northern hemisphere is June. Correspondingly, the bottom curve is drawn for the month when the Sun is at its lowest, for the northern hemisphere it is December. Dashed lines (slightly deviating from the vertical) that intersect the individual Sun paths are hourly curves (they represent consecutive full hours of a day). The curve denoting solar noon (12 o'clock) is a vertical straight line at the

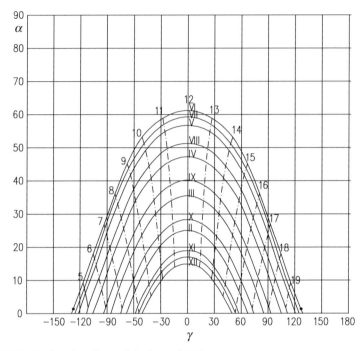

FIGURE 2.12 Sun chart for the latitude $\varphi = 52°$ N.

chart's center (azimuth angle zero). Sun path in the diagram goes left-to-right. Intersection with the x-axis determines azimuth angle of the sunrise (negative values) and sunset (positive values) on typical days of each month (Figure 2.12).

The Sun path for a specific location (latitude) may be used to read the solar altitude angle and azimuth angle for a selected hour and day. This may be useful for determining shading of buildings and solar systems, as described in [1,43−45].

For a given latitude φ and a given day n of a year, the solar altitude angle $\alpha(t)$ and azimuth angle $\gamma_s(t)$ are functions of the hour angle $\omega(t)$, and therefore also of solar time t_s (as defined with Eqn (2.2)). In order to determine shading, apart from the position of the Sun in the sky also the valid surroundings description is needed. Neighboring obstacles may have irregular shapes. They also may be opaque or only partially transparent for solar radiation and those properties may be also variable with time. Neighboring obstacles may be described in cylindrical coordinates. Then those coordinates are spherically projected on the Sun chart.

Layout of the surrounding objects specified in cylindrical systems is sometimes [50] called a polygon area. For a given shading obstacle, distinctive points of its shape are identified (number of points is i) and their coordinates in the chart are specified (α_i, γ_i). Those coordinates define contours of the projected shape. Number of points depends on the complexity of the obstacle's shape and expected calculation accuracy. Contour points of the projected obstacles should be connected in proper order to provide an approximate shape of the projected object. If at time t the apparent position of the Sun in the sky, described by coordinates (α_s, γ_s), is contained within a projected obstacle described by a set of points (α_i, γ_i), then, at that time, the discussed neighboring obstacle shades the observation point. Subsequently, if the Sun at time t is beyond the projected object areas, there is no shading.

Shading fully affects the beam radiation and partially also diffuse radiation (i.e., radiation incident from the sky obscured by the shading obstacle). If f_{sb} denotes the shading factor for beam radiation and τ is the transmittance of the shading obstacle, the f_{sb} factor may be expressed as follows:

$$f_{sb} = \begin{cases} 0 & when \ (\alpha_s, \gamma_s) \supset (\alpha_i, \gamma_i) \\ 1 & when \ (\alpha_s, \gamma_s) \subseteq (\alpha_i, \gamma_i) \\ 1 - \tau & when \ (\alpha_s, \gamma_s) \subseteq (\alpha_i, \gamma_i) \end{cases}$$

According to this description $f_{sb} = 0$, if there is no shading and the Sun's position in the chart is beyond the contours of a shading obstacle, then if $f_{sb} = 1$, shading is full and the Sun's position in the diagram is within the contours of an opaque shading obstacle or at the contour line. The last case applies to the partial shading situation for partially transparent obstacles characterized by transmittance τ.

It may be stated that determining radiation availability as a result of shading is relatively simple for the beam component, and therefore the mathematical model of the shading and its computer simulation are pretty straightforward, although time-intensity of calculations may be fairly high. In case of a partially transparent shading obstacle, the model gets considerably more complex, as usually the transmittance varies over time. Obstacles with partial transmittance include, for example, broadleaf trees, whose transmittance varies not only with seasons (related with changing condition of leaves; i.e., their number and size) but also with tree age. Trees also have variable contours because of their growth. Usually, an analysis is simplified by assuming total shading ($\tau = 0$) in summer, average transmittance ($\tau = \tau_{avg}$) in winter, and in fall or spring certain intermediate values, lower by $\Delta\tau$ from the average value. This difference depends on the tree species and age.

Calculation of the diffuse component of the solar radiation incident on a shaded object is a more complex task. Because of existence of obstacles in the surroundings of the investigated object, the diffuse component of radiation is always attenuated. If there are any shading obstacles nearby, then part of the diffuse radiation incident from the obscured part of the hemisphere will always be blocked. In order to determine the availability of diffuse radiation (lowered in reference to the hemispheric radiation because of presence of solid shading obstacle) it is assumed that the shading area is equal to the projection of the obstacle's shape onto the celestial hemisphere. The method for determining the area shaded by a solid obstacle in reference to diffuse radiation has been proposed by Quaschning [50].

It should be remembered that a tilted surface is also reached by reflected radiation. Shading may also affect this component. Sometimes it is proposed to simplify the analysis of radiation variations by assuming certain arbitrarily lowered value of the surrounding's reflectance. However, this simplification does not guarantee good results. A shading obstacle limits the availability of both diffuse and beam radiation, thus obscuring the sky, ground, and objects located behind it. It may happen that the surface of an obstacle has high reflectance, much higher than the shaded ground, for example, in case of a building with mirror glazing. As a result, the radiation reflected from a highly reflective obstacle surface is more intense than if it was coming from the not obscured area.

Shading analysis is required for design of low-energy buildings. For a specific location (latitude) using Sun charts, it is possible to read the solar altitude and azimuth angle for any hour of any day. This is very useful for designing buildings and solar systems. Sun charts, as already written, may be used to determine local shading on a vertical southern facade that is shaded with natural terrain features (e.g., nearby hills), neighboring buildings, or special architectural features (e.g., overhangs) used for intentional artificial shading. If a building is at its design stage, Sun diagrams allow the analysis of different variants of location of a building, its orientation, shape, and dimensions. If locations close to the ground level happen to be strongly shaded, it may be advisable to plan residential or working rooms on higher floors. More and more frequently, the modern architectural trends emphasize the so-called "rights for the Sun," particularly important in high-latitude countries during winter. Application of Sun charts allows considering a building as an object that may utilize available solar energy, adapt to local conditions, and, if desirable, reduce the energy gains from solar radiation.

As already mentioned, this discussion applies to shaded horizontal surfaces and tilted surfaces turned toward the south. Creating the Sun diagrams for other exposed locations (i.e., for surfaces oriented toward other direction than south), is much more complex. It requires using numerical simulations and drafting multiple diagrams in (γ_s, θ) coordinates (azimuth angle and radiation incident angle on investigated surfaces). Such diagrams are very rarely published in the literature. An example of such a diagram according to [45] is shown in Figure 2.13. The real sunrise and sunset time is pointed out in the figure.

Shading analysis is essential for designing roof-mounted solar systems (thermal and PV). Considerable challenges arise particularly in urban areas, where probability of a roof being shaded by neighboring structures is much higher than outside cities, and, in extreme cases, may even make a collector

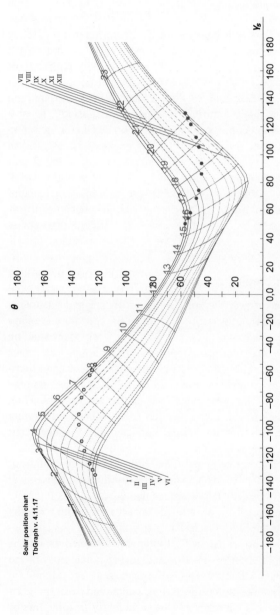

FIGURE 2.13   The Sun chart in (γ_s, θ) coordinates. Surface location: $\varphi = 52°$, $\gamma = 90°$, and $\beta = 45°$ [45].

system unfeasible. Collector systems should not be shaded by neighboring objects constituting opaque obstacles. A detailed analysis based on the Sun charts should always be carried out to estimate whether the actual shading will have an impact on the operation of the solar system and, if so, how large it will be.

It is also possible that there will be some shading with intensity variable in time, caused, for example, by broadleaf trees. During summer, this shading would be practically total, whereas during winter only partial. This kind of shading is usually desirable for buildings exposed to excessive solar radiation during summer. Utilization of Sun charts allows to plan the appropriate plants to create shading during the most sunny periods. Thus, it helps in maintaining thermal comfort inside the rooms. Obtaining information on exposure conditions for various building surfaces is essential in order to design a building with low-energy consumption (i.e., low demand for heat and not requiring cooling in summer). Until recently, optimization of the reduction of energy consumption focused on limiting heat demand only. It might lead to excessive heat gains during summer, because of unnecessary solar radiation gains. As a result, the need for air conditioning can increase quickly.

2.5.3 Reflectance of Elements in Surroundings

Upon reaching the Earth's surface, the solar radiation (beam and diffuse) is partially absorbed and partially reflected. Reflection phenomena are the subject of discussions and publications [1,58,59] less frequently than shading caused by intentional or accidental neighborhood of buildings and other objects of certain reflectance. These elements may, under certain conditions, cause additional gains related to an increased share of reflected radiation in the total solar energy incident on the investigated surface. Apart from reflection from objects located on the ground, also reflection from natural surrounding may occur, including reflection from the ground itself and its cover, and from surface waters, if there are any in the area.

Ground surface reflectance for solar radiation varies, depending on properties of reflecting surfaces, from 0.05 to values close to unity. Most solar energy studies assume the average value of reflectance to be 0.2 [1]. Table 2.2 contains solar radiation reflectance for ground surfaces depending on the surface type (sourced from the literature [60]).

Except for the type of a surface (its reflectivity), the reflected irradiance is also affected by the inclination of the exposed surface. It is possible to introduce special elements with highly reflective surfaces, like aluminum cladding or other reflectors of high reflectivity, which increase reflected radiation quite a lot. However, such reflectors are used for solar collectors of active thermal systems, rather not for building passive solutions (not to cause the glare effect). As it has been already mentioned, in case of using

TABLE 2.2 Examples of Ground Surface Solar Radiation Reflectance [60]

Surface	Solar Radiation Reflectance, ρ
Water at solar altitude $\alpha_s > 40°$	0.05
Water at solar altitude $\alpha_s < 40°$	0.05–1.0
Soil after rainfall	0.16
Dry soil	0.32
Vegetation after rainfall	0.15
Dry vegetation	0.33
Dry asphalt	0.07
Dry concrete	0.35
Fresh snow	0.87
Old snow	0.46

isotropic radiation model, as per Eqn (2.44), the correction factor R_r for the reflected radiation reaches its maximum for vertical surfaces. Lowering tilt results with lowering the share of reflected radiation in hemispheric radiation.

REFERENCES

[1] Duffie JA, Beckman WA. Solar engineering of thermal processes. New York: John Wiley & Sons, Inc; 1991.
[2] Gordon J. Solar energy the state of the art; 2001. ISES position papers, UK.
[3] Kambezidis HD. The solar resources. Comprehensive renewable energy, vol. 3. Elsevier; 2012. 27–83.
[4] Kalogirou SA. Solar thermal systems: components and applications − introduction. Comprehensive renewable energy, vol. 3. Elsevier; 2012. pp. 1–25.
[5] Benford F, Back JE. A time analysis of Sunshine. Trans Am Illumination Eng Society 1939;34:200.
[6] Braun JE. Mitchel: solar geometry for fixed and tracking surfaces. Solar Energy 1983;31:439.
[7] Mazria E. The passive solar energy book. Emmaus, PA: Rondale Press; 1979.
[8] EN ISO 9488 Solar energy − terminology.
[9] Willson RC. Active cavity radiometer. Applied Optics 1973;12:810–7.
[10] WMO. Guide to meteorological instruments and methods of observations. OMM No 8. 5th ed. Geneva, Swizerland: Secretariat of the World Meteorological Organization; 1983.
[11] Anderson B. Solar energy. Fundamentals in building design. New York: McGraw Hill Book Company; 1977.

[12]   Anderson EE. Fundamentals of solar energy conversion. Reading, MA: Addison-Vesley Publ. Co; 1982.

[13]   Coulson K. Solar and terrestrial radiation. Methods and measurements. New York: Academic Press Inc; 1975.

[14]   Gawin D, Kossecka E. Komputerowa fizyka budowli. Typowy rok meteorologiczny do symulacji wymiany ciepła i masy w budynkach [Computational physics in buildings, typical meteorological year for heat and mass transfer simulation in buildings]. Łódź: Politechnika Łódzka; 2002 [in Polish].

[15]   Guemard C. Prediction and performance assessment of mean hourly global radiation. Solar Energy 2000;68(2):285–303.

[16]   Lund H. Test reference years TRY, commission of the European communities. Technical University of Denmark, Department of Building and Energy; 1985. Report no. EUR 9765.

[17]   Klucher TM. Evaluating models to predict insolation on tilted surfaces. Solar Energy 1979;23(2):111–4.

[18]   Kondratyew K. Radiation in the atmosphere. New York: Academic Press; 1962.

[19]   Maxwell EL. METSTAT- the solar radiation model used in the production of the National Solar Radiation Data Base (NSRDB). Solar Energy 1998;62(4):263–79.

[20]   Miguel A, Bilbao J. Test reference year generation from meteorological and simulated solar radiation data. Solar Energy 2005;78(6):695–703.

[21]   Perez R, Ineichen P, Maxwell EL, Seals RD, Zelenka A. Dynamic global to direct irradiance conversion models. ASHRAE Trans 1992;98(1):354–69.

[22]   Perez R, Seals P, Zelenka A, Ineichen P. Climatic evaluation of models that predict hourly direct irradiance from hourly global irradiance: prospects for performance improvements. Solar Energy 1990;44(2):99–108.

[23]   Redmund J, Salvisberg E, Kunz S. On the generation of hourly shortwave radiation data on tilted surfaces. Solar Energy 1998;62(5):331–44.

[24]   Reindl DT, Beckman WA, Duffie JA. Evaluation of hourly tilted surface radiation models. Solar Energy 1990;45(1):9–17.

[25]   Sfeir A. A stochastic model for predicting solar system performance. Solar Energy 1980;25(2):149–54.

[26]   Wang Q, Tenhunen J, Schmidt M, Kolcun O. A model to estimate global radiation in complex terrain. Boundary Layer Meteorol 2006;119(2):409–29.

[27]   Wilcox S, Marion W. Users manual for TMY3. Data sets. Colorado: National Renewable Energy; 2006.

[28]   Liu BY, Jordan RC. The interrelationship and characteristic distribution of direct, diffuse and total solar radiation. Solar Energy 1960;4(3).

[29]   Collares-Periera M, Rabl A. The average distribution of solar radiation – correlation between diffuse and hemispherical and between dail and hourly insolation value. Solar Energy 1979;22:155.

[30]   Erbs DG, Klein SA, Duffie JA. Estimation of the diffuse radiation fraction for hourly, daily and monthly – average global radiation. Solar Energy 1983;28:293.

[31]   Hottel HC, Woertz BB. Performance of flat plate solar collectors. Trans ASME 64/91; 1942:91–104.

[32]   Hottel HC, Whiller A. Evaluation of flat plate solar collector performance. Trans Conf Use Solar Energy, II Arizona October 31–Noveber 1, 1955;2:74–104. Tucson.

[33]   Liu BY, Jordan RC. The long term average performance of flat-plate solar energy collectors. Solar Energy 1963;7:53.

[34]   Veziroglu TN, editor. Solar energy and conservation. Oxford: Pergamon Press; 1978.

[35] Hay JE, Davies JA. Calculation of the solar radiation incident on an inclined surface. In: Proceedings first Canadian solar radiation data workshop. Ministry of Supply and Service Canada; 1985. p. 59.

[36] Temps RC, Coulson KL. Solar radiation incident upon slopes of different orientations. Solar Energy 1977;19:179.

[37] Perez R, Stewart R, Arbogast C, Seals R, Scott J. An anisotropic hourly diffuse radiation model for sloped surfaces − description, performance validation, and site dependency evaluation. Solar Energy 1986;36. 481−98.

[38] Perez R, Seals R. A new simplified version of the Perez diffuse irradiance model for tilted surfaces. Solar Energy 1987;39:221−31.

[39] Karlsson J. Windows − optical performance and energy efficiency. Upsala: Acta Universitatis Upsaliensis; 2001.

[40] Chwieduk D. Modelowanie i analiza pozyskiwania oraz konwersji termicznej energii promieniowania słonecznego w budynku [Modeling and analysis of thermal conversion of solar energy in a building]. PRACE IPPT; 11/2006 [in Polish].

[41] Aitken DW. Frank Lloyd Wright's "solar hemicycle" revisited: measured performance following thermal upgrading. In: The 17th national passive solar conference, vol. 17; 1992. pp. 52−7.

[42] Akbari H, Pomerantz M, Taha H. Cool surfaces and shade trees to reduce energy use and improve air quality in urban areas. Solar Energy 2001;70(3):295−310.

[43] Balcomb JD, editor. Passive solar buildings. Cambridge, Massachusetts: The MIT Press; 1992.

[44] Chwieduk D. Zacienianie budynków. Wykorzystanie diagramów drogi Słońca przy określeniu zacienienia [Shading of buildings. Applying of the sun charts to determine shading]. Polska Energetyka Słoneczna 2004;(2−4):18−22 [in Polish].

[45] Kazmierski S. Szacowanie osłabienia promieniowania słonecznego na skutek zacienienia dzięki wykorzystaniu wykresów pozycji Słońca [Estimation of reduction of solar irradiance in result of shading through application of the Sun Charts]. Polska Energetyka Słoneczna 2011;(2−4):5−16.

[46] Humm O. NiedrigEnergie und PassiveHauser. Okobuch Verlag: Staufen bei Freiburg; 1998.

[47] Kapur NK. A comparative analysis of the radiant effect of external sunshades on glass surface temperatures. Solar Energy 2004;77:407−19.

[48] Kristl Ž, Krainer A. Energy evaluation of urban structure and dimensioning of building site using iso-shadow method. Solar Energy 2001;70(3):23−34.

[49] Olgyay A, Olgyay V. Solar control and shading devices, Chap. 8. Princeton: NJ Princeton University; 1957.

[50] Quaschning V, Hanitsch R. Irradiance calculation on shaded surfaces. Solar Energy 1998;62(5):369−75.

[51] Yezioro A, Shaviv E. Shading: a design tool for analyzing mutual shading between buildings. Solar Energy 1994;52(1):27−37.

[52] Holzberlein TM. Don't let the trees make a monkey of you. In: Proceedings of 4th national passive solar conference, Kansas City, MO; October 3−5, 1979. p. 416.

[53] Montgomery D, Keown SI, Heisler. Solar blocking by common trees. In: Proceedings 7th national passive solar conference, Knoxville, TN, August 30−September; 1982. p. 473.

[54] Motloch JL, Song KD. Approximating tree transmissivity using visual and image capture/interpretation methods. In: The 15th national passive solar conference, vol. 15; 1990. pp. 335−9.

[55] Utzinger DM, Klein SA. A method of estimating monthly average solar radiation on shaded surfaces. In: Proceedings 3rd national passive solar conference, San Jose, CA. Newark, De: American Section of the ISES; January 11−13, 1979. p. 295.

[56] Pereira FOR, Silva CAN, Turkienikz B. A methodology for sunlight urban planning: a computer based solar and sky vault obstruction analysis. Solar Energy 2001;70(3):217−26.

[57] Utzinger M. Sustainable communities: ecological design of the 21st century suburb. In: The 17th national passive solar conference, vol. 17; 1992. pp. 216−21.

[58] Kim J-J. Modeling of reflected solar radiation from adjacent building surface. In: The 17th national passive solar conference, vol. 17; 1992. pp. 14−8.

[59] Tsangrassoulis A, Santamouris M, Geros V, Wilson M, Asimakopoulos D. A method to investigate the potential of south-oriented vertical surfaces for reflecting daylight onto oppositely facing vertical surfaces under sunny conditions. Solar Energy 1999;66(6):439−46.

[60] ASHRAE. Standard method for determining and expressing the heat transfer and total optical properties of fenestration products. BSR/ASHRAE Standard 142P (Public Review Draft). Atlanta, Georgia: American Society of Heating, Refrigerating and Air-Conditioning Engineers (ASHRAE); 1996.

Chapter 3

Shaping Building Envelope with Regard to Incident Solar Radiation

3.1 GENERAL RECOMMENDATIONS

The problem of availability of solar radiation incident on a building envelope is usually discussed in publications on solar buildings, solar architecture, and solar passive systems [1−19], and it is considered with different levels of detail. Anderson's publication [1] is one of the first such studies. However, it needs to be noted that conditions discussed and conclusions presented by him apply mainly to low latitudes (below 40°), and only part of them is general. In the opinion of Anderson and other researchers [1−13,17,18], the best orientation of the main building facade is the southern one, with possible extension to the east or west by up to 30°. Exactly on this side, most windows and other glazed elements should be located. Key observations related to orientation and shape of a building are:

- rectangular building with larger eastern and western facades has poorer ability for using solar energy than a building constructed on a plan of square (during both seasons winter and summer);
- building with a plan of square is not the most effective solution for utilizing solar energy;
- the most appropriate solution for effective solar energy utilization is a rectangular building extended along the east-west direction (i.e., with the largest southern facade).

This general recommendation for the most effective building orientation for solar energy utilization (i.e., facing its main facade to the south, in the northern hemisphere) combined with extending it along the east-west axis, is also indicated by ASHRAE standards [14], Kalokotsa, and others [18]. Moreover, the standards recommend keeping the ratio between the southern wall and eastern/western walls as 2:1, although, in Anderson's opinion [1], the preferred ratio is 3:2.

Buildings designed to have very good solar radiation availability sometimes have a well-developed southern part in a form of terraced shape (stepped

Solar Energy in Buildings. http://dx.doi.org/10.1016/B978-0-12-410514-0.00003-7

FIGURE 3.1   National Renewable Energy Laboratory, Boulder, Colorado, USA.

design). There are a number of terraces with large window panes. The rest of the building is usually opaque. An example of such a building is shown in Figure 3.1, it is the National Renewable Energy Laboratory in Boulder near Denver, Colorado.

In the 1980s, it was thought [9] that altering the shape of the southern facade from flat into an almost cylindrical one would not boost the solar energy gains. However, in the early 1990s, a clear tendency to create such curved main facades appeared in global architecture of low-energy buildings, particularly in high-latitude countries. Figure 3.2 presents an example of such a facade.

FIGURE 3.2   A low-energy building, Environment Center, Cottbus, Germany.

The first research studies and demonstration projects for solar buildings were carried out primarily in the United States. As a result, analyses for solar energy availability mainly focused on the North American geographic conditions (below the latitude of 40° N). Recommendations for shaping the building envelope with regard to incident solar radiation formulated at that stage are therefore naturally different from recommendations appropriate for buildings located at high latitudes. Moreover, some researchers [9–11] thought that differences in wall tilt were insignificant for solar radiation availability. They claimed that irradiation differences between the wall most irradiated during a year (i.e., the wall appropriately tilted and oriented to the south), and a vertical south wall would not exceed 15%. However, in higher latitudes, those differences may be considerably larger. Actually, they grow together with the latitude. For example, for the latitude of 52°, the difference is around 40% (as shown in Figure 3.3).

In late 1980s, Sodha et al. [11] formulated recommendations for energy efficient room placement inside a building. A room layout was planned according to its function. It was pointed out that rooms occupied during daytime (e.g., living rooms) should be located at the south, whereas bedrooms and store rooms should be moved to the north. It was recommended to locate the kitchen at the first floor level (at the north) to maximize utilization of internal heat gains from the kitchen appliances. Also, a need of shading of the south facade was highlighted. Particularly important was blinding the windows to avoid overheating rooms during summer. External shading elements of a building structure placed over the window (e.g., overhangs), were proposed as the most effective form of a building shading element. Those elements should be designed in such a way as to block beam radiation during summer, while letting it reach the window in colder months. A number of fundamental recommendations concerning such shading elements were formulated. The most important are:

- the necessity of using external shading elements, which are more effective than internal ones (note: this is always true, regardless of latitude);
- external shading elements should be in dark color, as it increases shading effects (note: this recommendation is at least disputable, it is better to use light-colored shades of higher reflectivity; while the glazing itself may be dark as increased absorptivity decreases transmissivity);
- appropriate use of external shades allows reducing the beam solar irradiation, which could reach the interior of a building by even more than 90% (note: of course it is possible and often used, but usually also related to restricting window view).

It has also been pointed out that other elements of direct surroundings should be utilized to reduce energy consumption of a building [11–19,21–25]. Except for solar energy considerations, prevalent wind direction and typical wind speeds also play a crucial role. Depending on a season, high wind speed may be

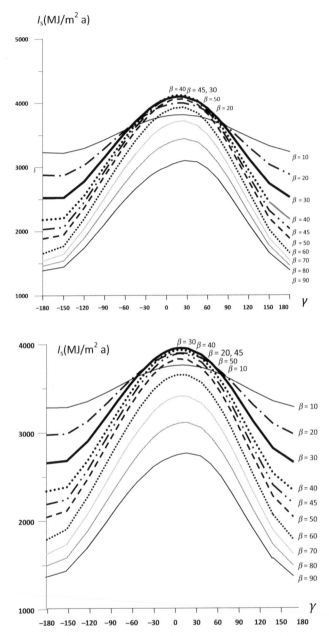

FIGURE 3.3 Comparison of annual hemispheric irradiation obtained with the anisotropic (top) and the isotropic (bottom) diffuse radiation models [20].

desirable or not. During summer, free air movement is beneficial. However, in winter, excessive wind speed increases heat losses from a building. Typically, wind directions during winter and summer are different. This should be taken into account when planning a building's layout. Knowledge of wind directions in different seasons may help to choose and arrange plants to ensure screening building from winter winds, while exposing it to the summer ones. It has been found out [11] that numerous (e.g., four) rows of plants can have better screening effect than a full concrete wall of the same height. Such "live barriers" should be placed at the places of highest wind speeds. Their height should be from two to five times smaller than their distance from the building.

Further, part of this chapter discusses the influence of solar radiation on a building envelope in a high-latitude country and present conclusions concerning recommended shape of such an envelope, as well as the layout of the interior and surroundings.

3.2 DESCRIPTION OF SOLAR RADIATION DATA USED FOR CONSIDERATIONS

Analyses of solar radiation availability and potential of solar energy utilization by solar passive and active elements of an envelope require determining the position of the investigated surface in reference to the sun. Location of a surface on the globe is described with geographic coordinates: latitude and longitude, and position coordinates: tilt (angle to the horizontal surface) and azimuth angle (position in reference to cardinal directions). Solar irradiation of a surface depends on the surface location. Position in reference to the Sun is determined using mathematical relations for spherical geometry of mutual position of the Earth and the Sun, as described in detail in Chapter 2.

Solar radiation data used for discussions in this publication refer to the solar time and represent information on global and diffuse radiation for central Poland (in approximation representative for high-latitude countries). These data have been obtained by arithmetic averaging of measured hourly sums of global and diffuse radiation (by actinometrical station in Bielany, district of the city of Warsaw) [26]. Averaging has been performed on the 30-year set of data (1971–2000). In the first step of averaging, the hourly sums of global and diffuse radiation were averaged for all days of each month of every year. This yielded average hourly irradiations of one averaged day of every month for every year of a 30-year period. Then, one averaged day with hourly irradiation of every month of the averaged year (averaged by 30 years) was created. This averaged data model has been used as a base for calculating irradiation of variously located surfaces.

Polish conditions, just like those of other high-latitude countries, are characterized by considerable variation of irradiance in time, both in daily and annual cycles. This situation is illustrated by Figure 3.4 for the averaged days of every month of the averaged year. The highest hourly irradiation in a day is observed

Gs (W/m²)

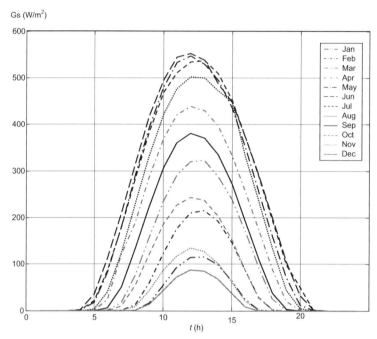

t (h)

FIGURE 3.4   Daily distribution of global solar irradiance for the averaged day of every month of the averaged year for Warsaw, Poland.

between 11 a.m. and 2 p.m. The highest daily irradiation is from June to August. The lowest irradiation occurs between November and January. Figure 3.4 shows also large differences in solar operation hours for different months. Days are the shortest in November, December, and January (approximately 8 h of solar operation per day), and the longest in May, June, and July (16 h).

Table 3.1 shows the averaged daily irradiation for global and diffuse radiation for every month of the averaged year. Characteristic high share of diffuse radiation may be noted. By average, the share of diffuse radiation over a year is around 55%. The lowest share of some 40–45% is in summer, and the highest of 80% is in December. Diffuse radiation dominates just after sunrise and just before sunset, regardless of the season (this cannot be seen in Table 3.1).

Radiation data presented in Table 3.1 apply to a horizontal surface and may be used to calculate irradiation of tilted surfaces of any orientation. It needs to be noted that during an analysis of solar radiation availability on a building envelope, the entire year should be considered (not only the heating season, as it is used to be in high-latitude countries). Any building is exposed to solar radiation influence during the entire year. Thus, knowledge of irradiation in all seasons is important for the design of the building envelope and internal layout of the building, regardless of its geographic location.

TABLE 3.1 Averaged Daily Sums of Global Radiation H_s and Diffuse Radiation H_d of Every Month of the Averaged Year (Actinometrical Station Bielany)

	I	II	III	IV	V	VI	VII	VIII	IX	X	XI	XII
H_s, kJ/m²	2091	4508	8058	12,766	17,395	18,566	17,789	15,473	10,019	5567	2529	1491
H_d, kJ/m²	1508	2734	4651	6704	8349	9421	8897	7407	5229	3183	1766	1187

3.3 COMPARISON OF RESULTS OF CALCULATION OF SOLAR IRRADIATION ON SURFACES DIFFERENTLY SITUATED USING ISOTROPIC AND ANISOTROPIC SOLAR RADIATION MODEL

Analysis of availability of solar radiation incident on an arbitrary tilted surface for location discussed in this section (of this chapter and in the next one) has been performed using computer simulation. Solar input data, used for calculation of irradiation of variously situated surfaces, are averaged hourly irradiations of global and diffuse radiation for averaged days of every month of the averaged year (methodology for averaging of solar data is described briefly in Section 3.2). Calculations have been performed using two solar radiation models: the Liu-Jordan isotropic model and the HDKR anisotropic model. Irradiation calculations have been performed for the whole range of inclinations β and azimuth angles γ. For the isotropic radiation model, Eqns (2.1)–(2.19) have been used, whereas for the anisotropic model calculations have been based on Eqns (2.57), (2.58), (2.61), and (2.62).

Simulation study results obtained with both solar radiation models: isotropic and anisotropic, have been compared. Comparative analysis of variations of averaged hemispheric annual irradiation as a function of the azimuth angle and the surface tilt (slope) (obtained with both models) confirms a well-known general rule: the use of an isotropic model results with underestimation of the diffuse irradiation, as illustrated in Figure 3.5.

Calculations with the anisotropic model have shown that the highest annual irradiation is obtained for the surface inclined at $\beta = 40°$ and oriented at $\gamma = +15°$, whereas with the isotropic model the maximum is observed for $\beta = 30°$ and $\gamma = +15°$. The difference is 270 MJ/(m^2 year). In case of a small scale active solar system (e.g., for a single family house) with a solar collector area of some 10 m^2 the difference would be small. However, for a large scale system it could be considerable. For example, for the total collector area of 1000 m^2, it would be 270 GJ/yr.

Figure 3.3 presents a comparison of the annual hemispheric irradiation calculated for selected slopes, from 10° to 90°, as a function of the azimuth angle γ (full range from $-180°$ to $+180°$).

FIGURE 3.5 Comparison of annual solar irradiation, calculated with two diffuse radiation models, in a function of the tilt angle (slope) of the exposed surface.

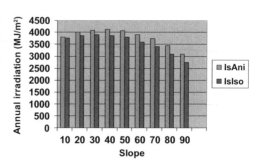

Figure 3.3 shows that the anisotropic model usually yields higher irradiation for identically situated surfaces. The largest differences between the annual hemispheric irradiation are observed in case of vertical surfaces. When the slope lowers, so do the differences in irradiation. It turns out that, in case of surfaces oriented toward directions from northeast to northwest, the irradiation obtained with the isotropic model is slightly higher (in case of vertical surfaces both models give the same results).

Results of the average annual irradiation of surfaces of different inclination and orientation are of very general character and may be used primarily for estimating insolation conditions of active solar systems. For passive solar systems, annual irradiation does not reflect the actual nature of solar radiation.

Analysis of calculation results obtained with both models in reference to variations of averaged hemispheric daily irradiation (as a function of the surface slope and orientation) for every month of the averaged year allows drawing conclusions, which confirm the main findings from comparing annual irradiations. More detailed conclusions may be formulated after analyzing variations of the averaged daily irradiation, taking into account its components (i.e., direct, diffuse, and reflected radiation). Differences are largest in case of surfaces exposed to the highest beam irradiance, which of course results from the impact of the circumsolar diffuse radiation component (high anisotropy). For the radiation data under consideration (central Poland), the largest differences are observed for surfaces facing south and southwest during the summer with a high anisotropy index A_i and high share of the beam radiation in global radiation (here in May and August). At high slope, especially in case of vertical surfaces, also the share of the brightening horizon diffuse radiation increases. Irradiation of a surface of the building envelope during the summer should therefore be investigated in detail because of possible room overheating. Lowest values of the anisotropy index are observed in winter (in December, then in January), and also during late fall (i.e., in November). For that reason, in winter (December) for small slope the solar irradiation obtained with both diffuse radiation models may be identical.

Results of the comparative analysis of two models under consideration with regard to daily irradiation for 12 months (of the averaged year), taking into account components of solar radiation, allow drawing the following conclusions.

- Isotropic model may be used for:
  - small slope in reference to the horizontal plane;
  - low values of anisotropy factor (<0.1);
  - surfaces oriented from the northeast (through the north) to the northwest, regardless of their inclination and season.
- Anisotropic model should be applied for:
  - large slope, in particular for vertical surfaces;

- not particularly low values of anisotropy index (>0.1);
- surfaces oriented toward east, south, and west (i.e., for azimuth angles from $-90°$ to $+90°$), regardless of their inclination and season.

These findings allow us to conclude that the anisotropic model should be used to analyze the influence of solar radiation on a building, as a building envelope is primarily composed of vertical walls and inclined elements of a roof.

The most detailed conclusions concerning differences between both considered models may be drawn upon a comparative analyses of variations of hourly averaged sums of solar radiation (taking into account their components) for the averaged days of every month of the averaged year. Those conclusions confirm and extend the ones just presented.

The largest differences between the two models are observed for vertical surfaces. In this case, the influence of the isotropic component of diffuse radiation is lowest, while the anisotropic component of the brightening horizon has a considerable impact.

Share of the circumsolar component in the diffuse radiation is highest at that time of day when the beam radiation is able to reach the considered surface (i.e., when the surface can "see" the Sun). The share of the circumsolar radiation grows when the anisotropy index increases. In case of the southeastern and eastern surfaces, the circumsolar radiation share is largest before noon, because of their orientation (when these surfaces can "see" the Sun). At that time of day, the diffuse irradiation is higher in the anisotropic model. Then in the afternoon, the higher diffuse radiation is given by the isotropic model. In case of south-western and western surfaces, the situation is opposite: before noon the diffuse irradiation is lower in the anisotropic model, and in the afternoon it is higher. The isotropic model underestimates irradiance, when the solar radiation is able to reach the surface directly (i.e., exactly when there is a risk of overheating of the building's interior). It turns out that, in case of applying both models for the summer period, differences in hourly irradiation of a vertical wall facing south or southwest for noon or afternoon may reach 20%. Shading of the building, and in particular of its glazing, is at that time essential.

As has been already mentioned, irradiation calculations have been performed for a latitude of 52° N. Obtained results and conclusions are applicable for insolation conditions characteristic of high-latitude countries. In these countries, during summer, the sunrise azimuth angles are much smaller than the exact east direction (lower than $-90°$), and those of the sunset much larger than the west direction (higher than $+90°$). For example, in Poland, in summer the sunrise can be seen at $-135°$ of azimuth angle (i.e., at northeast), and the sunset at $+135°$ of azimuth angle (i.e., at the northwest). This situation causes peculiar exposure of northern walls. For example, during summer, when the anisotropy index is relatively high and the northern wall is able to get the beam radiation, the share of the circumsolar component increases. At small slope (about 20°) a north-oriented surface can "see" the Sun during the entire day of the hottest month. As a result, hourly diffuse irradiation is higher in the anisotropic model. However, it should be noted that the level of the beam radiation is not very high (for most of a day solar radiation incident angles are very high). As the slope angle grows, the time of beam radiation availability gets shorter.

In case of a vertical surface facing north during mornings and evenings, when the surface can "see" the Sun the circumsolar radiation and the brightening horizon radiation make the hourly irradiation considerably higher than those obtained with the isotropic model. However, for the rest of the day, the beam radiation does not reach the surface in question, and it is only exposed to diffuse radiation. For that period and for small and medium slope angles (up to 45°), the isotropic model gives slightly higher irradiation. For larger slopes, both models provide very similar results. For the north vertical surface, the anisotropic model gives slightly higher irradiation (the influence of horizon brightening component).

It should be mentioned that, in summer, the north surface of low and medium slopes is exposed to the beam radiation for the longest time in comparison to other cardinal orientations. Of course, this only applies to the exposure time, not to the irradiation level. As can be expected, in other seasons, the longest exposure is observed on the south facade.

Summarizing, it can be written that the claim that anisotropic model increases values of diffuse radiation is too general, as it usually applies to values summed up over longer periods of time (e.g., for a day). Thus, it should be stated that analysis of solar radiation availability on a building envelope should be performed with hourly resolution. At this level of detail, as it has been just discussed, it turns out that there may be certain periods of time within a day (spans of a few hours) when the irradiation obtained with the isotropic model is actually higher than the irradiation obtained with the anisotropic model. It also needs to be emphasized that application of an anisotropic diffuse radiation model is necessary for irradiation conditions characterized by a high share of beam radiation (over 50%) in hemispheric radiation incident on a surface under consideration, when anisotropy indexes are not low (over 0.2), and when the slope of the considered surface is big (bigger than latitude), and always for vertical surfaces.

General recommendations to select suitable radiation models for analyses of solar radiation availability can be formulated as follows:

- if the analysis concerns operation of an active solar systems with low heat loads, an isotropic model may be used;
- if the analysis concerns preliminary estimation of solar radiation availability and general building concept, an isotropic model may be used;

- if the analysis concerns operation of medium- and large-scale active solar systems, an anisotropic model should be used;
- in case of analyses of solar radiation availability on a building envelope at the building design stage and planning of shading elements, an anisotropic model should be used.

3.4 SOLAR IRRADIATION OF SURFACES DIFFERENTLY SITUATED IN THE CONSIDERED LOCATION

This section presents selected results of numerical simulations of solar irradiation for variously situated (inclined and oriented) surfaces in Warsaw, Poland. Although the results are based on averaged irradiation data for a specific location, they still show certain features characteristic for high latitude countries. Figure 3.4 presents a daily distribution of global solar irradiation (of a horizontal surface) for every month of the averaged year in Warsaw. Of course, as it has been just mentioned, the averaged data do not give a full picture of insolation conditions for the selected location, as they do not reflect sudden changes in irradiance and its extreme values characteristic for stochastic parameters, like solar irradiance. Still, the averaged data well reflects certain tendencies in insolation conditions, which was one of the objectives of the simulation. Averaged data on global and diffuse hourly sums of solar radiation (of a horizontal surface) of every month of the averaged year (described in Section 3.2) were used as a basis for determining irradiation of variously situated surfaces. The HDKR anisotropic model (selection of an anisotropic model was justified in Section 3.3) was applied for calculations.

Thanks to analyzing irradiance distributions for all months of the averaged year, insolation conditions have been grouped into three periods, roughly corresponding to the seasons:

- period of best solar radiation availability: spring-summer (April−August);
- period of the moderate solar radiation availability: winter−spring (i.e., February−March); and summer-fall (i.e., September−October; there is analogy between February and October, and between March and September);
- period of worst (poor) solar radiation availability: late autumn−winter (i.e., November−January).

During the best solar radiation availability period, the most solar exposed are surfaces with small tilt ($\beta \leq 15°$), nearly horizontal and facing nearly to the southwest (azimuth angle close to $\gamma = +45°$). In June, the month with the best insolation conditions, averaged daily irradiation reaches 19 MJ/(m² day). As the tilt increases (in reference to the horizontal plane), the irradiation drops, reaching the lowest values in case of a vertical surface. For the vertical south wall ($\gamma = 0°$) the averaged daily hemispheric irradiation reaches 10.5 MJ/(m² day). It is almost twice as less than the most solar irradiated cases. During the summer, south-facing surfaces with high tilt, especially vertical ones, are irradiated in a relatively even way throughout a day and their daily hemispheric irradiation is considerably lower than that of eastern and particularly western surfaces. Western walls of a building envelope are most insolated throughout a summer.

In June, as well as during other months of the best solar availability period, characteristic high share of the beam component in hemispheric radiation is observed. Increasing tilt results with gradual decrease of hemispheric solar radiation and its beam component, but, at the same time, the share of reflected radiation and horizon brightening diffuse radiation increases. The highest hourly hemispheric irradiation is observed in June around noon, when at low tilts and southerly or south-westerly orientation ($0° < \gamma < +45°$) hourly irradiation exceeds $2\,MJ/m^2$, with the beam component constituting 60% of the total at around $1.2\,MJ/m^2$. With increased tilt angle not only hemispheric irradiation decreases, but so does the time of exposure to beam radiation during the relevant time of day. Therefore, if a building is used only in summer (e.g., summer houses), because of the time of exposure to beam radiation, as well as irradiation levels, in order to restrict availability of solar radiation, and especially the beam component share, it is recommended to turn the main facade of a building toward the east or southeast (not toward the west or southwest).

During the period of the worst solar radiation availability (winter), the best solar exposure is observed on surfaces of high tilt (some 10° larger than local latitude) facing south. In December (the month of worst insolation conditions), average daily hemispheric irradiation of the best situated surface ($\beta = 65°$, $\gamma = 0°$) in Warsaw is around $2.5\,MJ/(m^2\,day)$. Daily irradiation of a southern vertical surface at this period is around $1.9\,MJ/(m^2\,day)$. It is five times lower than in June. Daily global irradiation (of a horizontal surface) is just $1.5\,MJ/(m^2\,day)$ in December. In winter, solar irradiation is sensitive both to the tilt and orientation. Best insolated surfaces are turned toward the south (or slightly turned toward the east or west), as they are exposed to beam radiation during the entire day. In December, as well as during other months of the poorest availability of solar radiation, the irradiation level is very low and so is the number of daylight hours (e.g., in December only some 8 h/day). At this time, beam radiation is incident only for 6 h/day, and this time may be further shortened by certain orientation of a surface (if a surface is too much turned from the south). In case of the most favorable insolation (i.e., at tilt of the exposed surface of around $\beta = 60°$ and south orientation), at noon, the hourly hemispheric irradiation reaches level of $0.5\,MJ/m^2$, including approximately $0.3\,MJ/m^2$ of the beam component (60% of the total). In case of larger tilts, from $\beta = 60°$ up, shares of the brightening horizon diffuse radiation and reflected radiation grow, while the role of the isotropic component of diffuse radiation diminishes. For example, for a vertical surface with a typical reflectance of surroundings ($\rho = 0.2$), the share of reflected radiation exceeds 10%. Role of reflective surfaces intentionally installed nearby may play a noticeable role in increasing availability of solar energy during winter.

During moderate-intermediate periods (spring and fall) the level of hemispheric irradiation and its beam component is relatively high, especially at favorably situated surfaces. The highest hourly hemispheric irradiation is observed at noon. At favorable tilt ($\beta = 45°$) and orientation (due south), this hourly irradiation reaches some 1.7 MJ/m^2, with the beam component of 1.1 MJ. Further increase of tilt to $\beta = 60°$ (or even more, to $\beta = 90°$) gradually reduces solar energy availability (although for such tilts roles of reflected component and brightening horizon component increase, they do not make up for the loss of beam and isotropic diffuse radiation). Eastern and western surfaces are not exposed to beam radiation before sunset and after sunrise, respectively. As the tilt increases, the period when beam radiation is not available gets longer. In comparison to the summer, the period of daytime hours are decreased (12 in September against 16 in June). Knowledge of the beam radiation distribution is particularly important in cases in which its share is relatively high (over 60). Irradiation in moderate-intermediate periods is not as high as in the summer. However, if a building is used seasonally, the design of a building facade should take into account shading for the most exposed surfaces. Main facade should be turned toward the southeast. Then, in order to increase solar energy gains, the southern surfaces with tilt angles several degrees lower than local latitude should be used.

Figures 3.6 and 3.7 present daily distribution of hemispheric irradiance of surfaces differently situated for consecutive hours of the averaged day of every month of the averaged year. Surfaces with orientation most typical for external elements of a building envelope have been selected for further analysis and presentation. These are: vertical surfaces ($\beta = 90°$) (Figure 3.6) and surfaces sloped at $\beta = 30°$ (Figure 3.7), as the potential inclination of roof surfaces, oriented toward the four cardinal directions (corresponding to azimuth angles $\gamma = 0$, $-90°$, $+90°$, and $180°$). Obtained results have been used to draw conclusions of general and particular character. Graphic interpretation enables relatively accurate and transparent presentation of key characteristics of solar energy availability on considered surfaces, as hypothetical elements of a building envelope.

In case of vertical surfaces, the irradiance curves have clear maxima occurring at different times of a day for every cardinal direction. This depends on the period when a surface can "see" the Sun directly (i.e., when it is exposed to the beam radiation and circumsolar diffuse radiation). As it might have been expected, rooms located at the southern side of a building are most irradiated around noon, eastern rooms before noon, and western ones in the afternoon hours.

Particularly distinctive is the distribution of irradiance of a northern wall during summer. As already mentioned in Section 3.3, vertical surfaces are exposed to the beam radiation in the morning and late afternoon (as the Sun

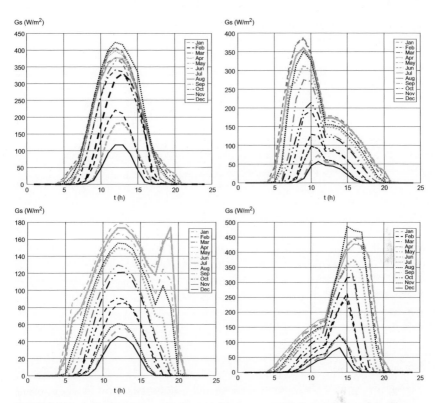

FIGURE 3.6   Daily distribution of solar irradiance on a vertical surface: southern, eastern, northern, and western (starting from the top, left-to-right) of every month of the averaged year.

rises and sets beyond the east-west line). Irradiance levels are particularly high around noon and in the afternoons, causing distinctive sharp rise of irradiance of a vertical northern surface, when in the afternoon it starts to "see" the Sun again (Figure 3.6; irradiance peak during summer around 6 p.m.). Sudden changes in irradiance distribution observed also for vertical eastern and western faces result from a sudden change of exposure to beam and circumsolar diffuse radiation at certain times of a day. During the entire year, the most insolated surfaces are southern and western walls of a building. It turns out that during late spring and summer (May to August), vertical western walls are much better insolated than southern ones. Unfortunately, it is often neglected, even if it is a clear tendency in some climate conditions (e.g., of Central Europe; and not only). This results from higher irradiance in the afternoon and higher share of beam radiation in the afternoon than in the morning. In summer, there are big ambient air temperature differences between night and day. In the late evening and at night at low temperatures, vapor contained in the air condenses and starts to re-evaporate in the morning as the

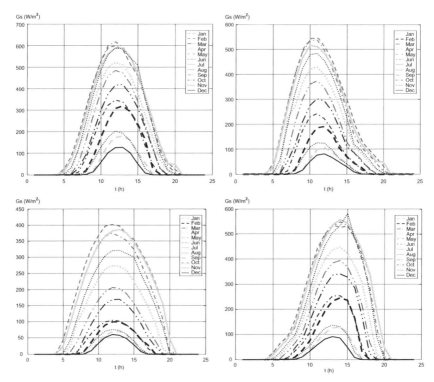

FIGURE 3.7 Daily distribution of solar irradiance on a surface tilted at 30°: southern, eastern, northern, and western (starting from the top, left-to-right) of every month of the averaged year.

temperature rises. This results with frequent morning fogs and mists that block access to the beam radiation. No such phenomena are observed during the afternoons.

As the surface tilt decreases (here to 30°), irradiance variations become smoother. Solar radiation reaches a surface in a more even way throughout a day, as surfaces with low tilt can "see" the Sun for a longer time, when compared to surfaces of high tilt and same orientation (in particular vertical ones). Lowering the tilt angle of the envelope surfaces to 30° increases the period of solar energy availability for all orientations during summer, early fall, and spring. During summer, the best insolated surfaces are those with small tilt angles of some 10°−15°. In winter, however, the best insolation is observed on southern walls with high tilt of 50°−60° (smaller incident angle; as mentioned before sunrise in the winter occurs in the southeast and sunset in the southwest). During late spring and summer (May to August), the irradiance of the southern surface of small tilt (30°) at noon is around 600 W/m². As the tilt increases, the irradiance decreases, and for vertical surfaces during the same period, it may only reach 400 W/m². In case of a vertical western wall, solar radiation availability is better than for a vertical southern one and during summer (August) irradiance may reach 500 W/m².

Analyzing Figures 3.6 and 3.7 allows us to note that the lowest irradiance is observed from November to February, with the lowest irradiation level and shortest daytime in December. In winter, the lowest irradiation is for surfaces with low tilt angles and horizontal ones. At that time, tilt increase results with increased solar energy gains. Irradiance of a surface tilted at 50° may reach 140 W/m² (diagrams for this tilt are not shown), whereas for a horizontal surface it is only 80 W/m². In case of a south-facing vertical surface, the irradiance may reach some 120 W/m², just like in case of a tilt of 30°. In January and November, insolation conditions are better than in December, whereas in February they are much better; then the maximum irradiance is two or even three times higher (depending on the tilt and orientation) than in December. In February, irradiance of vertical surfaces reaches 350 W/m² (compared to 210 W/m² on a horizontal surface). Daily irradiation for every month of the averaged year is represented by the area below the relevant curve.

Solar radiation availability on a surface is considerably affected by the beam radiation component. To illustrate this phenomena following Figures 3.8 and 3.9, present daily distributions of hemispheric radiation and its components for every month of the averaged year for vertical surface facing nearly south ($\gamma = -8°$) and nearly north ($\gamma = +172°$). South and north orientations were selected as significant to demonstrate differences between insolation conditions, taking into account radiation components for totally opposite situated surfaces. Results are presented for vertical surfaces, as they are most typical for building envelopes. Slight shift to the east of the southern surface (in reference to cardinal directions) and to the west of the north surface allows us to present clearly how sensitive the north vertical wall is to slight orientation changes (between March and September) from a point of view of availability of beam radiation.

3.5 IRRADIATION OF AN EXTENDED SOUTHERN FACADE

During general discussion concerning the shape of low-energy building facades in Section 3.1, it was pointed out that it is recommended to use extended curved facades in high-latitude countries. This kind of solution has been analyzed for Warsaw climatic conditions [27] and some results are presented in this section.

During winter in high-latitude countries, solar energy should influence on the energy balance of a building as much as possible in order to reduce the energy demand for space heating. In winter, as already mentioned, sunrise and sunset are characterized by relatively small azimuth angles (i.e., much smaller than those for the east and west direction). Therefore, extension of the main

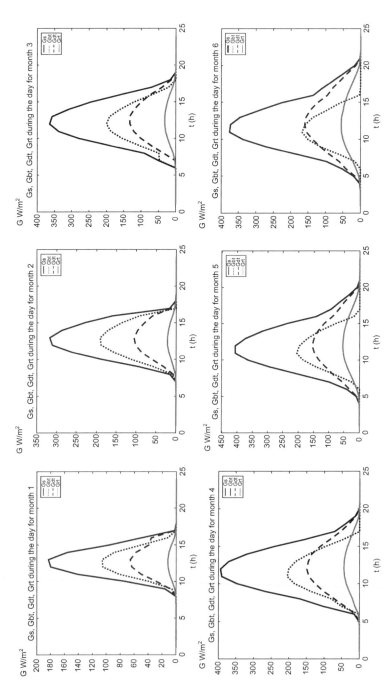

FIGURE 3.8 Daily distributions of hemispheric radiation and its components during every month of the averaged year for a vertical nearly south surface ($\gamma = -8°$).

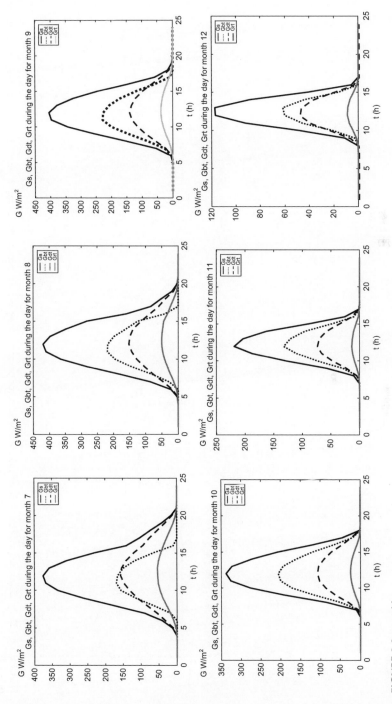

FIGURE 3.8 cont'd

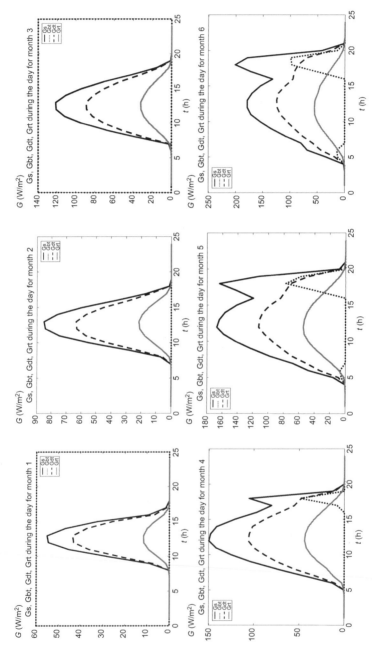

FIGURE 3.9 Daily distributions of hemispheric radiation and its components during every month of the averaged year for a vertical nearly north surface ($\gamma = +172°$).

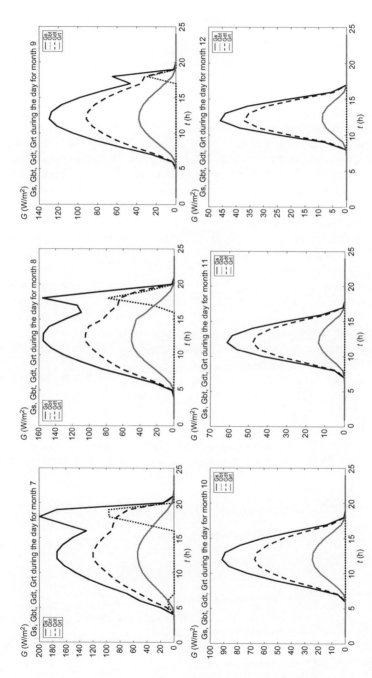

FIGURE 3.9   cont'd

curved facade should be suitable to the azimuth angle of the sunrise and sunset of the month with the worst insolation conditions. In case of geographic latitude of about 50° (for Warsaw it is 52°), the extension of the glazed facade should be at the maximum range of azimuth angles from −50° to +50°, which practically corresponds with directions from the southeast to the southwest (see the Sun chart in Figure 2.12).

It is assumed that the extended glazed facade has a shape close to a flattened ellipsis (i.e., it is curved in a horizontal projection) in order to increase solar radiation availability. Shape of the facade is actually curved by creating a polyline contour of straight sections. Azimuth angles of neighboring sections differ by more than 10° (as in Figure 3.2) spanning the range from −42° to +42°. This range is assumed as during winter irradiance for larger azimuth angles does not exceed 20 W/m² (i.e., negligibly small for practical use heating purposes). Detailed simulation enabled us to obtain daily distributions of hemispheric irradiance incident on a vertical surface of the discussed building's facade for consecutive hours of a day for every month of a year (the same averaged model of solar radiation has been used, as in previous sections).

Figure 3.10 presents distributions of maximum solar irradiance values incident on surfaces of the discussed extended facade (i.e., from a nearly southeast direction to nearly southwest for every months of the averaged year). Additionally, for the sake of comparison, maximum irradiance for eastern and western walls is also shown, although it is assumed that they do not constitute parts of the main glazed facade.

An analysis of obtained data (shown in Figure 3.10) reveals that the worst insolated is the east side, practically in every month of a year except June, and particularly so in winter. Another surface with low insolation in winter and fall (i.e., September−February) is the west one. However, this surface is most irradiated in late spring and summer (from May until the end of July). These observations confirm the previous assumption that such surfaces

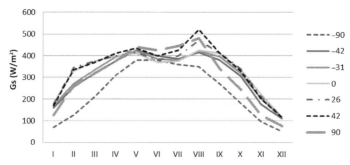

FIGURE 3.10 Maximum solar irradiance on differently oriented vertical surfaces of an extended main facade of a building throughout the entire averaged year.

FIGURE 3.11 Concept for the extended shape of the glazed facade in high-latitude countries.

should not be used as elements of the glazed facade. Creating a glazed building envelope with a contour following polyline inscribed in a flattened circle, from southeast to southwest, improves solar radiation availability in fall, winter, and early spring (in comparison to other extension angles). Because of increased radiation availability during winter, extending the main glazed facade from azimuth angle of −45° to +45° could be recommended. However, in summer, from June until September, the southwest wall, especially within the azimuth angles from +25° to +45°, is exposed to very intense solar radiation (with maximum in August). Therefore, it becomes necessary to shade this part of the facade and limit its glazing, or not to use any glazing at all, that is the best solution. In result, it is recommended to extend the glazed facade with azimuth angles from −45° to +30. Recommended facade shape and location of opaque and transparent parts is schematically presented in Figure 3.11.

3.6 RECOMMENDATIONS FOR SHAPING BUILDING ENVELOPE WITH REGARD TO SOLAR ENERGY AVAILABILITY

In this section, recommendations concerning shaping the building envelope in high-latitude countries with regard to solar energy availability are presented. Recommendations are based on simulation studies and analysis of their results.

According to the proposed classification of insolation conditions into three periods of different availability, certain characteristic recommendations for building shapes have been developed for each of those periods. Those recommendations, which are formulated for high-latitude countries, may be expressed as shown below.

For the period of best solar radiation availability (late spring and summer) recommendations are presented below.

- It is necessary to use as large tilt as possible ($\beta \rightarrow 90°$) of the building envelope elements; it is imperative to avoid horizontal surfaces and surfaces

FIGURE 3.12 Horizontal glazed roof in an office building in Warsaw, Poland.

with low tilt angles, which means to avoid flat or almost flat roofs. The biggest possible design error is a horizontal glazed roof, which, unfortunately, is frequently encountered in modern architecture in cities (see example in Figure 3.12).

- Main glazed facade should be oriented toward the south ($\gamma \to 0$). In summer, solar irradiation on vertical southern walls is much reduced when compared to eastern and especially western ones. At solar noon in summer (in fact from May until September), the solar altitude is big. In case of vertical surfaces, the incident angles of solar radiation are also big. At high incident angles, availability of solar radiation decreases and so does glass transmissivity (although this decrease becomes significant only for angles bigger than 60°). For that reason, it could be justified to use main glazed facades with tilts larger than 90°, for example 110° or more, thus ensuring in summer incident angles to be bigger than 60° around noon. As a result, there would be considerable reduction of solar radiation that passes through glazing. An example of such an "inversely tilted" facade is shown in Figure 3.13. In this case, the tilt angle is approximately 135°. Facade tilted at such a large angle is not exposed at all to beam radiation around the solar noon during the hottest months (from April until the end of August).
- It is necessary to avoid west orientation of the main facade, especially of azimuth angles in range of $+45° \leq \gamma \leq +90°$. If possible, glazing from

FIGURE 3.13   Example of an office building in Warsaw, with an upper part of "reversed facade."

southwest to west side of the building should be minimized or, if un-avoidable, they should be shaded. Seasonal shading (for summer periods) needs to be applied also for walls (facade) oriented from the southeast to the west ($-45° \leq \gamma \leq +90°$).

For the moderate (intermediate) period, because of relatively low solar irradiation, recommendations are very simple.

- It is necessary to orient the main facade toward the south ($\gamma \rightarrow 0$).
- The envelope elements should be tilted at an angle of at least 40° ($\beta \geq 40°$). Horizontal surfaces and surfaces with low tilt should be avoided.

For the period of worst (poor) solar radiation availability (winter), because of very low solar irradiation, recommendations are even more strict.

- The facade should be opened for solar radiation from the south ($\gamma \rightarrow 0$). Envelope of a building must not be shaded by elements of a building ar-chitecture or surrounding objects (neighboring buildings, trees, etc.). If the extended curved of the south facade is used, the glazed part should have azimuth angles from $-30°$ to $+30°$.
- It is recommended to use envelope elements with high tilt angle, sloped roofs with tilt in ranges from 50° to 70°. Horizontal surfaces and surfaces with small tilt angle should be avoided.

Simulation of solar irradiation and analysis of the results allow proposing general recommendations for building shapes, structures, and location of different rooms. General recommendations for types of walls (opaque, glazing with shading, non-shaded glazing) can be formulated as well. It may be pointed out that recommendations for envelope shaping with regard to solar energy availability are mostly consistent, although there are certain differences in detailed recommendations for individual seasons. Some buildings are only utilized seasonally, like summer houses, lodges, and seasonal hotels at summer resorts. Some others may be not used during summer months, for example, schools closed for summer holidays, or hotels and shuttles in winter resorts. In such cases, seasonal recommendations are particularly important.

Because of high solar irradiation in summer, buildings used only during this season may not be equipped with typical passive solutions, which facilitate solar energy utilization. The situation is quite the opposite, as they need elements that restrict solar radiation access. The main facade of such a building should not face the south or west (not to be in range of the azimuth angles from $0°$ to $+90°$). From these directions, such structures should be shaded, both with their own envelope elements (overhangs, wing walls) and natural surrounding elements (trees, pergolas). The facade should be oriented toward the east or possibly southeast. The roof should be sloped at a big angle to avoid overheating.

In case of schools, which are not used during summer, the main facade should be oriented toward the south and the roof should be sloped at an angle of $50°$ or even more. Shading elements of the envelope should only provide shading in June (possibly also May and September). Solar active system elements integrated with the envelope of the building should be tilted at $30°-40°$ and oriented at azimuth angle from $-45°$ to $+45°$ (preferably directly due south).

The majority of buildings is used continuously throughout the entire year. Basic conclusions for general architectural concept of a generic continuously used building with regard to solar radiation are provided below.

It is recommended to use one of the following shapes for the building's plan.

- Rectangular shape, stretched along the east-west axis (shorter eastern and western walls), with the longer side from 1.5 to 2-times longer than the shorter one (former ratio according to Anderson [1], the latter per ASHRAE standards [14]) is recommended.
- Nearly elliptical main facade, extended from the south side, preferably from the southeast to nearly southwest (azimuth angles from $-45°$ to $+30°$, see Figure 3.11), or from nearly southeast to nearly southwest (azimuth angles from $-30°$ to $+30°$), is recommended for solar energy availability.

For orientation and tilt of envelope elements, recommendations are presented next.

- It is recommended to orient the main facade with large glazed surfaces toward the south, or in case of a curved shape to extend it in range of azimuth angles from $-45°$ (southeast) or even $-30°$ to $+30°$ maximum (much less than southwest). In both cases (i.e., flat and curved) the facade may be conventionally vertical or tilted at an angle larger than 90° (inverse tilt) in order to increase insolation in winter (small incident angles) and decrease it in summer (big incident angles). In case of inversely tilted facade, glazing transmissivity for solar radiation is reduced in summer (in natural way) because of big incident angles of solar radiation. For incident angles bigger than 60°, sudden drop of transmissivity is accompanied by a similarly sharp rise of reflectivity. In case of large inverse tilt (around 135°), insolation conditions in winter are considerably improved and in summer are appropriately reduced, providing beneficial results.

- If solar collectors or photovoltaic (PV) cells are to be installed on the roof, the gable roof is recommended facing north and south. Tilt on the southern side should be 30°−45° and the tilt of the north can be bigger. Slope of gable roofs does not have to be symmetrical. If no solar receivers are to be used, south tilt may be higher than 45° to limit insolation in summer and increase it in winter.

Southern parts of a building, mainly south inclined roofs, have good conditions for collecting solar energy in the entire year and should contain solar energy receivers, like solar collectors or PV panels integrated with the roof. Solar passive elements should be incorporated into the main facade ensuring the best possible access to solar radiation during winter.

During the summer, south facades should be shaded with both architectural envelope elements and objects in the vicinity (trees, other vegetation). Solar passive elements incorporated in the front of the facade should be equipped with elements reducing solar energy impact in summer: external blinds or shutters, appropriate overhangs over glazing, pergolas with seasonal vegetation, etc.

Also, innovative material technologies may be proposed, including smart windows with transmissivity for solar radiation depending on the angle of the incident solar radiation, and length of the radiation wave. Transmissivity may also be decreased, or rather regulated with electro-chromic windows, already available on European markets. In this kind of glazing, transmissivity is reduced because of increased solar radiation absorptivity.

In case of multistory buildings, natural (traditional) structural elements of envelope should be used to reduce solar radiation influence in summer and increase it in winter. Such elements are balconies, especially their floors, as illustrated in Figure 3.14 (windows below balconies are clearly shaded). In case of using rows of balconies on every floor, balcony floors provide overhangs ensuring shading of the rooms on the floor below. Balconies can form interconnected rows of terraces above all the rooms on every floor. Shading elements of appropriate depth (ensuring shading in summer, but not in winter) should also be applied above the top story. Low-energy buildings are often provided with overhangs over windows made of PV panels, which is presented

FIGURE 3.14 Window shading by balcony floors in a residential building in Warsaw, Poland.

in Figure 3.15. PV panel shading is also often used within glazed roof structures over rooms that require daylight access, as illustrated in Figure 3.16.

As already mentioned, simulation results show that western parts of buildings are more exposed to solar radiation during summer than others. Therefore, it is required, just like in case of southern facades, to use elements providing shading on glazing. Shading may be provided by envelope elements (overhangs, external blinds, shutters, or sunshades) or external architectural features (primarily pergolas). Shading can also be provided by surrounding objects like trees or shrubs. During the summer, eastern parts of a building may be more open to solar radiation influence (i.e., less shaded than southern and western parts).

In case of factory buildings, assembly halls, or warehouses, envelope shapes should be much simpler, and better suited for industrial function. It is recommended to use box-shaped buildings extended along an east-west direction. In case of refrigerating rooms, the plan may be trapezoidal with the longer base on the north side. Refrigeration rooms (with refrigeration chambers) should be designed according to opposite principles than buildings that are supposed to collect, store, and use solar energy, "against" typical recommendations of solar architecture. Warehouses may have typical box

FIGURE 3.15 Low-energy building with overhangs made of PV panels, CIEMAT, Madrid, Spain.

FIGURE 3.16 PV panel roof on a conference hall, Brundland, Denmark.

FIGURE 3.17 Glazed pyramid roof
of the main building of the Warsaw
School of Economics, Warsaw, Poland.

shapes with flat (horizontal) roofs, if they are not used to store food. In case of
food storage, just like in case of refrigeration rooms (chambers), it is required
to provide very good insulation for flat horizontal roofs. However, it is much
better to use sloped roof designs, even if the tilt angle is not big. Attic space
created under such a roof creates a natural buffer space. In case of assembly
spaces (halls) manned by workers, it is recommended to use traditional saw-
tooth roofs with skylights and avoid flat horizontal ones. This solution enables daylight access and limits interior overheating. Remaining part of the
envelope (i.e., vertical external walls), should be simple (not extended).

To summarize, using horizontal surfaces (e.g., roofs or horizontal glazing
above atrium) is not recommended in high-latitude conditions. Horizontal or
almost-horizontal surfaces are particularly well insolated during summer, when
it is not desirable, whereas, in winter, insolation is much smaller. Instead, roof
surfaces should be inclined at $30°-60°$, as such angles limit radiation availability in hot seasons (in June it may even halve irradiation when compared to
horizontal surface). In case of using transparent roofs in order to provide
daylight access, saw-toothed design (see Figure 3.16) or step-pyramid shapes
(Figure 3.17) may be used, whereas horizontal glazing is not recommended.

Passive elements of the building structure should be installed on its
southern side. In spring and summer (April—September) in case of vertical
envelope walls, characteristic equalization of solar irradiation throughout a day
is observed.

During fall and winter (until the end of March) south facade should be "open" to solar radiation influence, while during the remaining part of the year it should be closed or strongly shaded. As already mentioned, vertical west walls and glazing are much more prone to overheating in summer than south ones and special care should be taken to shade them using structural elements and neighboring objects. Regardless of its tilt, any surface oriented toward the west is much better irradiated than a surface facing east with the same tilt angle. Also, southwest orientation provides much better exposure than a southeast one.

If the south facade does not contain passive architectural elements, also a solution with reverse slope (tilt angle of $110° - 140°$) may be considered.

Maximum annual irradiation can be used to formulate recommendations for tilt and orientation of solar active elements of energy systems (i.e., locations where solar collectors and PV modules should be installed). Installation of active solar systems in new buildings requires appropriately designing the roof surface or other envelope elements on which the devices are to be installed. This is of particular importance if the solar active systems are to be integrated with the envelope structure. This solution is currently becoming increasingly popular. In newly designed buildings, envelope shape may be quite freely designed and therefore all requirements for proper situation of individual surfaces may be addressed. It also needs to be noted that, besides purely energy-related considerations, also esthetic and functional questions are important factors, so a certain degree of freedom is possible. Because of solar energy gains in annual scale, it is recommended to place the solar active system elements in a position described by the tilt and azimuth of $\beta = 40°$ and $\gamma = 15°$, respectively. Of course, tilts within the range of $30° - 50°$ and orientation from $-30°$ to $+45°$ may also be considered as correct ones. However, in case of a building equipped with both solar active and solar passive system elements, orientation of individual envelope elements must be a compromise between the recommendation for active and passive solutions.

If the location imposes a situation with a longer wall facing east or west, then the main facade, but with only certain selected passive elements, should face east (or preferably southeast). Then the active solar devices should be placed on the west side or on both sides, although the larger number of receivers should still face west.

According to the obtained results, the best insolation conditions over a year for considered location occur on a south-facing surface tilted at 40°, then slightly worse results are obtained for tilts of 45° and 30°. Recommended orientation is $\gamma = +15°$. Surfaces tilted at 40° and facing nearly south ($\gamma = +15°$) can receive almost 420 MJ/(m² year) in average. Surface tilt of 40° providing for the maximum annual energy yield is not entirely in line with the general rule of thumb [28], which recommends tilts between 42° and 62° (for considered geographic location).

The main south facade of a building or previously discussed curved south extended facade (from $-45°$ to $+30°$, or just from $-30°$ to $+30°$) should be the main part of the envelope structure. At this side, main rooms occupied by people during the day should be located, as they require daylight access. Such rooms are living rooms, hotel guest rooms, hospital wards, offices, kindergarten day rooms, classrooms, lecture halls, teaching laboratories, and public utility rooms (hotel lobbies, dining rooms, exhibition halls).

Certain types of spaces, like conference rooms, may be situated on different sides, depending on their specific demand for daylight access or shading, and character of such rooms (designed for audiovisual presentation, exhibitions, workshops, seminars, etc.). Another important factor is expected time of day when the room in question is to be used. For example, if it is only to be occupied in evenings and no daylighting or solar radiation impact is needed, the room may be located even on the northern side, but not from the west (too much heat gains in summer).

Because of the fact (shown in simulation results) that west building parts are most exposed to solar radiation during summer, it may be proposed that during hot seasons some rooms on the west side are used only before noon. Therefore, the role of such rooms should be planned accordingly. No bedrooms should be placed at this side, especially in buildings without artificial cooling/air-conditioning systems. In case of public utility buildings, and particularly office buildings, duration of human presence in the west oriented rooms should be limited.

Northern parts of the envelope are obviously the least exposed to solar radiation, regardless of season. Therefore non-residential functions, including larders, storages, depots, and technical facilities, should be placed in the northernmost rooms. Those rooms are (or should be) so-called cold rooms or a cold buffer zones. The northern side of a building should not be designed for continuous human presence. The number of windows should be minimized to the level absolutely required to provide necessary daylighting. The northern part of a building should contain rooms used only periodically, first of all storages, maintenance rooms, or boiler rooms.

In case of public utility buildings, this is a good place to locate server rooms, HVAC units, and heating equipment (including boilers). Also, all kinds of service and maintenance rooms should be placed on the northern side, just like periodically used rooms that do not require sunshine (or even require shading) like screening rooms, concert halls, lecture halls, etc.

The eastern part of the building usually may be used around the clock. In residential buildings, it may contain bedrooms, living rooms, and kitchens. In public utility buildings, it is a good place to locate office rooms, meeting rooms, and all other facilities, which require good daylight access during morning hours. In any case, those rooms will be much less heated in summer than the ones located at the south and west sides of a building.

In case of residential and public utility buildings, attics should not be considered utility (living) space, so the occupants of a building are not exposed to the results of overheating in hot seasons and excessive cold during winter. Instead, attics may hold technical installations, HVAC systems, etc. They may also be used as storage spaces. In fact, this used to be a traditional application of attics. In the past, they would never be used as inhabited rooms, as opposed to many modern residential buildings.

While analyzing solar radiation availability, it is important to pay attention to factors related to external surroundings, which may cause shading or increase surrounding reflectivity.

REFERENCES

[1] Anderson B. Solar energy: fundamentals in building design. Harrisville (NH): Total Environmental Action, Inc.; 1975.

[2] Balcomb JD. Passive solar buildings. Cambridge (MA): The MIT Press; 1992.

[3] Crosbie MJ. The passive solar. Design and construction handbook. New York: Steven Winter Associates. John Wiley& Sons, Inc.; 1998.

[4] Feist W. The passive houses at Darmstadt/Germany. Darmstadt: Institute Housing and Environment; 1995.

[5] Humm O. NiedrigEnergie und PassiveHauser. Staufen bei Freiburg: Okobuch Verlag; 1998.

[6] Szokolay SV. Environmental science handbook for architects and builders. New York: John Wiley & Sons; 1980.

[7] Tombazis AN, Preuss SA. Design of passive solar buildings in urban areas. Sol Energy 2001;70(3):311−8. Elsevier Science Ltd., UK.

[8] Olgyay A, Olgyay V. Solar control and shading devices [Chapter 8]. Princeton: NJ Princeton University; 1957.

[9] Carter C, de Villiers J. Principles of passive solar building design. Pergamon Press; 1987.

[10] Johnson T. Solar architecture. The direct gain approach. MIT; 1981.

[11] Sodha MS, Bansal NK, Bansal PK, Kumar A, Malik MAS. Solar passive building. Science & design. Pergamon Press; 1986.

[12] Hsieh JS. Solar energy engineering. Prentice-Hall, Inc.; 1986.

[13] Meltzer M. Passive & active solar heating. Prentice-Hall, Inc.; 1985.

[14] ASHRAE. Handbook of fundamentals. Chapter: solar energy. Atlanta: American Society of Heating, Refrigerating and Air Conditioning Engineers, SI Edition; 1997.

[15] Athientis AK, Santamouris M. Thermal analysis and design of passive solar buildings. London (UK): James and James Publisher; 2002.

[16] Florides GA, Tassou SA, Kalogiru SA, Wrobel LC. Review of solar and low energy cooling technologies for buildings. Renewable Sustainable Energy Rev 2002;6(6):557−72.

[17] Yezioro A. A knowledge based CAAD system for passive architecture. Renewable Energy 2009;34(3):769−79.

[18] Kolokotsa D, Santamouris M, Synneta A, Karriesi. Passive solar architecture. Compr Renewable Energy 2012;3:637−65.

[19] Stathopoulou M, Synnefaa A, Cartalis C, Santamourisa M, Karlessia T, Akbarib H. A surface heat island study of Athens using high resolution satellite imagery and measurements of the optical and thermal properties of commonly used building and paving materials. Int J Sustain Energy 2009;28(1−3):59−76.

[20] Chwieduk D. Recommendation on modeling of solar energy incident on a building envelope. Renewable Energy. International Journal, Elsevier 2009;34:736—41.

[21] Akbari H, Pomerantz M, Taha H. Cool surfaces and shade trees to reduce energy use and improve air quality in urban areas. Sol Energy 2001;70(3):295—310.

[22] Mazria E. The passive solar energy book. Emmaus (PA): Rondale Press; 1979.

[23] Montgomery D, Keown SI, Heisler. Solar blocking by common trees. In: Proc. 7th national passive solar conference; 1982. Knoxville, TN, August 30—September 1, p. 473.

[24] http://www.wbdg.org/resources/psheating.php.

[25] Solar heating, Part 2 Solar technology. Trainer guidelines. EUREM. www.energymanager.eu; 2007.

[26] Chwieduk D, Bogdańska B. Some recommendations for inclinations and orientations of building elements under solar radiation in Polish conditions. Renewable Energy J 2004;29:1569—81.

[27] Chwieduk D. Availability of solar energy on a building envelope. Solaris 2011. In: Proceedings of the 5th international conference on solar radiation and daylighting. Brno University of Technology; 2011. pp. 51—6.

[28] Duffie JA, Beckman WA. Solar engineering of thermal processes. New York: John Wiley & Sons, Inc.; 1991.

Chapter 4

Photothermal Conversion in a Building

4.1 USE OF PHOTOTHERMAL CONVERSION IN A BUILDING

Every building is exposed to the influence of solar radiation. Photothermal conversion occurs in its elements in an uncontrolled manner. Solar energy is converted into heat in a natural way, in a process driven by the laws of physics. Photothermal conversion may also occur in a planned way, thanks to using the appropriate methods and equipment. There are two essential types of solar systems involved: active and passive.

An active solar system is a system that in an active way (i.e., utilizing appropriate equipment) converts solar energy into useful heat. The task of an active solar system is to absorb solar energy, convert it into heat, store heat gained, and then distribute it to consumers in a controlled manner. Devices that convert solar energy into useful heat are called solar collectors. Generated heat is transferred to a working fluid circulated in the collector's loop. Active systems operate thanks to mechanical devices, which force working fluid circulation. Those devices are circulation pumps (in systems with liquid working agent) or fans (when air is used). An active system may operate only when powered externally (e.g., with electricity). Typically, an active system is provided with a storage tank, which discharges heat toward consumers in a controlled manner.

Active systems usually operate at low temperatures, although solutions with medium temperature levels (90–200 °C) are becoming increasingly popular, too [1,2]. The latter, when used in buildings, are often coupled to sorption cooling systems (absorption and adsorption chillers) used for air conditioning [3–10]. Such combined systems may operate during the entire year, providing cooling in hot seasons, and heating in cold ones. In the process of photothermal conversion in low- or medium-temperature systems, solar energy is transferred (via different processes) into the receiving medium. Energy transfer is related to optical phenomena, as well as heat and mass transfer. Thanks to those processes, obtained heat, which is only part of the solar energy incident on an exposed surface, is stored within the volume of the

Solar Energy in Buildings. http://dx.doi.org/10.1016/B978-0-12-410514-0.00004-9

working fluid. This storage is usually based on heat capacity of the medium or, less frequently, on its ability to change phase at a predetermined temperature.

Passive solar systems are systems in which conversion of solar energy into heat occurs naturally because of the laws of physics, with no need for any mechanical drives. Such processes occur within the structure of a building, which is sometimes specially designed to use this process (but without special installations). Passive systems also include gravity installations, so-called thermosiphon systems. No driving energy (e.g., electricity) is needed for passive systems to operate (enforce fluid flow). Gravity systems are built in a similar manner to active systems, but they do not contain any mechanical equipment (pumps or fans).

In solar passive architecture, individual structural elements have functions of different elements of a solar heating system. Thus, envelope elements act as solar collectors, heating (heating/cooling) and ventilation ducts are used to distribute collected heat, and elements of internal structure (internal walls, floors, ceilings), and, to some extent, also envelope elements (walls) are used as heat stores. Some elements of a building, for example, specially designed solar-spaces together with their envelope, may fulfill several functions at the same time. Storage of energy captured from solar radiation may involve choosing the proper materials for the envelope elements and interior, and utilization of so-called phase change materials [11,12].

There are also semipassive systems, which are essentially passive systems utilizing certain control components and devices periodically forcing the flow of collected solar energy. Those are usually mechanical devices intensifying the flow of heated air within ducts. Such auxiliary drives are primarily used in high-latitude countries, where cloud formation and low irradiance do not allow development of natural convection.

4.2 FUNDAMENTALS OF RADIATION PROCESSES IN PHOTOTHERMAL CONVERSION

4.2.1 Absorption, Reflection and Transmission of Solar Radiation

Individual parts of this section describe shortly the fundamentals of photothermal conversion, including fundamental definitions and laws, which are presented in detail in publications on fundamentals in physics [13–15]. The description focuses on those problems that are important when analyzing the phenomena occurring during photothermal conversion. Special attention has been put on the phenomena related to the transmission of solar radiation through transparent media, as such media (i.e., glazing), allow the radiation to penetrate the building's interior directly, thus significantly affecting its energy balance.

Solar energy is carried by electromagnetic waves of different lengths. As a wave, solar radiation is subject to the same phenomena as any other electromagnetic wave (i.e., reflection, refraction, absorption, polarization, etc.)

[15,16]. Radiation incident on a body may be reflected by that body, absorbed within, and transmitted through the body and outside. Share of each of those processes depends on the type of body (material), properties of its surface, distance traveled by radiation within the body, wavelength, and angle of incidence.

For certain wavelength λ within a range $\Delta\lambda$, the monochromatic absorptance α_λ determines the share of absorbed radiation $\phi_\lambda \Delta\lambda$ in reference to the total incident radiation (where $\Delta\phi = \phi_\lambda \Delta\lambda$). Monochromatic absorptance therefore determines how large part of radiation of certain wavelength may be absorbed by the investigated surface, if wave of such length is contained in the spectrum of the incident radiation. Monochromatic reflectance ρ_λ and monochromatic transmittance τ_λ are defined analogically. According to the energy conservation law for certain wavelengths, the following relation is true:

$$\alpha_\lambda + \rho_\lambda + \tau_\lambda = 1 \tag{4.1}$$

In reality, the radiation incident on some surfaces contain a wide spectrum of different wavelengths, and therefore radiation absorption occurs at different wavelengths as well. Absorptance α is therefore defined as a ratio of absorbed radiation with irradiance ϕ_{abs} to the total incident energy flux ϕ_{tot}, which may be expressed as:

$$\alpha = \frac{\phi_{abs}}{\phi_{tot}} \tag{4.2}$$

Total absorptance of a surface depends on the spectrum of incident radiation. Thus, Eqn (4.2) may be converted to a more detailed form:

$$\alpha = \frac{\int_{\lambda=0}^{\infty} \alpha_\lambda \phi_{\lambda,tot} d\lambda}{\int_{\lambda=0}^{\infty} \phi_{\lambda,tot} d\lambda} \tag{4.3}$$

Analogical relations are true for reflectance and transmittance of a surface. Ultimately, just like in Eqn (4.1), total absorptance, reflectance, and transmittance of a given surface satisfies the equation:

$$\alpha + \rho + \tau = 1 \tag{4.4}$$

Values of each of those three radiation properties may vary from 0 to 1. Three perfect cases may be therefore listed as:

- white body, body that reflects all incident radiation, whose reflectance ρ is equal to 1, while $\tau = 0$ and $\alpha = 0$;
- black body, which absorbs all incident radiation, whose absorptance α is equal to 1, while $\tau = 0$ and $\rho = 0$;
- transparent body, which fully transmits the incident radiation, whose transmittance τ is equal to 1, while $\alpha = 0$ and $\rho = 0$.

A body may be considered black if it absorbs all incident radiation. This means that monochromatic absorptance value α_λ for every wavelength λ is equal to 1, and also total absorptance $\alpha = 1$. Thus, the absorptive properties of a black body are not dependent on wavelengths (and also on temperature). In reality, there are no ideally black bodies (the same is true for white bodies and transparent bodies).

A building consists of opaque and transparent media. Opaque media are walls, whereas transparent ones include glazing and contained air. Each surface of an external wall is characterized by certain absorptance and reflectance for solar radiation (i.e., $\tau = 0$), while glazing, except for absorptance and reflectance, also has certain transmittance value. Clean air is considered to be fully transparent for solar radiation (i.e., $\tau = 1$).

To sum up, solar radiation, which is an electromagnetic wave, when it reaches a body, may be absorbed, reflected, or transmitted through that body (if the body is transparent) [17]. The solar energy flux incident on a surface may therefore have three components:

- reflected,
- absorbed,
- transmitted (in case of transparent bodies).

Transmittance (i.e., ability to transmit radiation) it depends on the radiation of the angle of incidence, its spectrum, and polarization state. It is possible to distinguish directional transmittance (in reference to radiation transmitted directionally) and diffuse transmittance (referring to radiation transmitted in diffuse form). Total transmittance is a sum of directional transmittance and diffuse transmittance. The transmittance value is considerably affected by the harshness of a body surface. For example, for a 2 mm thick glass, whose one external face is grinded and the other polished, transmittance is 0.85 and 0.78, respectively, depending on which face is exposed to radiation [18].

The solar radiation incident on an interface between media may be refracted. If an electromagnetic wave is incident on a medium with lower optical density and the angle of incidence is larger than the boundary angle ($>$arc sin (n_2/n_1), where n_1 and n_2 are values of refractive index for the adjacent media), a total internal reflection occurs. If light reflection/refraction phenomena are analyzed, the first medium (through which the incident wave travels) must be transparent. Light reflection is classified according to the type of interface. If the surface is smooth (surface irregularities with diameters lower than the wavelength λ), the reflection is specular (mirror-like). In case of irregularities with dimensions comparable to the wavelength, then in case of irregular distribution of those irregularities, diffuse reflection may occur, whereas in case of regular distribution of surface irregularities the character of reflection becomes a special case similar to the reflection from diffraction gratings. These phenomena are very important for analysis of the application of special systems (materials) for glazing aimed at the proper use of daylighting.

About 99% of solar radiation spectrum is visible radiation and near-infrared radiation. Wave absorption is the conversion of wave energy into other forms of energy as a result of interaction with a medium or body, through which the wave propagates or which blocks wave propagation. In case of light absorption, the radiation energy is converted into different kinds of internal energy, or into energy of secondary radiation emitted in a different direction or characterized by different spectrum. Light absorption may cause the body heating, ionization, or atom excitation. The medium or the body may also disperse waves. Another possibility is induced emission. This is emission of waves with direction and spectrum identical to the transmitted light. In case of near-infrared radiation, absorbed energy increases the internal energy of a body, causing its temperature to rise (which may be used in solar heating systems) and may be further emitted away.

As the main focus of this discussion is photothermal conversion, further analysis covers those phenomena the lead to conversion of solar radiation into heat. Solar energy absorbed by a body, or transmitted by a transparent body and then absorbed by another body or a medium, is converted into heat. Solar radiation may be reflected from a certain body and may reach another body, and gets absorbed there and converts into heat. Generated heat may be further transferred because of heat transfer [19−21] (i.e., heat conduction, convection, or radiation).

Heat conduction is energy transmission within one medium or between two media in direct contact from higher temperature region to lower temperature ones. Individual particles of the medium or media that conduct heat display practically no change of their position. Heat conduction in pure form is observed in solids.

Convection occurs if particles of a medium that transfers heat change their position. Convection occurs in fluids and is always related to their flow, which may be either natural or forced. In case of free flow, the process is called natural convection and is driven by density differences caused by temperature gradients. In case of forced convection, the fluid flow is forced by either natural external causes (e.g., wind) or appropriate devices.

Heat radiation is heat transfer through an electromagnetic wave with wavelengths in a range of $0.74-200$ μm. Heat exchange by thermal radiation does not require a medium and may be carried out in a vacuum. Radiation energy, as the energy of an electromagnetic wave, is transmitted at the speed of light. In low temperature, thermal radiation is often neglected and its influence rises together with the temperature increase of bodies that exchange heat.

The processes of reflection, absorption, and transmission of solar radiation lead to heat generation and then transfer of generated heat within one medium or between media, through mechanisms of conduction, convection, and thermal radiation. Those processes are utilized in both active and passive solar systems. All those phenomena may have either a beneficial or adverse effect on solar energy conversion and utilization, as they can influence both energy gains and losses, and then the amount of generated heat.

4.2.2 Thermal Radiation Emission

Monochromatic emissive power $E_{\lambda,T}$ ($E_{\nu,T}$) characterizes emission of radiation from that body (other types of emission are: field emission, thermionic emission, secondary emission, and induced emission [17]). All bodies, except for white bodies and perfectly transparent ones, emit radiation corresponding to their temperature.

Monochromatic emissive power $E_{\lambda,T}$ ($E_{\nu,T}$) is the energy flux of electromagnetic radiation $d\phi_e$ emitted by a unit of surface over a unit of time within wavelength range from λ to $\lambda + d\lambda$ (frequencies from ν to $\nu + d\nu$), called monochromatic irradiance of emitted radiation $E_{\lambda,T}$ ($E_{\lambda,T} = d\phi_e/d\lambda$) (or $E_{\nu,T}$: $E_{\nu,T} = d\phi_e/d\nu$). Irradiance $E_{\lambda,T}$ is a function of wavelength and temperature. Additionally, it depends on the chemical composition of the body, characteristics of its surface, and its shape.

Planck has formulated a law describing the distribution of energy as a function of wavelengths of radiation emitted by a black body. According to the Planck law, for certain wavelength λ, monochromatic radiant flux density $E_{o\lambda,T}$ (monochromatic irradiance) of a black body reaches its maximum. As a body temperature grows, so does the radiant intensity, which also moves toward shorter wavelengths. Mathematical expression of the Planck law for a black body provides the relation between black body monochromatic irradiance $E_{o\lambda,T}$ or emissivity $\varepsilon_{o\lambda,T}$ and wave frequency ν or wavelength λ, as well as temperature T:

$$E_{ov,T} = \frac{2\pi \nu^2}{c^2} \frac{h\nu}{\exp\left(\frac{h\nu}{\sigma T}\right) - 1} \qquad (4.5)$$

or:

$$E_{o\lambda,T} = \frac{2\pi c^2}{\lambda^5} \frac{h}{\exp\left(\frac{hc}{\sigma \lambda T}\right) - 1} = E_{ov,T} \frac{c}{\lambda^2} \qquad (4.5a)$$

Total energy emitted by an elementary surface toward the interior of a solid angle of 2π during 1 s over the entire range of frequencies from zero to infinity, is described by the Stefan–Boltzmann law and equal to:

$$E_o = \int_0^\infty E_{o\lambda} d\lambda = \sigma_o T^4 \qquad (4.6)$$

Symbols used in the equations above are:

h = Planck constant equal to 6.62491×10^{-34} (Js);
σ = Boltzmann constant for a black body equal to 5.67×10^{-8} (W/(m^2 K^4));
c = speed of light in vacuum (m/s);
T = absolute temperature (K).

If a spectrum of the real body's radiation is continuous and its characteristic is similar to that of a black body, it is called a gray body. A perfect gray body has constant absorptance (independent on wavelength), its transmittance τ is zero, and reflectance ρ is larger than zero.

The Stefan–Boltzmann law, while defined for a black body, remains reasonably accurate for gray bodies, too. In case of gray bodies, the constant σ value is multiplied by emissivity ε of a specific surface of a body, which does not depend on wavelength (frequency), but only on body structure, surface condition, and temperature.

According to the Wien displacement law, which describes energy distribution in a black body spectrum, wavelength corresponding to the maximum emissivity changes in inverse proportion to the absolute temperature (function of $E_{\lambda,T}$ monochromatic irradiance at certain temperature reaches the maximum value for certain wavelength λ_{max}), and the product of λ_{max} and T is constant equal to 2.898×10^{-2} (m K).

The Kirchhoff law specifies the relation between the ability to absorb energy and irradiance of emitted radiation. According to Kirchhoff's theory, emissivity of a body is equal to its absorptance, which means that a black body may emit all the absorbed radiation.

Total emissivity ε of a certain surface of a gray body is defined as a ratio of radiant intensity E of radiation emitted by the analyzed gray body surface to the intensity E_0 of radiation emitted by a black body at the identical conditions (same temperature). Emissivity of a gray body's absorbing surface may be expressed as:

$$\varepsilon = \frac{E}{E_0} \tag{4.7}$$

Monochromatic emissivity ε_λ for certain wavelength λ is defined analogically as:

$$\varepsilon_\lambda = \frac{E_\lambda}{E_{\lambda 0}} \tag{4.7a}$$

According to Kirchhoff's law, emissivity of a body is equal to its capacity to absorb at certain wavelengths:

$$\alpha_\lambda = \varepsilon_\lambda \tag{4.8}$$

This law has been formulated assuming that the analyzed system is a thermally insulated system composed of a gray body and a black body located close to each other with heat-exchanging surfaces being equal in size and parallel. The system is assumed to remain in thermal equilibrium, which means that energy radiated by each of those bodies must be equal to the energy absorbed by it. The energy balance for a gray body may be expressed as:

$$\alpha_\lambda dq = FE_\lambda d\lambda \tag{4.9}$$

This means that the gray body absorbs part α_λ of energy dq incident on it in the form of radiation of wavelength λ, which is equal to the energy radiated by that body during the same time (F is the heat exchange surface).

Energy balance for the black body (index 0) is analogical to Eqn (4.9), except for the fact that the absorptance is equal to 1:

$$dq = FE_{0\lambda}d\lambda \qquad (4.10)$$

If we use the formula for dq given in Eqn (4.10) and in the Eqn (4.9), and transform it appropriately, we may obtain:

$$\frac{E_\lambda}{E_{0\lambda}} = \alpha_\lambda \qquad (4.11)$$

Equation (4.11) is analogical to Eqn (4.7a) and is a confirmation of the Kirchhoff's law expressed by Eqn (4.8). A discussion similar to the one above may be carried out for the entire wavelength spectrum. This would lead to another form of the Kirchhoff's law:

$$\alpha = \varepsilon \qquad (4.12)$$

Emissivity of a gray body is equal to its ability to absorb energy, which means that the ratio of irradiance of energy emitted by a gray body to its ability to absorb energy is a constant value equal to the irradiance of energy of a black body.

Two bodies of different emissivity ε_1 and ε_2 may interact with each other by exchanging a certain amount of heat Q_{1-2} between their surfaces (F_1 and F_2) through radiation. The amount of exchanged heat Q_{1-2} in relation to the surface F_1 (assuming that $F_1 \ll F_2$), is described by the relation:

$$Q_{1-2} = \varepsilon_{12}\sigma F_1\left(T_1^4 - T_2^4\right) \qquad (4.13)$$

Value of the factor ε_{12} is calculated with appropriate equations, whose form depends on the mutual position of two surfaces that exchange heat by radiation.

4.2.3 Radiation Transmission Through Transparent Media

This part will present a detailed discussion of physical fundamentals of solar radiation's transmission through transparent media like glazing. As already mentioned, glazing plays a huge role in determining the magnitude of the impact of solar energy on an energy balance of a building. In the future, type and structure of glazing will become decisive for controlling solar radiation availability inside a building. That is why it is so important to learn the physical fundamentals of the phenomena occurring within glazing, which are mostly related to the optics [16,22].

Fresnel has formulated a law describing the reflection of the unpolarized light incident on a smooth surface and passing from a medium 1 with refractive index n_1 to medium 2 with refractive index n_2. He derived a relation describing reflection coefficients r, separately addressing the component with perpendicular polarization (s-polarized) r_s and that with parallel polarization (p-polarized) r_p. Graphic illustration of the phenomena occurring at the interface of two media is presented in Figure 4.1.

Radiation I_0 strikes the interface between two media at angle θ_1 and is reflected at the same angle (I_r, reflected radiation) and refracted at angle θ_2. Reflection coefficient r, which is the ratio of reflected radiation I_r to the incident radiation I_0 striking the surface, is mathematically expressed by Eqn (4.14). This relation is a mathematical expression of the Fresnel law, which expresses the total reflection coefficient r as an arithmetic mean of coefficients for both components r_p and r_s, where the former has perpendicular polarization and the latter has parallel.

$$r = \frac{I_r}{I_o} = \frac{1}{2}\left(r_s + r_p\right) = \frac{1}{2}\left(\frac{\sin^2(\theta_2 - \theta_1)}{\sin^2(\theta_2 + \theta_1)} + \frac{\tan^2(\theta_2 - \theta_1)}{\tan^2(\theta_2 + \theta_1)}\right) \qquad (4.14)$$

Relation of the refractive indices of both media (ratio of the value for medium 1 to that of medium 2) is equal to the ratio of the sinus of refraction angle to the sinus of angle of incidence (reflection). This is the Snell's law described by Eqn (4.2).

$$\frac{n_1}{n_2} = \frac{\sin \theta_2}{\sin \theta_1} \qquad (4.15)$$

If the angle of incidence θ_1 and angle of refraction θ_2 are known (calculated from Eqns (4.14) and (4.15)), then it is possible to determine the reflection coefficient for a single interface between two media.

If the radiation strikes interface between two media as normal radiation (i.e., along the normal to the interface in question), then both components (s- and p-polarized) yield the same refractive index, because the angle of

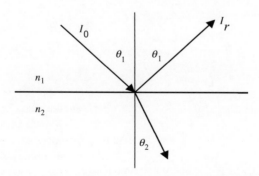

FIGURE 4.1   Passage of an unpolarized light beam through interface between two media.

incidence/reflection θ_1 and the angle of refraction θ_2 are both equal to zero. Thus, Eqn (4.14) simplifies into:

$$r(0) = \left(\frac{n_1 - n_2}{n_1 + n_2}\right)^2 \qquad (4.16)$$

If one of the media is air, the refractive index n_2 is close to one, and Eqn (4.16) transforms into:

$$n_2 \rightarrow 1$$

$$r(0) = \left(\frac{n_1 - 1}{n_1 + 1}\right)^2 \qquad (4.17)$$

The following section describes a phenomenon of radiation transmission through a long transparent plate (e.g., glass), with air on both sides. Radiation attenuation on both sides is identical. Adding up the shares of radiation repeatedly reflected by media interfaces yields the total transmittance of the discussed plate.

Radiation transmission through a transparent cover is schematically illustrated in Figure 4.2 (this and the following drawings in this section present the transmission in a simplified way, without showing the difference between the angle of incidence and angle of refraction).

As shown in Figure 4.2, a transparent cover is struck by radiation, with the share $(1 - r)$ of the incident radiation penetrating its interior, and r getting reflected. The share of radiation that emerges on the other side is $(1 - r^2)$, while $(1 - r) r$ gets reflected back from the second interface. Of that part $(1 - r) r^2$ is again reflected by the upper interface, while $(1 - r)^2 r$ emerges outside. The phenomena of reflection and transmission is further repeated.

Mathematically, the phenomena of radiation transmission and repeated reflection in a transparent barrier are described by the equation for the cover's ability to transmit and reflect radiation (i.e., equation for transmittance and

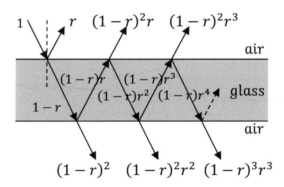

FIGURE 4.2   Transmission and repeated reflection of radiation within a transparent plate.

reflectance of the discussed transparent cover). Those properties in reference to the s-polarized radiation (τ_s and ρ_s) are expressed with the following infinite sums:

$$\tau_s = (1 - r_s)^2 + (1 - r_s)^2 r_s^2 + (1 - r_s)^2 r_s^4 + \ldots$$

$$= (1 - r_s)^2 (1 + r_s^2 + r_s^4 + \ldots) = (1 - r_s)^2 \sum_{n=0}^{\infty} r_s^{2n} \qquad (4.18)$$

$$\rho_s = r_s + (1 - r_s)^2 r_s + (1 - r_s)^2 r_s^3 + (1 - r_s)^2 r_s^5 + \ldots$$

$$= r_s + \left((1 - r_s)^2 r_s\right) \sum_{n=0}^{\infty} r_s^{2n} \qquad (4.19)$$

As the addends of sums presented above constitute convergent geometric series, they may be expressed as:

$$\tau_s = \frac{(1 - r_s)^2}{1 - r_s^2} = \frac{1 - r_s}{1 + r_s} \qquad (4.18a)$$

$$\rho_s = r_s + \frac{\left((1 - r_s)^2 r_s\right)}{1 - r_s^2} \qquad (4.19a)$$

Relations for the p-polarized component are derived analogically and for transmittance it may be stated that:

$$\tau_p = (1 - r_p)^2 \sum_{n=0}^{\infty} r_p^{2n} = (1 - r_p)^2 \left(1 + r_p^2 + r_p^4 + \ldots\right) = \frac{(1 - r_p)^2}{1 - r_p^2} = \frac{1 - r_p}{1 + r_p}$$

$$(4.20)$$

Total transmittance, which in the discussed situation is an arithmetic mean of s-polarized transmittance (Eqn (4.18a)) and the p-polarized one (Eqn (4.20)) is described with the following equation:

$$\tau_r = \frac{1}{2}\left(\frac{1 - r_p}{1 + r_p} + \frac{1 - r_s}{1 + r_s}\right) \qquad (4.21)$$

In case there are n separate parallel plates of identical properties, their total transmittance may be expressed as:

$$\tau_{r,n} = \frac{1}{2}\left(\frac{1 - r_p}{1 + (2n - 1)r_p} + \frac{1 - r_s}{1 + (2n - 1)r_s}\right) \qquad (4.22)$$

The discussion just presented concerns the ability to transmit radiation without absorption within the cover's structure. The ability to absorb radiation within a partially transparent medium is addressed by the Bouguer Law (Bouguer–Labert–Beer). Transmission factor τ_a is defined to express the ratio of radiation transmitted by the medium in question I_{trans}, which gets attenuated because of absorption processes within that medium to the total incident radiation I_0:

$$\tau_a = \frac{I_{trans}}{I_0} = e^{(-KL)} \qquad (4.23)$$

The coefficient K in the equation above is assumed to be constant for the entire spectrum of solar radiation. The L value is the length of the radiation's path within the transparent medium (cover) after refraction. This path is longer than the distance between both interfaces (i.e., cover's thickness δ) and is equal to:

$$L = \frac{\delta}{\cos(\theta_z)} \qquad (4.24)$$

Transmittance, reflectance, and absorptance of a single transparent cover, taking into account losses on reflection and absorption, may be determined in a way similar to the one already described (Eqns (4.18)−(4.22)), taking into account the repeated reflections of the radiation incident on media interface. Components with p- and s-polarization are considered separately. Phenomena occurring in a transparent cover (analogy to Figure 4.2), taking into account radiation absorption, are presented in Figure 4.3.

Transmittance, reflectance, and absorptance of the plate for the s-polarized component may be described by the following infinite sums:

$$\tau_s = (1 - r_s)^2 \tau_a + (1 - r_s)^2 r_s^2 \tau_a^3 + (1 - r_s)^2 r_s^4 \tau_a^5 + (1 - r_s)^2 r_p^6 \tau_a^7$$
$$+ \ldots = \left((1 - r_s)^2 \tau_a \right) \sum_{n=0}^{\infty} \left(r_s^{2n} \tau_a^{2n} \right) \qquad (4.25)$$

$$\rho_s = r_s + (1 - r_s)^2 r_s \tau_a^2 + (1 - r_s)^2 r_s^3 \tau_a^4 + (1 - r_s)^2 r_s^5 \tau_a^6$$
$$+ (1 - r_s)^2 r_s^7 \tau_a^8 + \ldots\ldots = r_s + \left((1 - r_s)^2 r_s \tau_a^2 \right) \sum_{n=0}^{\infty} \left(r_s^2 \tau_a^2 \right) \qquad (4.26)$$

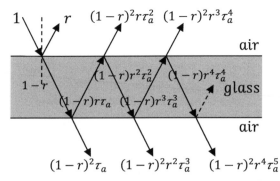

FIGURE 4.3 Repeated reflection in a transparent plate, including absorption.

$$\alpha_s = (1 - r_s)(1 - \tau_a) + (1 - r_s)(1 - \tau_a)r_s\tau_a + (1 - r_s)(1 - \tau_a)r_s^2\tau_a^2$$
$$+ (1 - r_s)(1 - \tau_a)r_s^3\tau_a^3 + (1 - r_s)(1 - \tau_a)r_s^4\tau_a^4 + \ldots\ldots$$
$$= ((1 - r_s)(1 - \tau_a)) \sum_{n=0}^{\infty} \left(r_p\tau_a\right) \tag{4.27}$$

Addends in Eqns (4.25)–(4.27) constitute convergent geometric series, and may be expressed as:

$$\tau_s = \frac{(1 - r_s)^2\tau_a}{1 - r_s^2\tau_a^2} \tag{4.28}$$

$$\rho_s = r_s + \frac{(1 - r_s)^2 r_s\tau_a^2}{1 - r_s^2\tau_a^2} \tag{4.29}$$

$$\alpha_s = \frac{(1 - r_s)(1 - \tau_a)}{1 - r_s\tau_a} \tag{4.30}$$

Relations similar to the ones just presented may be obtained for the p-polarized radiation, and then transmittance, reflectance, and absorptance for the cover may be calculated as the arithmetic mean of relevant values for both polarized components.

The phenomena of radiation's passage through two transparent covers may be analyzed in a similar manner, as presented in Figure 4.3, taking into account the repeated reflection on all interfaces which is schematically presented in Figure 4.4. Two transparent covers may be made of different materials characterized by different values of refractive index, absorptance, and the

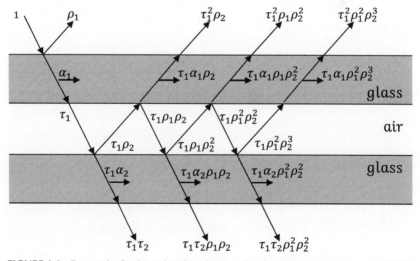

FIGURE 4.4 Repeated reflection at interfaces between the air and two transparent covers, taking into account absorption.

possibility of different thickness. Because of that transmittance, reflectance and absorptance are first calculated for every cover separately using Eqns (4.18)–(4.22) for s- and p-polarized components. Then, for every cover, the total transmittance, reflectance, and absorptance for unpolarized radiation are calculated as arithmetic averages of s- and p-polarized components.

Symbols ρ_1, τ_1, α_1, and ρ_2, τ_2, α_2 stand for transmittance, reflectance, and absorptance of both covers, respectively. Just like in case of a single cover, Figure 4.4 may be used to derive relations describing transmittance, reflectance, and absorptance of a system of two covers for s- and p-polarized radiation. Equations for transmittance, reflectance, and absorptance of each cover separately, taking into account their arrangement (order) in reference to the incident radiation and their mutual interactions, are then used to determine the optical parameters of the entire system for both polarized components of solar radiation.

Transmittance, reflectance, and absorptance of a system of two covers for the s-polarized radiation are written below in the form of infinite sums, which create relevant convergent geometric series:

$$
\begin{aligned}
\tau_s &= \tau_{s,1}\tau_{s,2} + \tau_{s,1}\tau_{s,2}\,\rho_{s,1}\rho_{s,2} + \tau_{s,1}\tau_{s,2}\,\rho_{s,1}^2\rho_{s,2}^2 + \tau_{s,1}\tau_{s,2}\rho_{s,1}^3\rho_{s,2}^3 \\
&\quad + \tau_{s,1}\tau_{s,2}\,\rho_{s,1}^4\rho_{s,2}^4 + \ldots = \left(\tau_{s,1}\tau_{s,2}\right)\sum_{n=0}^{\infty}\left(\rho_{s,1}\rho_{s,2}\right) = \frac{\tau_{s,1}\tau_{s,2}}{1 - \rho_{s,1}\rho_{s,2}}
\end{aligned}
\tag{4.31}
$$

$$
\begin{aligned}
\rho_s &= \rho_{s,1} + \tau_{s,1}^2\rho_{s,2} + \tau_{s,1}^2\rho_{s,1}\rho_{s,2}^2 + \tau_{s,1}^2\rho_{s,1}^2\rho_{s,2}^3 + \tau_{s,1}^2\rho_{s,1}^3\rho_{s,2}^4 \\
&\quad + \tau_{s,1}^2\rho_{s,1}^4\rho_{s,2}^5 + \ldots = \rho_{s,1} + \left(\tau_{s,1}^2\rho_{s,2}\right)\sum_{n=0}^{\infty}\left(\rho_{s,1}\rho_{s,2}\right) \\
&= \rho_{p,1} + \frac{\tau_{s,1}^2\rho_{s,2}}{1 - \rho_{s,1}\rho_{s,2}}
\end{aligned}
\tag{4.32}
$$

$$
\begin{aligned}
\alpha_s &= \alpha_{1,s} + \tau_{s,1}\alpha_{s,2} + \tau_{s,1}\alpha_{s,1}\rho_{s,2} + \tau_{s,1}\alpha_{s,2}\rho_{s,1}\rho_{s,2} \\
&\quad + \tau_{s,1}\alpha_{s,1}\rho_{s,1}\rho_{s,2}^2 + \tau_{s,1}\alpha_{s,2}\,\rho_{s,1}^2\rho_{s,2}^2 + \tau_{s,1}\alpha_{s,1}\rho_{s,1}^2\rho_{s,2}^3 \\
&\quad + \tau_{s,1}\alpha_{s,2}\,\rho_{s,1}^3\rho_{s,2}^3 + \tau_{s,1}\alpha_{s,1}\,\rho_{s,1}^3\rho_{s,2}^4 + \ldots \\
&= \alpha_{1,s} + \tau_{s,1}\left(\alpha_{s,2} + \alpha_{s,1}\rho_{s,2}\right) + \tau_{s,1}\left(\alpha_{s,2} + \alpha_{s,1}\rho_{s,2}\right)\rho_{s,1}\rho_{s,2} \\
&\quad + \tau_{s,1}\left(\alpha_{s,2} + \alpha_{s,1}\rho_{s,2}\right)\rho_{s,1}^2\rho_{s,2}^2 + \tau_{s,1}\left(\alpha_{s,2} + \alpha_{s,1}\rho_{s,2}\right)\rho_{s,1}^3\rho_{s,2}^3 + \ldots \\
&= \alpha_{1,s} + \frac{\tau_{s,1}\left(\alpha_{s,2} + \alpha_{s,1}\rho_{s,2}\right)}{1 - \rho_{s,1}\rho_{s,2}}
\end{aligned}
\tag{4.33}
$$

Transmittance, reflectance, and absorptance for three covers may be determined likewise:

$$
\tau_s = \frac{\tau_{s,2}\tau_{s,3}}{1 - \rho_{s,2}\rho_{s,3}}
\tag{4.34}
$$

$$\rho_s = \rho_{s,2} + \frac{\tau_{s,2}^2 \rho_{s,3}}{1 - \rho_{s,2}\rho_{s,3}} \tag{4.35}$$

$$\alpha_s = \alpha_{s,2} + \frac{\tau_{s,2}\left(\alpha_{s,3} + \alpha_{s,2}\rho_{s,3}\right)}{1 - \rho_{s,2}\rho_{s,3}} \tag{4.36}$$

Then, for the general case of n covers for the s-polarized component we get:

$$\tau_s = \frac{\tau_{s,n-1}\tau_{s,n}}{1 - \rho_{s,n-1}\rho_{s,n}} \tag{4.37}$$

$$\rho_s = \rho_{s,n-1} + \frac{\tau_{s,n-1}^2 \rho_{s,n}}{1 - \rho_{s,n-1}\rho_{s,n}} \tag{4.38}$$

$$\alpha_s = \alpha_{s,n-1} + \frac{\tau_{s,n-1}\left(\alpha_{s,n} + \alpha_{s,n-1}\rho_{s,n}\right)}{1 - \rho_{s,n-1}\rho_{s,n}} \tag{4.39}$$

Using Eqn (4.28) with appropriate transformations, the transmittance of a glazed part of the envelope for the s-polarized component may be described with the following equation:

$$\tau_s = \frac{\left(1 - r_s\right)^2 \tau_a}{1 - r_s^2 \tau_a^2} = \tau_a \left(\frac{1 - r_s}{1 + r_s}\right)\left(\frac{1 - r_s^2}{1 - r_s^2 \tau_a^2}\right) \tag{4.40}$$

A similar relation exists for transmittance for the p-polarized component:

$$\tau_p = \frac{\left(1 - r_p\right)^2 \tau_a}{1 - r_p^2 \tau_a^2} = \tau_a \left(\frac{1 - r_p}{1 + r_p}\right)\left(\frac{1 - r_p^2}{1 - r_p^2 \tau_a^2}\right) \tag{4.41}$$

Equations (4.40) and (4.41) may be simplified, as the last addend in practice is very close to unity [5]. Therefore, in simplified analyses, the total transmittance τ of a single transparent cover may be expressed as:

$$\tau = \tau_a \frac{1}{2}\left(\left(\frac{1 - r_p}{1 + r_p}\right) + \left(\frac{1 - r_s}{1 + r_s}\right)\right) \tag{4.42}$$

This equation describes total transmittance τ of a single transparent cover, as a product of transmittance τ_r, which takes into account the phenomena of repeating reflection within the cover (Eqn (4.20)), and transmittance τ_a, which addresses absorption (Eqn (4.23)). After introducing simplifications, the transmittance in question may be expressed as:

$$\tau = \tau_r \tau_a \tag{4.43}$$

For n parallel and identical covers, the total transmittance τ_n is described by equation:

$$\tau_n = \frac{1}{2}\left(\frac{1 - r_p}{1 + (2n - 1)r_p} + \frac{1 - r_s}{1 + (2n - 1)r_s}\right)e^{(-K\sigma/\cos(\theta_2))} \tag{4.44}$$

Duffie and Beckman [17] propose certain simplifications for analyzing properties of transparent covers. According to them, absorptance for a single cover (described by Eqn (4.30) for the s-polarized component) may be approximated by the following equation:

$$\alpha \cong 1 - \tau_a \tag{4.45}$$

Reflectance for a single transparent cover, where $\rho = 1 - \alpha - \tau$ may also be (using Eqns (4.45) and (4.43)) approximated by:

$$\rho \cong \tau_a(1 - \tau_r) = \tau_a - \tau \tag{4.46}$$

In case of two nonidentical transparent covers Eqns (4.32) and (4.33) in reference to nonpolarized radiation transmittance, reflectance, and absorptance are described by the following approximate relations:

$$\tau = \frac{1}{2}\left[\left(\frac{\tau_1\tau_2}{1 - \rho_1\rho_2}\right)_p + \left(\frac{\tau_1\tau_2}{1 - \rho_1\rho_2}\right)_s\right] \cong \left(\frac{\tau_1\tau_2}{1 - \rho_1\rho_2}\right) \tag{4.47}$$

$$\rho = \frac{1}{2}\left[\left(\rho_1 + \frac{\rho_2\tau_1^2}{1 - \rho_1\rho_2}\right)_p + \left(\rho_1 + \frac{\rho_2\tau_1^2}{1 - \rho_1\rho_2}\right)_s\right] \cong \left(\rho_1 + \frac{\rho_2\tau_1^2}{1 - \rho_1\rho_2}\right) \tag{4.48}$$

$$\alpha = \frac{1}{2}\left[\left(\alpha_1 + \frac{\tau_1(\alpha_2 + \alpha_1\rho_2)}{1 - \rho_1\rho_2}\right)_p + \left(\alpha_1 + \frac{\tau_1(\alpha_2 + \alpha_1\rho_2)}{1 - \rho_1\rho_2}\right)_s\right]$$
$$\cong \left(\alpha_1 + \frac{\tau_1(\alpha_2 + \alpha_1\rho_2)}{1 - \rho_1\rho_2}\right) \tag{4.49}$$

Indices 1 and 2 refer to the first and second cover, respectively, counted along the radiation direction of incidence. When determining transmittance, reflectance, and absorptance of a double nonidentical cover on the base of Eqns (4.47)–(4.49), it is important which cover is struck by radiation first. If both covers are identical, then Eqns (4.47)–(4.49) may be simplified to the following forms:

$$\tau_c = \left(\frac{\tau^2}{(1 - \rho^2)}\right) \tag{4.50}$$

$$\rho_c = \left(\rho + \frac{\rho\tau^2}{1 - \rho^2}\right) \tag{4.51}$$

$$\alpha_c = \left(\alpha + \frac{\tau\alpha(1 + \rho)}{(1 - \rho^2)}\right) \tag{4.52}$$

The denominator from Eqn (4.50) is usually close to unity, as reflectance of typical glass is low (e.g., around 0.04), and becomes negligibly low if squared.

For the sake of estimations, it is assumed that, in case of two identical standard glass panes (like a double-glazed window), transmittance is equal to:

$$\tau_2 = \tau^2 \tag{4.53}$$

In case of triple glazing, we may approximate that $\tau_3 = \tau^3$, therefore, for N layers glazing we get:

$$\tau_N = \tau^N \tag{4.54}$$

Simplified Eqn (4.54) is often used for a few covers and enables quite easy estimation of radiation attenuation caused by penetrating consecutive covers. It is also possible to estimate cover absorptance using formula $\alpha = 1 - \tau - \rho$.

When discussing solar radiation transmission through transparent covers, it is needed to consider absorptance of each cover separately in order to determine the share of solar radiation absorbed in every cover. Using Figure 4.4, it is possible to formulate the absorptance equation (for both s- and p-polarized radiation) for every cover, taking into account their arrangement and interactions. Thus, appropriately for the first outer and second inner covers we get:

$$\alpha_{s,out} = \alpha_{1,s} + \tau_{s,1}\alpha_{s,1}\rho_{s,2} + \tau_{s,1}\alpha_{s,1}\rho_{s,2}^2\rho_{s,1} + \tau_{s,1}\alpha_{s,1}\rho_{s,2}^3\rho_{s,2}^2 + \ldots$$

$$= \alpha_{1,s} + \tau_{s,1}\alpha_{s,1}\rho_{s,2}\sum_{n=0}^{\infty}(\rho_{s,2}\cdot\rho_{s,1}) = \alpha_{1,s} + \frac{\tau_{s,1}\alpha_{s,1}\rho_{s,2}}{1 - \rho_{s,1}\rho_{s,2}} \tag{4.55}$$

$$= \alpha_{1,p}\left(1 + \frac{\tau_{s,1}\rho_{s,2}}{1 - \rho_{s,1}\rho_{s,2}}\right)$$

$$\alpha_{s,in} = \tau_{s,1}\alpha_{s,2} + \tau_{s,1}\alpha_{s,2}\rho_{s,1}\rho_{s,2} + \tau_{s,1}\alpha_{s,2}\rho_{s,1}^2\rho_{s,2}^2 + \ldots$$

$$= \tau_{s,1}\alpha_{s,2}\sum_{n=0}^{\infty}(\rho_{s,1}\cdot\rho_{s,2}) = \frac{\tau_{s,1}\alpha_{s,2}}{1 - \rho_{s,1}\rho_{s,2}} = \alpha_{2,s}\frac{\tau_{s,1}}{1 - \rho_{s,1}\rho_{s,2}} \tag{4.56}$$

Relations for the p- and s-polarized radiation are identical, whereas absorptance for every transparent cover struck by unpolarized light is calculated as an arithmetic mean of both components. If both glasses are identical, so are their relevant optical properties:

$$\alpha_1 = \alpha_2 = \alpha; \quad \tau_1 = \tau_2 = \tau; \quad \rho_1 = \rho_2 = \rho$$

Then absorptance of the outer and inner cover is, respectively:

$$\alpha_{out} = \alpha\left(1 + \frac{\tau\cdot\rho}{1 - \rho^2}\right) \tag{4.57}$$

$$\alpha_{in} = \alpha\left(\frac{\tau}{1 - \rho^2}\right) \tag{4.58}$$

Depending on the available source data, analyses of spectral properties of transparent covers may utilize three basic models of multilayer covers and methods for determining their transmittance, reflectance, and absorptance. "External" and/or "internal" radiometric properties or optical indexes are used. Models based on "external" and "internal" radiometric properties utilize the description of the repeated reflection phenomena occurring in a system of n covers, as already described. There are at least three equivalent concepts for solving the repeated reflection problem: Rubin's ray tracing method, described above (1982), Edwards's energy balance method (1977), and Harbecke's matrix transformation method (1986). Sometimes, also an optical indexes model is used. It is based on relations between the refractive index values of media involved and application of the Fresnel law and Snell law in the following form:

$$\cos(\theta_2) = \left[1 - \left(\frac{n_1}{n_2}\right)^2 \sin^2(\theta_1)\right]^{\frac{1}{2}}$$

However, in practice, formulas already discussed often prove to be difficult because of poor availability of spectral parameters of transparent materials. In many cases, only data for the radiation incident along the normal are available. Solar energy potential analyses use approximate methods for determining optical parameters, primarily transmittance, which is seen as the most important optical parameter for a building's glazing elements.

Transmittance of a single transparent cover may be calculated with Rivera's relation:

$$\tau = 1.018 \cdot \tau_o \left(\cos\theta + \sin^3\theta \cos\theta\right)$$

where:

τ_0 = transmittance of the cover for the radiation incident along the normal direction.

As already mentioned, the presented discussion is of key importance for analyzing the transparent elements of a building envelope, which will be further investigated in the following parts of this publication, including Section 4.3.2 of this chapter.

4.3 ANALYSIS OF PHENOMENA OCCURRING IN SOLAR ENERGY RECEIVERS

4.3.1 Simplified Heat Balance of Any Solar Energy Receiver

The essential purpose of every solar thermal system, active or passive, is to capture solar energy and convert it into heat with the highest possible efficiency, and then transfer the generated heat to consumers in an effective and relatively simple way [23,24]. Because of the periodical character of solar energy availability, the key role in every system utilizing that energy for heating is played by heat storage. In case of solar thermal systems, solar energy is collected in solar collectors where the photothermal conversion occurs.

Then the heat carrier transports the collected energy to the heat storage. From the storage, heat is supplied to consumers. Two main energy balance equations may be proposed for every solar system. One concerns only the solar radiation receiver, the other describes the energy storage tank. This section is focused on the energy balance for the solar energy receiver.

Solar radiation with irradiance of G_s reaches the surface of a receiver (e.g., solar collector). In case of a typical collector, its absorbing surface is covered with a transparent protective cover shielding it from potential damage and negative (thermal) influence of the environment. This cover has certain transmittance $\tau < 1$ and attenuates radiation to the value of τG_s. Radiation, which passes through that cover, is subsequently absorbed by the absorbing surface characterized by certain absorptance α (for solar radiation), while part is also reflected.

Usually the absorbing surface (plate), because of photothermal conversion, is heated up to a temperature T_p higher than the ambient air temperature T_a. For simplicity of consideration (at this stage), it has been assumed that the absorbing plate temperature and transparent cover temperature is the same, and the thermal state of the transparent cover, including its role in reduction of heat losses from the absorbing surface, can be neglected. Therefore, the heat losses from the absorbing surface are proportional to the temperature difference between the absorbing surface and its surroundings, and inversely proportional to the thermal resistance of the absorbing surface R_L. Heat exchange between the flat surface of an energy receiver and its surroundings is described by the Hottel–Whillier equation [25]. This relation shows that the net energy gain of the absorbing surface that is:

$$Q_{net} = \tau \alpha A_p G_s - \left[A_p \left(T_p - T_a\right)/R_L\right] = \eta_{sp} A_p G_s \qquad (4.59)$$

where η_{sp} is the efficiency of solar energy capture by a flat absorbing surface (plate). According to the equation above, this efficiency may be described as:

$$\eta_{sp} = \tau \alpha - \left[\frac{T_p - T_a}{R_L G_s}\right]$$

Part of the net energy gain is transferred to the working fluid with a temperature T_f. It is desirable to make the difference between the absorbing surface temperature T_p and fluid T_f as small as practicable in order to maximize the efficiency η_{pf} of heat exchange between those media (make it close to 1). Then it is possible to derive an equation for useful energy gained by the medium (working fluid of the receiver), which transports heat captured from solar radiation and stored within the medium's volume:

$$Q_u = \eta_{pf}\left[\tau \alpha A_p G_s - A_p\left(\left(T_p - T_a\right)/R_L\right)\right] = \rho c V \frac{dT_f}{dt} \qquad (4.60)$$

It is an energy balance equation for the solar radiation receiver (a solar collector) [26]. A solar collector is usually used as a component of an active solar thermal system. Typically, used working fluid is water (possibly with antifreezing agents), sometimes air is also used [27]. The role of a solar collector may also be played by a solar collector wall of a passive system. In such a case, the balance equation is identical to Eqn (4.60), and the medium used to capture and store heat is air.

In some passive building envelope solutions in countries located at lower latitudes (mainly southern states of the United States), collector-storage walls are used. In some of these walls, water tanks are placed inside the structure of the external walls to act as heat storage tanks. However, most storage walls follow the Trombe or Trombe–Mitchell design. In this case, the energy balance equation has a more complex form. Besides being transferred to a heat carrier (i.e., air in an air gap between the cover and wall), heat gained is also transferred to the wall itself, which acts as a main storage. Phenomena occurring in such a system together with relevant heat balances are presented in Chapter 5.

One more commonly used solar energy receiver is the building itself. If we neglect the influence of solar radiation on opaque covers, we may assume windows to be transparent envelopes with certain transmittance and describe the interior with appropriate absorptance. Energy balance for a building understood as a solar collector and heat storage system is the balance for air contained in that building, with a temperature T_{in} variable over time. In general form, it may be expressed as:

$$Vc_p\rho \frac{dT_{in}}{dt} = \dot{Q}_{in}(t) - \dot{Q}_{out}(t) + \dot{Q}_{qv}(t) \tag{4.61}$$

Solar energy Q_{sol} is always an energy gain, as it flows into the building (during the daytime). Heat transferred through external walls, windows, roof, ground, and ventilation may be directed outward (heat losses) or inward (heat gains). Heat transfer direction depends on ambient air temperature, interior air temperature, and heat capacity of the building. Interior air temperature may also be affected by internal heat sources. Bearing all that in mind, we may transform Eqn (4.61) into a general form of energy balance equation. Thus, taking into account heat losses through the envelope and ventilation, and heat gains because of internal heat sources and solar energy gains, the energy balance equation for a building itself is:

$$Vc\rho \frac{dT_{in}}{dt} = \sum \dot{Q}_{walls}(t) + \sum \dot{Q}_{win}(t) + \sum \dot{Q}_{roof}(t)$$
$$+ \sum \dot{Q}_{gt}(t) - \sum \dot{Q}_{sol}(t) - \sum \dot{Q}_{qv}(t) \tag{4.62}$$

Mathematical expression of individual energy flows may be more or less complex, depending on the aim of calculations. Detailed formulas for all flows are presented in Chapter 6.

4.3.2 Transparent Covers

The share of glazing in modern building envelopes has been increased a lot recently. People staying indoors like to have a view of the outside. This provides them with visual contact with external surroundings and is important both in urban and rural conditions. The type of glass used, understood as its optical and radiative properties, depends on the role it is to play in a building structure.

The type of glass most recommended for increasing solar radiation availability indoors is glass with low iron content. Such glass should contain less than 0.02% of iron oxide (Fe_2O_3) [18]. This kind of glass looks totally transparent, even for an observer looking at it at a very large angle. The opposite solution is absorptive glass with significantly increased iron content. It is used to restrict access of solar radiation (transmittance) and protect rooms from excessive irradiation. This kind of glass may absorb roughly 50% of incident radiation. The glass surface may have the appropriate finish increasing reflection of radiation incident at small angles, which allows blocking more radiation around noon. Glass may be covered with different substances in order to restrict access of radiation at certain wavelengths (i.e., to provide certain spectral properties). More information about glazing, especially windows, is given in the chapter dealing with energy balance of a building (Chapter 5).

Optical properties of a transparent body are described through its transmittance, reflectance, and absorptance. These parameters are decisive for the amount of losses (i.e., attenuation of radiation passing through such transparent bodies). Transparent bodies, like covers used in active and passive solar energy systems in buildings, should preferably be perfectly transparent ($\tau = 1$), although, as already mentioned, transmittance of transparent bodies is always lower. In addition, increasing the cover thickness results with lowered transmittance for solar radiation. However, because of the structural strength of this material, the cover thickness cannot be too small.

In case of transparent covers used in solar systems, an important role is played by the interface between the two media: glass and air. In most favorable conditions (i.e., for the solar radiation angle of incidence equal to zero), the refractive index of a typical glass is 1.526 (for a polycarbonate cover it is 1.60), while for air it is equal to 1. Therefore, according to Eqn (4.17), reflectance is 0.0434. As there are two interfaces on a cover, the total reflected radiation is $2 \times 4.34\% = 8.68\%$.

Variations of glass transmittance depending on radiation wavelength are very favorable for solar energy applications. Glass covers of solar collectors are almost totally transparent for wavelengths of solar radiation (by average some 91%) and usually opaque for long waves of thermal radiation (totally opaque for ultraviolet radiation). Nevertheless, glass does absorb longwave radiation emitted by absorber surfaces. This leads to increasing glass temperature and, thus, also increased convective heat losses. The clear disadvantage of glass is its fragility. Glass may be easily damaged and is not particularly resistant to thermal stress. For those reasons, transparent covers are also made of plastics, whose role is currently increasing, especially in solar active systems (new generation plastics are getting increasingly popular). Plastics are usually more transparent than glass, their transmittance for solar radiation is around 0.92—0.94, and, in some cases, may even reach 0.97. Unfortunately, they are also partially transparent for longwave radiation.

Covers of both collectors and passive elements may also be made of transparent insulating materials (polycarbonate or aerogels), which are described in more detail in Chapter 5. However, one of their disadvantages is the low melting point around 120 °C. This level of temperature may occur in solar collector covers, leading to degradation or even destruction of such a cover.

Another possible solution is using loosely packed fiberglass materials, which have much higher heat resistance than typical pure-glass covers, as their structure contains small pores filled with air. Nevertheless, thermal resistance of a material with closed small air cells drops sharply, if the material gets wet or if the air-filled pores become larger causing intensified heat exchange through convection. It may be added that thermal resistance for conduction always drops in wet and humid materials.

The highest transparence for radiation occurs at angles of incidence equal to zero (incidence along the normal). If the angle of incidence grows beyond 60°, transmittance drops, while reflectance increases (absorptance gets only slightly higher). At angles close to 90°, practically all radiation is reflected. Those parameter variations are shown in Figure 4.5 [28].

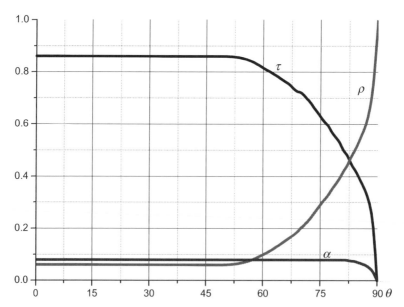

FIGURE 4.5 Transmittance, reflectance, and absorptance for solar radiation as a function of radiation's angle of incidence.

In simplified calculations, it is assumed that for angles of incidence of radiation lower than or equal to 70°, glass transmittance is of some 0.91. According to the Maxwell law, reflectance depends on the refractive index of the material in question and radiation's angle of incidence. It turns out that for glass, at angles of incidence <40° reflectance of visible light is 0.08. This means, that if there was no absorption in the glass, its transmittance would be 0.92. However, in reality, some small part of radiation is always absorbed in glass.

Descriptions of transparent cover's ability to transmit solar radiation often involve the extinction coefficient K used in Eqn (4.23). The value of K depends on the type of material and is usually assumed to be constant for the entire spectrum of solar radiation. It varies from 0.04/cm in case of white glass of good quality to some 0.30/cm for typical window glass that contains iron. Exemplary parameters of glass used to make covers are shown in Table 4.1 [6]. It needs to be pointed out that these parameters do not take into account the possible additional surface layers, which may alter properties of a cover according to its function.

Each additional glass layer within a cover decreases transmittance for solar radiation because of the phenomena described in detail in Section 4.2 of this chapter.

The discussion just presented is true for beam radiation. However, the solar radiation incident on a transparent cover consists of beam, diffuse, and reflected components. Beam component is strictly directional, while diffuse and reflected radiation is incident from the entire range of angles from 0° to 90°. Solar radiation transmitted by a transparent cover may be approximately determined by integrating that radiation over the whole range of the angles of incidence. However, the usually detailed angular distribution of diffuse radiation is unknown and also the spectral properties of covers are not available.

In case of an isotropic model of diffuse radiation, a simplified method may be used to determine the cover's transmittance. It is assumed that the discussion refers to an equivalent angle of incidence for beam radiation,

TABLE 4.1 Key Parameters of Glass Used to Make Covers

Characteristic Parameter	Common Glass (Pane)	Glass with Lowered Iron Content	Low-Iron Glass (High-Transparency Glass)
Iron oxide content (%)	0.12	0.05	0.01
Refractive index	1.52	1.50	1.50
τ (along normal) (%)	79–84	88–89	91–92
Thickness (cm)	0.32–0.64	0.32–0.48	0.32–0.56

approximately equal to 60° [29]. This means that transmittance of any transparent cover for the diffuse radiation incident from all directions in a uniform manner (isotropic model) is equal to the transmittance (the same is true for reflectance and absorptance) for the beam radiation incident at an angle of 60°. Increasing sky cloudiness brings the actual character of diffuse radiation closer to the isotropic model.

According to Brandemuehl and Beckman [29], the equivalent angle of the incidence of diffuse radiation on a tilted surface, for which transmittance is determined using an isotropic model, varies from 55° to 60° and the latter value refers to both the horizontal and vertical surface. The equivalent angle of incidence of diffuse radiation on a tilted surface may be calculated with the following equation:

$$\theta_d = 59.68° - 0.1388\beta + 0.0011497\beta^2$$

The equivalent angle of the incidence for radiation reflected from the surroundings incident on a tilted surface used to determine transmittance varies from 60° to 90°. The angle of 60° is the value for the vertical surface, and 90° for the horizontal one. This angle may be calculated with the following equation:

$$\theta_r = 90° - 0.5788\beta + 0.002693\beta^2$$

In case of clear sky, the equivalent angle for diffuse radiation varies from 52.9° to 56.6° and is correlated to the angle of incidence of beam radiation through the Gueymard equation:

$$\theta_{d,b} = 56.6° \left[1 - \exp\left(- 8.88 + 10.45 \cos \theta - 4.30 \cos^2 \theta\right)\right]$$

In case of a partly cloudy sky, another formula may be used to calculate the equivalent angle of incidence $\theta_{d,p}$, as a function of cloudiness index c_c, which varies from 1 (for overcast) to 0 (clear sky):

$$\theta_{d,p} = 59.4° c_c + \theta_{d,b}(1 - c_c)$$

In case of using an anisotropic model of diffuse radiation, it is assumed that the circumsolar diffuse radiation has the same angle of incidence as beam radiation. Horizon brightening diffuse radiation is usually only a small part of the hemispherical radiation incident on a tilted surface and therefore is assumed to arrive at the same equivalent angle of incidence as the isotropic diffuse radiation.

In solar energy calculations aimed not on analyzing the very process of solar energy transmission through a transparent cover, but at more global description of solar energy conversion in passive and active elements and in building's envelope, it is assumed [17] that transmittance of one or several glass panes is equal to the transmittance for a normal angle of incidence, as in practical terms, it remains constant within angles 0°−60°, as shown in Figure 4.5. It needs to be noted that collectors, and therefore also their covers, are tilted in reference to the horizontal plane. For tilted surfaces, radiation angles of incidence are smaller than for horizontal ones. Moreover, in high-latitude countries in case of a vertical surface oriented toward the south, which is typically true for the main glazed facade of a building, angles of incidence are small. The smallest values are observed in winter, but also during summer the values do not exceed the solar altitude angle (for Warsaw they change from $0°$ to the maximum of $61°$ observed only at noon in June). Transmittance for glazing of windows located in vertical walls and tilted roof windows may be assumed to be equal to the transmittance to the angle of incidence of $0°$.

4.3.3 System of Transparent Cover and Absorbing Surface

The question of using covers over surfaces of solar energy absorbers is important in case of key elements of active solar systems (i.e., solar collectors), as well as in passive systems and traditional buildings. In typical solar active and passive systems, a transparent cover is installed at a small distance (a few or several centimeters) from the absorbing surface.

Most active and passive solar systems are equipped with transparent covers to protect the actual absorbing surface (this does not apply to swimming pools absorbers, which do not have covers). Installation of a transparent cover over the absorbing surface results with a certain effect on transmission and absorption, which affects the optical efficiency of the system.

After passing the transparent cover, solar radiation reaches the absorbing surface. Part of the radiation is absorbed, while the rest is reflected and returns to the inner face of the cover. At that interface (air-cover), some radiation is reflected again and returns to the absorbing surface, where it is partially absorbed and partially reflected again toward the cover. It needs to be noted that radiation reflected from the absorbing surface is diffuse radiation. The phenomena of repeated reflection between a cover and the absorbing surface and absorption within that surface are graphically illustrated in Figure 4.6. These phenomena are similar to the repeated reflection in a system of two covers shown in Figure 4.4, although, now the second cover is replaced by an opaque medium with high absorptive capabilities. As a result, incident radiation cannot pass to the other side of this medium. The phenomena of repeated reflection

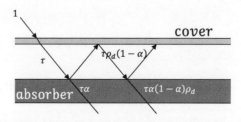

FIGURE 4.6 Repeated reflection between absorber and its cover.

between the absorbing surface and its cover are mathematically described by equation for the total transmittance-absorptance product [17]:

$$(\tau\alpha) = \tau\alpha \sum_{n=1}^{\infty} [(1-\alpha)\rho_d]^n = \frac{\tau\alpha}{1-(1-\alpha)\rho_d} \tag{4.63}$$

Reflectance ρ_d used in this equation is the reflectance of the cover valid for diffuse radiation scattered by the absorbing surface and incident at the inner side of the cover. Value of this reflectance may be calculated with Eqn (4.48), assuming the transmittance value for the angle of incidence is equal to 60° (we are dealing with diffuse radiation).

The amount of energy that may be absorbed by the absorbing surface depends on the optical properties of the media transmitting and absorbing solar radiation, which are represented in a combined form by the transmittance-absorptance product, also known as optical efficiency of solar radiation receiver. This parameter $(\tau\alpha)$ is slightly higher than the product of cover transmittance and absorber absorptance. In many simplified analyses, it is assumed that its value (for solar collectors with low iron glass) is equal to:

$$(\tau\alpha) \cong 1.01 \cdot \tau\alpha \tag{4.64}$$

Use of a cover over (in front of) the absorbing surface attenuates the solar energy gains. However, the main task of using a cover is to reduce heat losses from the absorbing surface of the receiver. In thermal analyses of solar systems operating in transient conditions, the so-called effective transmittance-absorptance product $(\tau\alpha)_{\text{eff}}$ is used. Its value is higher than that of $(\tau\alpha)$, which only depends on the steady-state optical properties of the involved media. The effective value takes into account the fact that energy absorbed in the glass increases its temperature, thus decreasing the temperature difference between the absorber and cover and, therefore, also reducing heat losses from the absorbing surface.

When analyzing radiation absorption on tilted surfaces separated from surroundings with covers, it may be taken into account that individual components of radiation are attenuated by the cover-absorber system to a different extent. Each component gets attenuated according to the transmittance-absorptance product $(\tau\alpha)$ valid for that component. If using the Liu-Jordan diffuse radiation model, hemispherical irradiation of a tilted surface is described by Eqn (2.31). If we take into account the cover-absorber system, the solar energy absorbed by a tilted surface during 1 h is equal to:

$$I_{sa} = I_b R_b (\tau\alpha)_b + I_d (\tau\alpha)_d \left(\frac{1+\cos\beta}{2}\right) + \rho_g (I_b + I_d)(\tau\alpha)_g \left(\frac{1-\cos\beta}{2}\right) \tag{4.65}$$

Coefficient $(\tau\alpha)_b$ in Eqn (4.65) is variable over time and refers to the beam radiation angle of incidence on the analyzed tilted surface. The value of this parameter for certain angles may be individually determined for the absorptive component (α/α_n) with Eqn (4.68) and for the transmittance component (τ) with Eqn (4.43). Transmittance τ_c for unpolarized radiation, taking into account repeated reflection effect, is described by Eqn (4.22) and transmittance τ_a, which addresses absorption, by Eqn (4.23).

As for the diffuse radiation described with an isotropic model, an equivalent angle of incidence of diffuse radiation on a tilted surface is determined. The value $(\tau\alpha)_d$ is assumed to be constant, with a value equal to $(\tau\alpha)_b$ for the beam radiation incident at $60°$. The value of $(\tau\alpha)_d$ may also be calculated with Eqn (4.63), and that of $(\tau\alpha)_g$ with Eqn (4.64).

In a special case of a vertical absorber with a transparent cover analyzed with an isotropic solar diffuse radiation model, the following is true: $(\tau\alpha)_d = (\tau\alpha)_g = (\tau\alpha)_{b60}$. Then Eqn (4.65) may be transformed to:

$$I_{sa}(t) = I_b(t)R_b(t)(\tau\alpha)_b + (\tau\alpha)_{b60}\left\{0.5\left[I_d(t) + \rho_g(I_b(t) + I_d(t))\right]\right\} \quad (4.65a)$$

If the HDKR anisotropic model of diffuse radiation is used to determine irradiation of a tilted surface, circumsolar and horizon brightening components need to be taken into account for cloudy skies. Anisotropic HDKR model is expressed by Eqn (2.50). If radiation is incident on a tilted surface separated from surroundings with a transparent cover, then the amount of absorbed solar radiation according to the HDKR model may be calculated as:

$$I_{sa}(t) = (I_b(t) + I_d(t)A_i(t))(\alpha\tau)_b(t)R_b(t) + I_d(t)(1 - A_i(t))(\alpha\tau)_d\left[\frac{1 + \cos(\beta)}{2}\right]$$

$$\left[1 + f(t)\sin^3\left(\frac{\beta}{2}\right)\right] + I(t)\rho_g(\alpha\tau)_g\left(\frac{1 - \cos(\beta)}{2}\right)$$

$$(4.66)$$

Many simplified studies utilize averaged values of $(\tau\alpha)_{mean}$. It is defined as a ratio of radiation absorbed by the absorbing surface to all radiation incidents on a transparent protective cover:

$$(\tau\alpha)_{mean} = I_{sa}/I$$

If the share of beam radiation in hemispherical radiation is high, then the value of $(\tau\alpha)_{mean}$ is close to the corresponding value for beam radiation. If, however, the role of diffuse radiation is dominating, the value of the mean coefficient will be closer to the value for diffuse radiation.

If we consider the whole building as a solar collector, then its transparent covers (i.e., windows) are much farther from the absorbing surfaces and the internal wall surfaces, acting also as heat storage systems. In such a case, the photothermal conversion and utilization of captured heat depend on transmittance of glazing, absorptance of individual surfaces indoors, and heat capacity of walls and objects inside the building.

After passing through a transparent cover (glazing) of a building's envelope, solar radiation reaches various internal surfaces, particularly walls, ceilings, and floors. Absorptance of any surface is lower than 1. Part of the radiation is reflected and returns to the inner face of the transparent cover (window). At the media interface (cover—air), radiation is partially reflected and returns to the absorbing surface where its part is absorbed and the rest gets reflected again and yet again reaches the cover. When it reaches the cover, part of the radiation is reflected again and part may pass through the transparent cover and "get out" of a building. These phenomena are repeated multiple times. The amount of energy, which may be absorbed by a surface absorbing solar radiation, depends on the optical properties of the media that transmit and absorb solar radiation.

Analyses of solar radiation penetration inside buildings use the concept of effective absorptance. Effective absorptance of a cavity (without transparent cover) is a ratio of solar radiation absorbed by a surface of internal walls (and floors, ceilings) of the cavity to all radiation incidents on that surface. Effective absorptance is a function of absorptances of all internal absorbing surfaces, as well as the ratio of the aperture area of all transparent surfaces (glazing) to the total area of the internal absorbing surfaces. It is expressed by the following equation [17]:

$$\alpha_{eff} = \frac{\alpha_i}{\alpha_i + (1 - \alpha_i)\frac{A_a}{A_i}} \tag{4.67}$$

where:

α_i = absorptance of the surface of an internal absorbing element,
A_a = aperture area of the transparent surface (glazing) (m^2),
A_i = total area of the internal absorbing surfaces (m^2).

Usually the relation between the angle of radiation incidence on a surface of a building and the absorptance of solar radiation for this surface is unknown. Duffie and Beckman [17] have proposed an approximate relation between the ratio of absorptance α_i for the solar radiation incident at any angle θ to absorptance α_r along the normal direction. This relation is expressed as a function of angle of incidence θ i.e., ($\alpha_i/\alpha_r = f(\theta)$) and is valid for irradiation of flat black surfaces. For angles of incidence $0° < \theta < 80°$, the approximate relation is:

$$\frac{\alpha}{\alpha_r} = 1 + 2.0345 \times 10^{-3}\theta - 1.990 \times 10^{-4}\theta^2 + 5.324 \times 10^{-6}\theta^3$$
$$- 4.799 \times 10^{-8}\theta^4 \tag{4.68}$$

In simplified calculations of effective interior absorptance, value of absorptance α_i is related to an effective angle of incidence of diffuse radiation, which is 60°. Using Eqn (4.68) and knowing the absorptance for the normal direction, it is possible to determine the absorptance α_i and then the effective absorptance with Eqn (4.67).

Window glass is characterized by certain transmittance τ_c for incident solar radiation. As already noted, the transmittance of glass for isotropic diffuse radiation is defined for an equivalent angle of the incidence of $60°$. The interior of a room is defined by its structural boundaries: walls, floor, and ceiling, with a total area of A_i. Their surfaces are characterized by certain absorptance for diffuse radiation α_i. The share of solar energy, which passes through the glazing and is absorbed by the interior (according to Eqn (4.67)), is described by the effective transmittance–absorptance product:

$$\tau_c\alpha_{eff} = \tau_c \frac{\alpha_i}{\alpha_i + (1 - \alpha_i)\tau_d \frac{A_a}{A_i}} \tag{4.69}$$

Solar radiation absorptance α_i may be different for different surfaces of the interior: walls, floor, and ceiling. In such a case, the average value may be used:

$$\overline{\tau_c \ \alpha_{eff}} = \tau_c \frac{1}{N} \sum_{n=1}^{N} \frac{\alpha_{ni}}{\alpha_{ni} + (1 - \alpha_{ni})\tau_d \frac{A_a}{A_{ni}}} \tag{4.70}$$

Transmittance of a transparent cover is a function of the wavelength of incident radiation. However, for most transparent covers, this dependence is insignificant within the spectrum of solar radiation and is normally omitted (except for the glass with high iron (Fe_2O_3) content around 50%, which has strong absorptive properties). Therefore, if a cover is made of a non-glass material, its transmittance may be considerably dependant on wavelength. In such a case, the monochromatic transmittance is used and then integrated for the entire spectrum. If the monochromatic transmittance also displays dependence on the angle, then the total transmittance for certain angles of the incidence θ is expressed as:

$$\tau(\theta) = \frac{\int_{\lambda=0}^{\infty} \tau_\lambda(\theta)I_{\lambda i}(\theta)d\lambda}{\int_{\lambda=0}^{\infty} I_{\lambda i}(\theta)d\lambda} \tag{4.71}$$

where $I_{\lambda i}(\theta)$ is the monochromatic irradiation incident on a transparent cover at an angle θ.

4.3.4 Influence of the Absorbing Surface Type on Solar Energy Collection on Selective Surfaces

When analyzing the conversion of solar energy into heat, it is important to determine the conditions that would enable the increased share of solar energy that may be collected. One of the possibilities is using an absorbing surface with a structure boosting solar energy absorption, while simultaneously decreasing losses through thermal radiation.

When referring to the phenomena of radiation absorption and emission, described in Section 4.2.3 to solar systems, it is necessary to note the specificity of the situation. The solar energy receiver receives radiation that is emitted by the Sun at a temperature of some 5800 K and in the spectrum below 3 μm, with the maximum for wavelength of some 0.5 μm. On the other hand, temperature of the solar radiation receiver (e.g., solar collector), is roughly 350 K. Radiation emitted from the surface of a receiver at such temperature is placed within the thermal radiation spectrum, with the maximum for wavelength of about 10 μm, as illustrated by Figure 1.6. In order to effectively utilize solar energy (increase solar gains and restrict thermal losses by radiation) it is necessary to increase absorptance of the absorbing surface for wavelengths at which solar radiation is absorbed, with the maximum monochromatic absorptance for solar radiation at $\lambda = 0.5$ μm. At the same time, the aim should be to decrease emissivity for the wavelength range at which the thermal radiation is emitted, especially for wavelengths where the maximum monochromatic emission of the absorber occurs, which, in this case, is $\lambda = 10$ μm. This condition may be generally presented as:

$$\alpha_{shortwave} \gg \varepsilon_{longwave} \qquad (4.72)$$

A surface fulfilling this condition is called wavelength selective, and usually selective. Nevertheless, it is quite difficult to fulfill this requirement in case of a regular black surface absorbing solar radiation, as it goes against the natural behavior of a typical gray body.

For a selective surface, the total absorptance and emissivity are weighted average values (Eqn (4.3)) for appropriately different wavelength ranges. The terms "equivalent absorptance" and "equivalent emissivity" are frequently used. Figure 4.7 presents the spectral characteristics of different surfaces, among them a perfect selective surface labeled with "1." A perfect selective surface would have absorptance equal to 1 for all wavelengths between 0.3 and 3 μm, corresponding with the solar radiation reaching the Earth's surface, and, at the same time, emissivity equal to 0 for waves longer than 3 μm, in range of

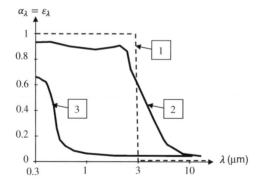

FIGURE 4.7 Spectral characteristics of different radiation absorbers. 1 = ideal selective surface, 2 = metal, and 3 = semiconductor.

thermal radiation (emitted by the solar radiation receiver). Such a perfect surface would totally absorb solar radiation, and would not emit any thermal radiation; therefore, there would be no thermal losses through radiation.

The selective surface is also called the semi-grey surface because it may be considered "gray" within the solar radiation spectrum (up to 3 μm) and also "gray" but with different properties within infrared spectrum (above 3 μm). In case of an opaque semi-grey selective surface, monochromatic absorptance α_λ is very high for wavelengths up to the 3 μm threshold (for the ideal case, it would be 1). Thus at:

$$\alpha_\lambda = (1 - \rho_\lambda) \to 1$$

reflectance ρ_λ is very low (for the ideal surface, it would be zero). Then for wavelengths over 3 μm reflectance ρ_λ is very high (for the ideal surface, it would be 1), and absorptance α_λ, and, therefore, also emissivity ε_λ are very low (for the ideal surface, they would be zero), because:

$$\alpha_\lambda = \varepsilon_\lambda = (1 - \rho_\lambda) \to 0$$

In real cases, variation of surface absorptance (emissivity) as a function of wavelength is different than those defined for the ideal selective surface. This is schematically presented in Figure 4.7 [18]. Except for the spectral characteristic of an ideal selective surface, it shows the absorptance changes of selective surfaces made of metal and semiconductor.

Figure 4.7 shows the spectral characteristics of a semiconductor material (Cu_2O)—curve 3, and a metal (Cu)—curve 2. Semiconductors have high absorptance within the visible radiation spectrum. For example, Cu_2O has high absorptance for wavelengths up to 2 μm. Some metals, like copper, have their absorptance high for shortwave radiation, up to 0.5 μm. Analyzing both of the spectral distributions presented in Figure 4.7, it may be noted that in order to receive a highly selective surface, it would be necessary to combine the properties of two such materials. This is possible through covering the metal with a thin layer of semiconductor. Energy absorbed on the semiconductor layer would be transferred deeper, to the metal layer, through conduction. Thermal conductivity of semiconductors is low, therefore, it would be recommended to keep the semiconductor layer as thin as practicable. This layer cannot be too thin though, as otherwise part of the radiation reaching the metal would be reflected back toward the surroundings.

The selectivity of an absorbing surface is defined as a ratio of shortwave absorptance to longwave emissivity:

$$Selectivity = \frac{\alpha_{shortwave}}{\varepsilon_{longwave}}$$

Highly selective surfaces are characterized by the high value of the numerator of this fraction (i.e., high absorptance for shortwave radiation), and at the same time very low emissivity in the shortwave (thermal) spectrum.

Selective surfaces for solar absorbers may be created in a variety of ways. The description of various methods of obtaining selective surfaces and their physical fundamentals are provided by Wackelgard, Nicklasson, and Granqvist [30]. Fundamentals of selective surfaces are also presented by Duffie and Beckman [17], Twidell and Weir [18], and Yianoulis, Giannouli, and Kalogirou [31].

Selective surfaces are currently commonly used in solar collectors. In case of a building envelope, using selective surfaces on external walls could be recommended, if their only task was collecting solar energy (e.g., in case of collector or collector-storage walls). However, during the summer, such solutions create a significant risk of overheating the building. In summer, the external wall surfaces should have low solar absorptance (i.e., high reflectivity for solar radiation) and high emissivity for thermal radiation. Thus, it might be advisable to consider a solution with changing the radiation properties of the wall surfaces depending on the season. It also needs to be added that collector walls in high-latitude countries must be insulated. This considerably restricts the application of collector-storage walls, as their very concept contradicts using insulation. A collector-storage wall during a cold season and low irradiance period does not fulfill its function, and may even act as a thermal bridge.

Further information on selective surfaces used to cover the solar collector absorbers are given in Chapter 7, which deals with active solar systems.

4.4 HEAT TRANSFER BETWEEN THE SOLAR ENERGY RECEIVER AND AMBIENT SURROUNDING

Except for collecting solar energy, a solar receiver placed outdoor interacts with its surrounding. There is always heat transfer between the two media of a different thermal state, in this case between the solar receiver and its surrounding. The influence of the surroundings usually causes heat losses from the solar receiver. However, it could happen that the ambient air temperature could be higher than solar collector's working fluid temperature. In such a situation, the solar receiver gains solar energy (when it is available), as well as heat gains from the surroundings (because of the temperature difference). Thermal efficiency of the photothermal conversion depends directly on the ambient conditions.

Heat exchange between the external face of a receiver and its surroundings is carried out by convection and radiation. Natural convection is caused by the temperature difference between the receiver's external surface and the surroundings. Forced convection is related to forced air flow around a surface, usually caused by wind. Radiative heat transfer occurs between the receiver's external surface and other objects in its surroundings of different temperature, as well as the celestial hemisphere.

Heat exchange through convection occurs between the face of the receiver, transparent cover or absorbing surface directly (if no cover is used), and ambient air. In case of a flat plate (surface of a receiver) in contact with air flow (wind) at a speed of $u < 20$ (m/s), the following equation may be used [18]:

$$\alpha_c = a + bu \qquad (4.73)$$

where coefficients a and b are, according to Mac-Adamson, equal to 5.7 (W/m²K) and 3.8 (J/m³K), respectively (it is noticed [17] that apart from convection also thermal radiation heat transfer is included).

Morrison [27] proposes to determine the convective heat transfer coefficient with a simple formula:

$$\alpha_c = 5.8 \ \mathrm{W(m^2K)}^{-1} \ \text{for} \ u = 1 \ (\mathrm{m/s}) \ \text{or} \qquad (4.73a)$$

$$\alpha_c = 8.8 \ \mathrm{W(m^2K)}^{-1} \ \text{for} \ u = 2 \ (\mathrm{m/s}) \qquad (4.73b)$$

Publications by the American Society of Heating, Refrigerating, and Air Conditioning Engineers [32] recommend to use Eqn (4.73) following a and b values:

for glass: 5.50 and 2.70, respectively;
for bricks: 10.21 and 4.57, respectively.

Mitchel proposed to use a correlation empirical equation showing the Nusselt number as a function of the Reynolds number, applied for the entire building subjected to the influence of wind [18]:

$$Nu = 0.42 \ Re^{0.6} \qquad (4.74)$$

The characteristic length of the Nusselt number ($Nu = \alpha L/\lambda$) used in Eqn (4.74) is assumed to be equal to a cube root of the total volume of the analyzed building. The wind speed u is contained within the Reynolds number: $Re = uL/\mu$.

In case of low wind speeds or total lack of wind, heat exchange with the surroundings occurs through natural convection. In this case, it is possible to calculate the heat transfer coefficients with correlation empirical equations valid for natural convection in unlimited space (external surroundings). Typically, empirical equations for the Nusselt number Nu as a function of a product of the Grasshof number ($Gr = [(g\beta'\Delta T^3)/\nu^2]$) and the Prandtl number ($Pr = \nu/a$) (for natural convection) are used. Using the Mikheev equation, the Nusselt number may be expressed in the following general form:

$$Nu = C(Gr \ Pr)^n \qquad (4.75)$$

In case of flat horizontal surfaces (e.g., roofs), it is possible to use the Lloyd and Moran equation. This equation [19] has been formulated for horizontal hot flat surfaces (with the length-to-width ratio of 7:1):

$$Nu = 0.76 \ Ra^{1/4} \ \text{for} \ 10^4 < Ra < 10^7, \qquad (4.76a)$$

$$Nu = 0.15 \ Ra^{1/3} \ \text{for} \ 10^7 < Ra < 3 \times 10^{10} \qquad (4.76b)$$

In this case, the characteristic length L included in the Nu is equal to the ratio of the quadrupled area of the analyzed surface to its circumference. The Rayleigh number used in these equations is defined as $Ra = (g\beta'\Delta TL^3)/(va)$.

In case of a vertical surface—typical case of a building envelope—it is possible to use the MacAdams equations:

$$Nu = 0.59\,Ra^{1/4} \text{ for } 10^4 < Ra < 10^9 \tag{4.77}$$

$$Nu = 0.13\,Ra^{1/3} \text{ for } 10^9 < Ra < 10^{12} \tag{4.77a}$$

It is also possible to use a more complex Shewen's formula from 1996 [33]:

$$Nu = \left[1 + \left(\frac{0.0665\,Ra^{\frac{1}{3}}}{1 + (9000/Ra)^{\frac{1}{4}}}\right)\right]^{\frac{1}{2}} \tag{4.77b}$$

Correlation empirical equations just presented are used in solar energy calculations when analyzing solar collector surfaces, passive walls, and typical building walls.

Radiative heat transfer occurs between the face of a receiver with a temperature of T_p and farther surroundings, represented by an apparent sky with effective temperature T_{sky}. Apparent sky is considered to be a black body. Heat flux exchanged through radiation with apparent sky (taking into account the view factor F_s) is described by equation:

$$\dot{Q}_{rpsky} = F_s \varepsilon_p \sigma A_l \left(T_p^4 - T_{sky}^4\right) \tag{4.78}$$

Heat flux exchanged through radiation with the nearest surroundings (taking into account the view factor value, which for this case is $F_g = 1 - F_s$) is described by equation:

$$\dot{Q}_{rpg} = F_g \varepsilon_p \sigma A_l \left(T_p^4 - T_a^4\right) \tag{4.79}$$

It may be noted that heat fluxes exchanged between the receiver's surface and the surroundings are functions of two different temperatures: ambient air temperature T_a and apparent sky temperature T_{sky}, which considerably complicates the calculations.

Sometimes, certain simplifications are proposed. For example, Morrison [27] assumes that the coefficient of radiation heat exchange with the sky may be assumed as constant, and proposes to utilize the value of 5−6 W/(m^{2}K) for clear sky conditions.

REFERENCES

[1] Weiss W, Bergmann I, Faninger G. Solar heating worldwide. Markets and contribution to the energy supply 2006. IEA, International Heating & Cooling Programme; 2008.
[2] A technology map for the European strategic energy technology plan, presentation at AGE meeting. Brussels: EC; October 2007.

[3]   CEN, EN 16316-4-3 Heating systems in buildings — Method for calculation of system energy requirements and system efficiencies — Part 4.3 Heat generation systems, thermal solar systems, 2007.

[4]   Weiss W, Bergmann I, Stelzer R. Solar heating worldwide. Market and contribution to the energy supply 2007. IEA, International Heating & Cooling Programme; 2009.

[5]   Dalenback JO. Take off for solar district heating in Europe. Pol Energ Słonecz 2010; 1—4/2009, 1/2010:9—13. Polskie Towarzystwo Energetyki Słonecznej — ISES.

[6]   Crema L, Bozzoli A, Cicolini G, Zanetti A. A novel retrofittable cooler/heater based on adsorption cycle for domestic application. Pol Energ Słonecz 2010;1—4/2009, 1/2010:43—53. Polskie Towarzystwo Energetyki Słonecznej — ISES.

[7]   Weiss W, Wittwer V. Contribution of solar thermal to the EU SET plan. Consolidated position of ESTTP. Presentation at the AGE Hearings, Brussels, V; 2007.

[8]   Maidment GG, Paurine A. Solar cooling and refrigeration systems. Compr Renewable Energy 2012;3:481—94. Elsevier.

[9]   Critoph R. Solar thermal cooling technologies. Pol Energ Słonecz 2010;1—4/2009, 1/2010: 14—9. Polskie Towarzystwo Energetyki Słonecznej — ISES.

[10] RETD Renewable Energy Technology Deployment. OECD/IEA Renewables for heating and cooling. Untapped potential; 2007.

[11] Kuznik F, Virgone J, Johannes K, Roux J-J. A review on phase change materials integrated in building walls. Renewable Sustainable Energy Rev 2011;15(1):379—91.

[12] Zhang Y, Zhou G, Lin K, Zhang Q, Di H. Application of latent heat thermal energy storage in buildings. State-of-the-art and outlook. Build Environ 2007;42:2197—209.

[13] Resnick R, Halliday D. Fizyka, tom 2. Warsaw: PWN; 1972.

[14] Jeżewski M. Fizyka. Warsaw: PWN; 1970.

[15] Feynman RP, Leighton RB, Sands M. The Feynmann lectures on physics, vol. 1, part 2. Pearson/Addison-Wesley; 1963.

[16] Frank S, Crawford jr. Waves. Berkeley physics course, vol. 3. New York: McGraw-Hill; 1968.

[17] Duffie JA, Beckman WA. Solar engineering of thermal processes. New York: Wiley; 1980.

[18] Twidell J, Weir T. Renewable energy resources. E&FN SPON, University Press Cambridge; 1996.

[19] Holman JP. Heat transfer. McGraw-Hill Higher Education; 2002.

[20] Kaviang M. Principals of heat transfer. New York: John Wiley & Sons, Inc.; 2002.

[21] Mikielewicz J. Modelowanie procesów cieplno — przepływowychIn Maszyny Przepływowe, tom 17. IMP PAN, Wydawnictwo PAN; 1995.

[22] t.1—3 Encyklopedia Fizyki. Warsaw: PWN; 1974.

[23] Faninger G. Solar hot water heating systems. Compr Renewable Energy 2012;3:419—48. Elsevier.

[24] Kalogirou SA, Florides GA. Solar space heating and cooling systems. Compr Renewable Energy 2012;3:449—80. Elsevier.

[25] Hottel HC, Whiller A. Evaluation of flat plate solar collector performance. In: Trans. conf. on use of solar energy, II Arizona; 1955.

[26] Hottel HC, Woertz BB. Performance of flat plate solar collectors. Trans ASME 1942; 64:91.

[27] Morrison G,L. Solar collectors. In: Gordon J, editor. Solar energy: the state of the art. ISES position papers. UK: James & James; 2001.

[28] Hollands KGT, Granqvist CG, Wright JLL. Glazings and coatings. In: Gordon J, editor. Solar energy: the state of the art. ISES position papers. UK: James & James; 2001.

[29] Brandemuehl MJ, Beckman WA. Transmission of diffuse radiation through CPC and flat-plate collector glazings. Sol Energy 1980;24:511.

[30] Wackelgard E, Niklasson GA, Granqvist CG. Selectively solar-absorbing coatings. In: Gordon J, editor. Solar energy: the state of the art. ISES position papers. UK: James & James; 2001.

[31] Yianoulis P, Giannouli M, Kalogirou SA. Solar selective coatings. Compr Renewable Energy 2012;3:301—12. Elsevier.

[32] ASHRAE: handbook of fundamentals. American Society of Heating, Refrigerating and Air-Conditioning Engineer; 1977.

[33] Shewen E, Hollands KGT, Raithby GD. Heat transfer by natural convection across a vertical air cavity of large aspect ratio. J Heat Transfer 1996;118:993—5.

Chapter 5

Passive Utilization of Solar Energy in a Building

5.1 REDUCTION OF BUILDING ENERGY CONSUMPTION

Actions aimed at limiting buildings' energy demand have been observed for many years, both in the case of newly built structures and renovated ones.

In the case of traditional energy systems (heating systems, electrical systems), energy-saving analysis is usually carried out separately for issues related to generation and those related to final consumption. The energy producer is focusing on energy supply. When purchasing fuel, the producer is mainly interested in getting a good price. Due to strong energy pro-efficiency and environmental regulations, the producer also aims to utilize this fuel as efficiently as possible in the energy-generation process. This requires minimizing the producer's own consumption of energy in the generation unit: power plant, Combined Heat and Power (CHP) plant, or heating plant. Its operator endeavors to improve efficiency of the process converting primary (fuel) energy into the final energy, and to reduce emissions of pollutants. On the other hand, a final consumer creates an energy demand. The consumer is primarily interested in reducing costs of energy used by himself or herself. It means the consumer reduces consumption of energy, which is purchased as a product, having no influence on its quality and supply methods. Consumer efforts are therefore focused on various forms of energy saving within the building, including its systems. First, this involves reducing energy demand, particularly by improving thermal insulation of a building's envelope. Second, the consumer focuses on reducing energy demand from internal systems of a building, particularly by switching off some electric appliances and heaters (radiators) for some time. This also reduces final energy consumption.

However, in the case of using solar energy and other renewable energy sources, a holistic approach to energy-saving matters is applied. Equal attention is given to saving energy in the process of capturing it, converting it, transmitting it, and finally, consuming it. This is because a user of a building becomes an energy producer and consumer at the same time. All energy-related processes occur locally, and can be locally managed and optimized. Moreover, reduction of energy intensity of a building is starting from decreasing energy demand. In case of heat, this requires not only using sufficiently good

Solar Energy in Buildings. http://dx.doi.org/10.1016/B978-0-12-410514-0.00005-0

insulation and planning appropriate heat capacity of a building, but also well-thought-out architectural and civil concepts enabling good utilization of energy available in the environment, including solar energy. Appropriate elements of building's envelope may increase or decrease the influence of solar radiation on building's interior. Appropriate installations are selected and dimensioned for the planned, significantly reduced demand for heating and cooling.

One more significant fact needs to be pointed out here: introduction of active solar systems (as well as heat pumps) has changed the approach to planning heating systems. Standard heat sources and heating systems are designed (according to standards) to cover peak load of heat (extreme weather conditions, i.e., minimum ambient air temperatures). For example in central Poland this requires dimensioning indoor heating systems for conditions occurring at ambient air temperatures of some −20 °C, and in eastern Poland even −25 °C. Solar system, however, may not be designed for extreme conditions with respect to the installed capacity (as it is done for traditional heat sources), as this would result in a very large system and therefore a big increase of the investment cost, while it would still remain impossible to cover 100% of the heat demand anyway, especially in high-latitude countries. This approach would result in the installation being considerably overdimensioned with reference to the heat demand in the summer and create problems with excess heat. For that reason, in case of solar thermal systems for domestic hot water (DHW), they should be designed (dimensioned) to cover 100% of water heating demand in summer. In case of solar combination systems (DHW + space heating), they may be based on the assumption of covering the entire heat demand in spring (e.g., March), although usually the actual dimensioning process is more complex. Prior to construction, dimensioning, and installation of a solar heating system, preliminary energy efficiency measures are always implemented in order to restrict the energy demand in a newly built structure or to reduce the heat load in case of existing buildings through appropriate thermal refurbishment.

It needs to be emphasized that utilization of solar energy for heating purposes was the reason behind developing comprehensive analyses of all energy-related phenomena occurring in a building, starting from collecting solar energy as the "primary fuel" to its final utilization involving demand side management (DSM). Thus prior to designing an energy system that utilizes solar energy, it is necessary to carry out a wide range of actions aimed at limiting energy demand during operation of the facility. This involves not only reducing heat demand for space heating but also reducing cooling demand and electricity demand.

The tendency to reduce energy demand is continuous and typical for modern energy-efficient buildings [1], and it is expressed by decreasing energy consumption indicators that describe final and primary energy consumption per square meter of usable (living) area over an entire year [2]. Development of a study of a building's energy characteristics [1] is aimed at decreasing useful energy demand and reducing energy consumption, both in reference to final and primary energy. In case of typical new energy-efficient design, as well as in case of thermal refurbishment of already existing structures, reduction of heat demand for space heating [3—6] involves:

- Ensuring good thermal characteristics of building elements by utilizing energy-efficient engineering solutions and materials, especially appropriate structural materials, insulation, and windows.
- Energy-efficient design of heating and cooling (air-conditioning) systems.
- Utilization of automatic control systems for thermal or microclimatic conditions of rooms and for controlling operation of various systems and their parameters.
- Monitoring consumption of energy and its carriers.

A building designed and built using innovative technical solutions may consume much less energy than a typical structure built according to common standards and technologies. Correct shape of a building alone may considerably reduce demand for heating energy during a heating season, and at the same time create appropriate microclimatic conditions during summer [4]. Modern, energy-conscious, especially "solar-conscious" design of a building, should address the following issues:

- Location and shape of a building ensuring proper insolation and daylighting conditions.
- Concept of building's interior, with the layout of rooms according to their function and time of use.
- Creation or adaptation of building structure for passive utilization of solar energy by using passive solar systems in the form of buffer spaces, solar collector walls, and storage walls.
- Selecting a building's structural elements and their materials to provide protection from excessive solar radiation.
- Utilization of the neighborhood of a building to enhance or reduce solar energy availability; layout of vegetation.
- Utilization of the neighborhood of a building for natural cooling and ventilation, complementary to automatic control of window opening ("ventilating" windows), precooling or preheating of ventilation air in structural elements of a building and below the ground level.
- Adaptation of structural and installation solutions for solar energy utilization, including active solar systems, heat pumps, ground storage systems, etc.
- Adaptation of structural, material, and installation solutions for waste energy utilization, including energy recovery from ventilation systems, wastewater, etc.

To summarize, utilization of solar energy requires using a number of energy-saving solutions listed above, which allows reducing energy demand for heating [4]. Solar energy utilization enforces appropriate design and construction of a building involving specific architectural, civil, and installation solutions or implementation of appropriate upgrades in the case of already existing structures. It should be added that solar heating systems in buildings usually require application of low-temperature heating systems, like floor and wall heating systems, or other unconventional solutions (e.g., using the chimney effect in room ventilation). In the case of thermal refurbishment of an existing building, its heating system needs to be adapted to new operating conditions. For instance, lowered temperature of the heating medium might require increased heat exchange area of radiators (e.g., achieved by increasing the number of radiators in individual rooms). Often however, old equipment is totally replaced by a new low-temperature system.

It needs to be noted that for utilization of solar energy and other unconventional solutions during design and construction of a building, collaboration of architects, civil engineers, and installation technicians with experts and energy consultants specialized in such unconventional solutions becomes essential. Close collaboration between engineers of various disciplines and solar energy specialists guarantees that a building will be a low-energy structure enabling efficient utilization of energy available in the surroundings and effective cooperation between conventional installations and unconventional systems based on solar energy.

Thermal comfort conditions inside heated rooms depend on changing climate conditions, architecture and design of a building, materials used, and energy efficiency of systems operation [7]. Individual elements of a building—depending on their geographic orientation, shape, location within a building structure, and in reference to the neighborhood and the properties of material they are made from—may differently react to changing ambient conditions related to weather changes. Selection of construction materials, architecture of a building, and civil engineering solutions are particularly important in the case of using passive systems. Passive solutions allow utilizing building structure to absorb, store, and distribute collected solar energy. Carefully planned architecture of a building and its location allow utilizing solar energy and ensuring required thermal comfort conditions indoors. Buildings should be "open" on their southern side (in the northern hemisphere), i.e., faced toward the sun, while from the north they should be "closed," i.e., well insulated and tight. This means that amount of transparent elements of the envelope, especially windows, on the northern side should be minimized (while ensuring required lighting). Using passive solutions decreases operational energy demand, considerably reducing energy consumption for heating. In many cases, use of passive solutions is combined with daylighting systems. It is particularly important in the case of large building structures: residential, industrial, and especially in public utility buildings.

Passive solutions play an increasingly important role in architectural designs. Use of passive solar systems leads to a wider use of glazing, verandas, and atria, as well as closer links between some utility rooms and the neighboring environment [7–13]. At the same time, application of passive solar systems enforces not only introduction of additional transparent elements (windows, glazed facades) but also use of appropriate elements of building structure for heat storage, and construction of appropriate ducts for distribution of heated air. Some solutions, like so-called solar walls, combine passive and active functions. A solar wall with air ducts within its structure plays a role of an air solar collector (passive or active, or semi-passive). Utilization of passive solutions also requires designing appropriate shading elements of a building envelope, e.g., overhangs, wingwalls, and of its external neighborhood, e.g., layout of green areas. Introduction of passive solutions requires careful analysis of energy demand and conditions for collecting solar energy for specific climate conditions.

In higher-latitude countries, direct solar gain systems are not recommended. Such systems include large-area windows and glazed façades used primarily on the southern elevation. This kind of solution, especially in climate conditions observed in higher latitudes, often results in large indoor temperature variations, both during summers and winters, leading to overheating or overcooling rooms adjacent to such glazed panels [14,15].

In most climatic conditions it is recommended to use indirect solar gain systems. In case of single-family housing those might include full or ventilated collector walls (depending on climate), or sometimes also combined solar collector–storage walls [16]. If climate conditions are characterized by frequent and prolonged periods of overcast, it is recommended to enforce airflow in ventilation ducts (this is a so-called semi-passive solution). In the case of higher latitudes, it is reasonable to use indirect solar gain systems with buffer spaces. A buffer space is a link between the neighborhood and heated room, enabling utilization of available solar energy. At the same time it protects heated rooms from sudden unpredicted weather changes, thus improving thermal protection of a building. Utilization of indirect solar gain systems with buffer spaces in practice involves adding additional glazed space (room) adjacent to the southern façade of a building. In single-family housing, this kind of a buffer space is usually created as glazed verandas, loggias, greenhouses, or winter gardens adjacent to a building. In multistory buildings, passive solutions with buffer space stretch across several stories and have a form of additional transparent envelope in front of the main one. These solutions have a form of multilevel winter gardens, so-called solar spaces, glazed loggias, balconies, atria, etc. A buffer space must be provided with natural ventilation, or—if needed—forced ventilation. In higher-latitude countries, it is recommended to locate the buffer space within the building's interior, at its southern side, as already discussed in Chapter 4.

The aim of using passive systems is reduction of energy demand for room heating and cooling. Each building structure exchanges heat with its surrounding (heat losses in the cold period, heat gains when the ambient air temperature and solar irradiation are high). Intensity of this heat transfer depends (among other factors) on structural materials used to make opaque walls and windows. Utilization of advanced and innovative materials, design solutions, and installations makes it possible to reduce the amount of energy needed for heating and cooling to maintain thermal comfort inside a building at a required level. In some special conditions it might be even possible to achieve energy self-sufficiency (in this case, self-sufficiency with respect to heating and cooling).

Passive solutions in architecture improve energy balance of a building. At the same time they affect the external look of a structure, creating very aesthetic and functional solutions. For instance, introduction of buffers in the form of solar spaces, winter gardens, etc. brings the inhabitants closer to nature and their environment. Passive architecture solutions such as overhangs, extended roofs, or wingwalls, which in many cases are traditional for regional architecture, allow limiting excessive influence of solar radiation during summer. Modern solar architecture, not only in small residential houses but also in large city buildings, should combine the ability to use energy available in the environment with the possibility to restrict energy interactions when required. At the same time it should be based on proven solutions and construction traditions of local geographic and climatic conditions.

Utilization of solar active systems, passive solutions, PV systems, and daylighting solutions in a building taken together realizes the solar building concept. The impact of buildings in the global energy demand of developed countries is high; therefore modern methods for reducing energy consumption are becoming increasingly important. Rules and methods used in solar buildings should be seen as significant elements of comprehensive energy-saving measures. They should be incorporated in both new buildings, and in upgrades of existing ones.

5.2 PASSIVE SOLAR SYSTEMS

5.2.1 Classification of Passive Solar Systems

Thermal comfort conditions inside heated rooms depend on changing climatic conditions, the building's architecture and design, utilized materials, and installation. Individual components of a building react differently to changing ambient conditions (weather changes) depending on their geographic orientation, shape, location within the building's structure, environment, and finally the materials used and their properties. Solutions recommended for warm and cold climates could be different. Some architectural and civil solutions of passive systems recommended for applications in southern Europe should not be used in central or northern European countries. Moreover, each building design involving passive utilization of solar radiation must be adapted not only to climate conditions but also to the local environment and the building's exact location.

Essentially, passive solar systems are divided into systems with direct and indirect solar gains. Here are a few main types of indirect solar gain systems.

- System with a solar collector—storage wall, including:
  - Full walls
  - Ventilated walls
- System with a solar collector wall
- System with a buffer space, including:
  - System coupled to a heated room with a transparent partition
  - System coupled to a heated room with a storage partition

Names of individual systems are related to those components of a building that act as solar energy receivers. Those systems also include other elements of a building that play roles similar to those in active solar systems, i.e., store and distribute collected energy. A further explanation is given for the concepts of applying the phenomena occurring in individual system types and character of those phenomena.

5.2.2 Direct Solar Gain Systems

A direct solar gain system utilizes solar energy that enters rooms directly. This kind of a system is the cheapest and simplest solution. In practical terms any room with a window is a simple passive solar system that utilizes solar energy in a direct although uncontrolled way. Solar radiation passes through glazing, which acts as a transparent cover of the solar energy receiver (the room). Then, depending on solar radiation absorptance of surfaces of internal walls and equipment (e.g., furniture) present within the room, solar radiation is absorbed (part of it is reflected). Absorbed radiation increases the internal energy of the material of the surface of a wall or equipment, and is converted into heat. The heat is then stored for a shorter or longer period within the element, depending on its thermal capacity. Then heat is exchanged between media (elements) of different temperatures. As a result, air temperature within the room increases. Heat stored in the walls affects the thermal comfort inside the room.

Flow of energy supplied to the room's interior through glazing may be reduced by using shades, blinds, shutters, etc. or increased by enlarging glazed areas, especially on the southern side. However, enlarging glazing area increases heat losses at low ambient temperature. It needs to be noted though, that in case of modern windows with compact glass panes, more heat is transferred through frames than through glazing, and this needs to be taken into account when designing windows—the frame area needs to be reduced. When using large glazed areas, the temperature inside a room may change considerably and sharply. As a result, in extreme weather conditions, thermal comfort significantly deteriorates. Indoor temperature variations follow the pattern of ambient temperature variations, which is often undesirable and onerous for inhabitants. A direct solar gain system is schematically shown in

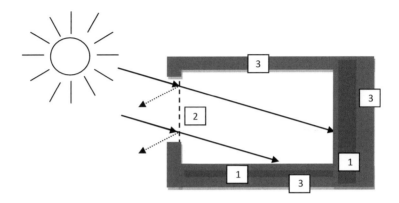

1 – Storage elements
2 – Transparent cover-glazing
3 – Insulation elements

FIGURE 5.1 Concept for application of a direct solar gain system.

Figure 5.1. The figure presents roles played by individual elements of rooms. Walls, floor, and ceiling, which are used for solar energy storage, are in direct contact with air inside heated rooms.

As already mentioned, windows are the main components of a direct solar gain system. During winter, their transmittance for solar radiation should be as high as possible, but at the same time they should provide relatively good thermal insulation. This latter characteristic not only depends on the glazing material but also on the entire window design, including the number of glass layers, their fastening method, material and thermal performance of frames, window mounting technology, etc. On the other hand, excessive solar energy flows during summer are undesirable and should be avoided.

In modern windows, heat transfer coefficients through the so-called central part of window panes are relatively low. It is the glazing (pane) edges, which remain crucial due to their contact with the frame. Also, the thermal effect of spacers between window panes can be significant. Metal spacers act as heat leakage bridges; therefore it is very important to choose a material with high thermal resistance for such elements. Recently so-called "warm" spacers have been introduced. They are made of insulating materials.

Thermal performance of windows is determined through experimental testing and numerical simulations. In 1989, Lawrence Berkeley National Laboratory introduced Windows 3.1 software simulating one-dimensional heat transfer phenomena occurring in vertical glazing. It was used for calculating coefficients of heat transfer through windows. Further research at Berkeley carried out in early 1990s brought a realization that while heat transfer through the central part is clearly one-dimensional, at the pane edges presence of spacers and a frame make the phenomena to be described at least two-dimensional ones. Accordingly, software was extended with a module handling two-dimensional heat transfer at pane edges and within the frame. It was later used for developing formulas for heat transfer coefficients through windows recommended by American ASHRAE standards [16]. The total heat transfer coefficient is calculated from separate components, with the central pane part, glass edge, and frame taken into account with appropriate weights.

Transparent facades and windows are essential elements of solar archi-tecture. Therefore glazing materials and window design or glazed façades are intensively researched all over the world. A separate discipline called "solar materials" was created within solar energy engineering to focus on applica-tions of transparent and opaque materials in solar technologies. The literature provides information on multiple interesting solutions related to application of windows and other glazed surfaces in passive solar systems. Some interesting solutions are discussed in studies by Hollands [17], Granqvist [18], Millburn and Hollands [19], Eames [20], Balcomb [21], and Papaethimiou [22].

Development of architectural concept for any structure aimed at adapting its envelope for active and passive solar energy utilization requires knowledge of the fundamentals of solar energy transmission through transparent bodies, which are essential envelope components of solar gain systems. Glass, espe-cially so-called solar glass with lowered iron content, is characterized by high transmittance for solar radiation (and low transmittance for thermal radiation), regardless of season. Solutions that could provide variable solar radiation transmittance depending on season (high during winter, low during summer) are being investigated nowadays. One such solution is electrochromatic glazing [18,22]. Windows and other glazed elements of a building's envelope (glazed façades) considerably influence the energy balance of a building.

5.2.3 Indirect Solar Gain Systems

A key disadvantage of a direct solar gain system is high variation of indoor temperature, often larger than normally tolerated by humans. It is possible to attenuate rapid temperature variations and also provide energy storage ca-pacity by isolating the interior of the building from the direct influence of solar radiation. The interior may be separated from the external environment by a storage wall acting as an intermediate element between two media. Such a wall is in fact a combination of a solar collector and storage system. This solution is known as the Trombe wall (Trombe was the Italian architect who proposed this concept) [10] and is schematically presented in Figure 5.2.

An indirect solar gain system may either include a massive collector—storage wall or a light collector wall. A collector—storage wall can be made as a full (Figure 5.2) or ventilated (Figure 5.3) structure. The latter solution may be either equipped with circulation openings above the floor and below the ceiling or with

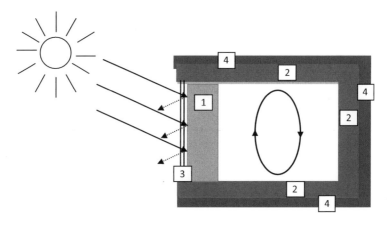

1 – Storage wall
2 – Storage elements
3 – Transparent cover
4 – Insulation elements

FIGURE 5.2   Concept for using an indirect solar gain system. Solar collector—storage wall.

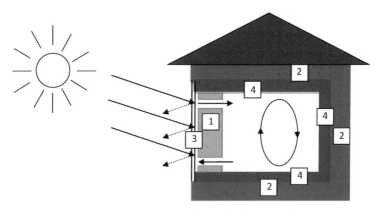

1 – Storage wall
2 – Insulation elements
3 – Transparent cover
4 – Storage elements

FIGURE 5.3   Concept of an indirect solar gain system with ventilated solar collector–storage wall.

a system of internal air ducts. From the outside, such wall is protected with a glass cover. A storage wall is often black to increase solar radiation absorptance. Collected solar energy, after passing through a transparent cover, is absorbed by the wall surface, where thermal conversion occurs. Generated heat is transferred into rooms thanks to conduction through the wall and then through convective and radiative heat transfer into the interior. Heat transfer in the bottom and top sections of the wall may be intensified by openings through which air is circulated. Cooler air is sucked in through the bottom openings, and then it is heated up inside an internal gap (between the wall and its cover), rises thanks to reduced density, and returns to the room through the upper opening. The flow of circulated air may be controlled by opening and closing circulation openings, changing the diameter of flow ducts, or external shading of the wall.

Reduced variation of indoor temperature is the key advantage of an indirect solar gain system. Heat conduction through a storage wall is a relatively slow process, so daily gains from solar radiation are transferred into the room during night. The time shift is at least a few hours, depending on the heat capacity of the wall (often time shifts are separately considered for day and night phases). However, an indirect solar gain system also has disadvantages. The outer part of a solar collector—storage wall is heated up to relatively high temperature, so heat losses from such a wall increase. The wall cannot have thermal insulation, as it would defeat its key purpose (to transmit the heat gained into a room). However, during days with low irradiation and low ambient temperature, the wall becomes de facto a thermal bridge and thermal losses become high.

It needs to be noted that in indirect solar gain passive systems, the solar collector—storage wall with air circulation openings, or so-called Trombe—Michel wall (Figure 5.3), should be relatively high. This is due to the necessity of perhaps obtaining a large density difference between air heated inside the wall gap and cool room air supplied from the bottom. High density difference creates a driving force for thermodiffusive airflow.

The main type of an indirect solar gain system, the Trombe wall, is an effective solution for countries with good insolation conditions and a warm climate with mild winters. Key disadvantages of this solution are the long time needed to heat up the storage wall, poor circulation of heated air in the gap and air supply (circulation) ducts, and high heat losses at low temperatures. For those reasons, such systems (Trombe wall, Trombe—Michel wall) are not a good solution for countries with poor insolation with climate characterized by long and harsh winters, and prolonged overcast periods. During operation of an indirect solar gain system, a large and rapidly changing temperature gradient may be formed within the storage block itself. Another problem is potential condensation of water vapor within the wall's structure during longer overcast periods characterized by low solar radiation flux, when airflow through the gap between the glass and the wall is stopped.

Figure 5.4 presents an indirect solar gain system with a solar collector wall. In this case, part of external elevation is built of collectors with natural, gravity-driven air circulation. A collecting wall is a light structure, protected with an external cover like any other solar collector, and its main component is an absorber insulated from the back side. Collected solar energy is transferred

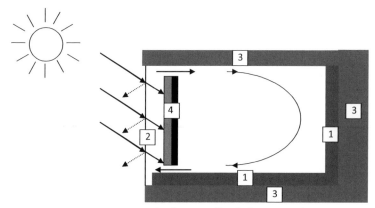

1 – Storage wall and other storage elements
2 – Transparent cover
3 – Insulation elements
4 – Solar collector wall

FIGURE 5.4 Concept for using an indirect solar gain system with a solar collector−wall.

to the heated room via convection. Solar energy accumulation takes place within internal structural elements, usually in internal walls.

Collector walls are often built in modules that are used as components or even entire walls of a building. Collector walls may absorb solar radiation directly, without glazed covers (so-called solar walls). Inside such walls, air is flowing beneath the absorbing surface through the entire space (cavity) or in ducts. Natural thermal diffusion occurs. Collector walls may have slightly perforated absorbing surfaces, which make them permeable for air. In addition, such a structure can be put on a regular insulated wall. In this case the heat is taken by air circulating in ducts or a gap (because of insulation, nearly no heat is transferred directly through the wall itself). The heated air flows to the ventilation center and is used directly or through a heat exchanger to release the gained heat. Solar walls can be used for heating in winter and cooling in summer. In some systems, forced heat flow is applied.

In some solar air-collector systems, heat is stored in a pebble bed beneath the heated building. Such a system in fact becomes an active air solar system. Air collectors mounted on external surfaces of elevations provide an additional thermal insulation layer and provide additional protection against heat losses. Solar energy storage in a pebble bed with a daily cycle (charged during the day, discharged at night) is a solution used effectively in lower-latitude countries. This kind of solar air-heating system appeared in the 1980s in the southern part of the United States. High solar irradiance allows storing heat during the daytime to be used for space heating during the nighttime. It needs to be noted that while ambient air temperature in such locations drops during the nights, it is usually only a few degrees lower than the required indoor temperature.

However, in higher-latitude countries, short-term storage of generated heat in a pebble bed is not effective and not recommended. Key problems could be the low solar irradiation and low ambient temperature at night. In such areas, temperature may drop below the freezing point even in May (for latitudes of about 50°). During the warmest periods of winter, average irradiation does not exceed 320 W/m^2 (March), and even if it does, it only occurs for a very short time (as illustrated by Figure 3.3). A few hours of daylight are insufficient for collecting enough heat to provide heating energy during the night, especially considering that during nighttime the ambient air temperature is typically much lower than the required indoor temperature. In winter at night it is usually below the freezing point. In early spring and late autumn at night, very often it may be about at the freezing point.

In 1980s in the United States, so-called water walls were proposed as a solution for increasing the heat capacity of collector–storage walls. In practical terms these walls were in the form of large, flat, box-shaped or cylindrical water containers made of metal. The tanks were painted with high-absorptance paint from the outside and protected with a transparent cover. Sometimes a water wall was made of semitransparent materials with water contained inside. Nowadays such solutions are no longer used. It may be added that this kind of wall may not be used in higher-latitude countries due to the risk of water freezing. For this reason, other solutions for passive solar systems need to be developed for countries with poor insolation conditions.

Special materials that would be transparent for solar radiation, but at the same time provide reasonable thermal insulation, have been developed in such countries. This research gave birth to an idea of transparent insulations (described in detail in Section 5.3.3). Other studies focus on improving conditions of heat accumulation in walls of various structures: slotted, porous, or with internal ducts. Intensification of heat exchange within walls is achieved by installation of small fans forcing heat flow; the resulting systems are described as semi-passive. Another innovative solution is integrating phase-change materials Phase Change Materials (PCMs) into building envelope elements. PCMs are used for short-term storage of gained solar energy in daily cycles (as discussed in Section 5.3.4).

5.2.4 Indirect Solar Gain Systems with Buffer Spaces

Indirect solar gain systems with buffer spaces will be discussed in the most detailed manner, because this kind of solution is most suitable for climatic conditions at higher latitudes. This is not very surprising, as buffer spaces have been used in traditional buildings in those areas for many years, first of all in form of glazed verandas.

As the name implies, a buffer space acts as a thermal buffer. On one hand it is an interconnector between the outside environment and heated (or cooled) indoor space. On the other hand it protects the heated (or cooled) space from effects of sudden and stochastic weather variations. Application of a buffer space allows collecting solar energy and improving thermal protection of a building. This combination of functionalities may not be achieved with the previously described indirect solar gain systems, and of course is also impossible with direct solar gain solutions. Moreover a buffer space separates the interior from the negative impact of the environment during both summer and winter.

In practical terms, application of an indirect solar gain system with a buffer space requires constructing an additional space adjacent to a building, covered with glazing (or incorporating such space within the main structure). Examples of buffer spaces adjacent to a building are shown in Figures 5.5 and 5.6. Figure 5.5 shows a typical buffer space added to a single-family house in the form of a so-called winter garden. Figure 5.6 presents a larger residential building with attached solar spaces that also play the roles of entrances and wind vestibules (entrance halls). Figure 5.7 shows an example of a buffer space incorporated into a main building structure. Also, additional external spaces formed by adding an additional glazed cover to existing elements of an envelope may be considered to be buffer spaces. The most typical solution of this type is glazing on loggias and balconies, as shown in Figure 5.8.

Modern solar systems with a buffer space were first proposed by Balcomb [9,10] in the form of a completely glazed veranda. A veranda built on the southern side is separated from the heated room by a massive storage wall. Solar radiation reaches interior of the veranda directly. Therefore inside verandas, temperature variations are sharp and large, as in any other direct gains system with large glazed areas. However, in this layout the inner room receives energy through the storage wall, i.e., indirectly. Such a solution has a

FIGURE 5.5 Typical example of a buffer space attached to a single-family house in form of a winter garden, Ulm, Germany.

FIGURE 5.6   Example of a buffer space attached to a residential building, Denmark.

number of practical and thermal advantages. Glazed verandas may be used for growing plants and, during certain seasons, for leisure, providing an additional useful area in a building. Such a structure provides a way for indirect transfer of solar energy into the building's interior and at the same time contributes to reduction of external environmental influence, including reduction of heat losses and gains. As a result, it helps maintaining desirable microclimate conditions inside a room. Additional elements providing thermal insulation (shades, blinds) may be added to enhance thermal effectiveness of entire system, particularly at nights.

FIGURE 5.7   Example of a buffer space incorporated into the structure of a single-family house, Warsaw, Poland.

FIGURE 5.8 Example of a buffer space in form of a glazed loggia, Warsaw, Poland.

Systems with buffer zones may utilize different types of walls between a veranda and heated room. It may be a storage wall, as in the Balcomb system. Figure 5.9 schematically presents an indirect solar gain system with a buffer space and a storage wall. In this solution, the veranda acts as a solar system composed of a solar collector and a heat accumulator. Its glazing acts as a transparent cover of the solar energy receiver. After passing through it, solar radiation is incident on surfaces of internal opaque components, including the storage wall, where the energy is partially reflected, but mainly absorbed and then accumulated. Part of the absorbed energy is transferred to the next room due to conduction through the storage wall. Solar energy converted into heat increases air temperature within the veranda. Thus the veranda becomes an element of thermal protection of the living spaces inside a building. Heat transfer may be controlled by regulating the airflow (between the veranda and heated room) through circulating ducts (schematically shown in Figure 5.9) or using shading devices as needed. A key advantage of this solution lies in the

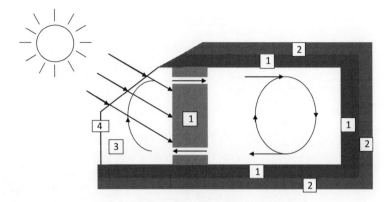

1 – Storage wall and other storage elements
2 – Insulation elements
3 – Buffer space
4 – Transparent cover of a buffer space

FIGURE 5.9 Concept of using of a buffer space in form of a glazed veranda coupled with a storage wall.

fact that it combines the functionality of a solar energy collector with the thermal protection of a building, and not only limits heat losses but also protects the building from unpredicted ambient air temperature variations, especially in extreme weather conditions. Buffer zones give extra thermal resistance for the interior of a building. Temperature variations within heated rooms for systems with a buffer space are much lower than in other types of indirect solar gain systems.

Another form of a buffer system is schematically shown in Figure 5.10. In this case, a veranda is separated from the heated room with a transparent barrier (e.g., additional windows). In this configuration the interior of the building plays the role of the solar energy receiver, which is protected by two transparent covers. A veranda may be very roughly compared to a gap between two transparent covers of a solar collector. External glazing of the veranda is acting as an outer cover (No. 4). Additional glazing (window) between the veranda and the internal rooms (No. 2) plays the role of the inner partition. Internal walls, floor, and ceilings act as storage elements (No. 1). Figures 5.11 and 5.12 show a real case of a buffer space seen from the outside and Figure 5.13 from the inside of a single-family house.

At higher latitudes, if a buffer space is placed within the building, then the time during a year when a buffer space may be used is usually longer than in case of buffer space attached to the building. This solution also increases thermal effectiveness of the entire system. When buffer space is incorporated into a building, then no glazed horizontal or sloped surfaces—which would be much more irradiated during summers than vertical ones—are needed (see Chapter 3). This often leads to excessive solar energy gains. Moreover, in the case of external buffer spaces attached to the building, glazed surfaces

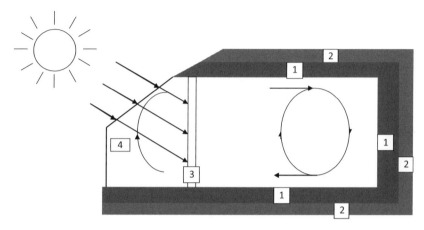

1 – Storage wall and other storage elements
2 – Insulation elements
3 – Transparent internal partition of a buffer space
4 – Transparent cover of a buffer space

FIGURE 5.10 Concept for using of a buffer space in form of a glazed veranda with transparent internal partition.

in direct contact with ambient air are much larger than for internal buffer spaces incorporated into the building. It means that during the cold season of low ambient air temperature, heat losses through external walls of attached buffer spaces are much higher than for walls of buffer spaces incorporated into the interior of a building. Besides, in an external attached buffer space, horizontal or sloped transparent glazing surfaces can "see" the sky much better than vertical ones. In result, heat exchange by thermal radiation with the sky is enhanced and heat losses increase, especially at nights. It may be noted that

FIGURE 5.11 System with a buffer space in the form of a glazed veranda with an internal glazed partition incorporated into a building.

FIGURE 5.12   Buffer space seen from the outside of a single-family house.

glazed verandas that play the role of heat buffers used in traditional architecture of higher-latitude countries (e.g., in Poland) have always been roofed,
as shown in Figure 5.14. Just like in case of modern buffer spaces, their
interior was used seasonally.

Another case of a buffer system is a solution with a double buffer space. In
this case, the first (outer) buffer space is separated from the room located
behind with a transparent barrier (windows). This next intermediate room is
separated from the living space of the building either with another transparent
barrier (windows) or with a storage wall. Figure 5.15 shows a solution with a
double buffer space with an internal storage wall.

FIGURE 5.13   Buffer space seen from the inside of a single-family house.

FIGURE 5.14   Buffer space in a traditional building.

Analogically to active solar systems, we may say that in case of a double buffer space with two transparent (glazed) internal barriers, we increase the number of transparent covers in order to reduce heat losses to the environment while retaining the possibility to use solar energy directly. This kind of solution is shown schematically in Figure 5.16. It is usually more expensive and requires more space than a system with a single buffer space incorporated into a building.

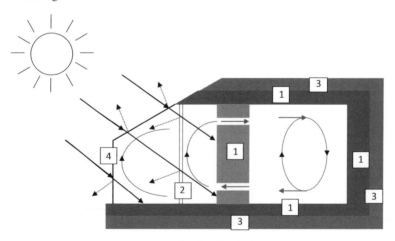

1 – Storage ventilated wall and other storage elements
2 – Transparent cover and interior transparent partition of buffer spaces
3 – Insulation elements
4 – Buffer zones

FIGURE 5.15   Concept for using a double buffer space in the form of a glazed veranda with two internal barriers: a transparent cover and a storage wall.

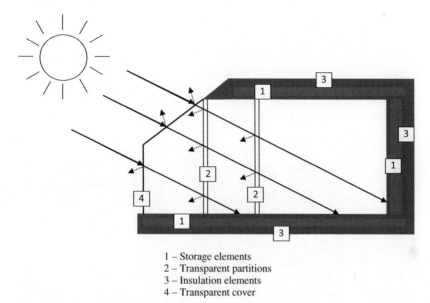

1 – Storage elements
2 – Transparent partitions
3 – Insulation elements
4 – Transparent cover

FIGURE 5.16   Concept for using a double buffer space in form of a glazed veranda with two glazed internal barriers.

A very important advantage of a system with a buffer space is the possibility of attaching it to an existing building during thermal refurbishment in a relatively easy way. Installation of glazing at a balcony or loggia, or construction of an adjacent greenhouse, is simple in terms of design and material selection. Also, labor intensity of such an extension is low, making it relatively inexpensive. Attaching a buffer space to a building improves thermal comfort, reduces heat losses, and provides additional living space. It is also preferred over other passive solutions by users who carry out thermal refurbishments. A very important aspect of installing glazing in a balcony or loggia is the necessity of providing ventilation. If this is neglected, then during cloudy weather with low solar irradiation, humidity will increase, and at low temperature below the dew point, water vapor will condense within the buffer space, leading to the growth of mold and fungi during prolonged overcast and low-temperature periods.

5.3 SELECTED BUILDING SOLAR TECHNOLOGIES

5.3.1 Research of Processes in Passive Solar Systems in Buildings

Research concerning the planned use of solar energy in passive systems of buildings commenced in late 1970s and for many years concentrated on theoretical studies and experimental works on passive systems operation [7,9,12,21]. During the 1990s, mathematical modeling of such systems' operation was also developed, as demonstrated by studies published at that time [10,11,23,24]. With time, the mathematical models became increasingly complex and detailed thanks to possible utilization of more and more accurate calculation tools [14,24−29]. In simplified studies, analysis usually concerns steady or quasi-steady states, and phenomena occurring within the envelope and surroundings of a building are described by constant total (equivalent) thermal resistance and heat transfer coefficient values. In the case of building's envelope, one-dimensional models are typically used.

It needs to be noted that research of heat transfer through an opaque wall in a one-dimensional model does not considerably distort the character of phenomena in simple and relatively homogenous wall structures (without thermal bridges). However, a transparent cover should be analyzed in a more sophisticated way. A window distorts spatial layout of an envelope due to its geometry, as well as its material and structural properties, and should be investigated in two or three dimensions.

When analyzing the impact of solar energy on a building, it is very important to consider shares of all individual components of solar radiation. In experimental studies using indoor laboratories and solar radiation indoor simulators, full verification of modeling results is not possible. Commonly used solar radiation simulators give the solar radiation spectrum for a given air mass. This means that total attenuation of solar radiation in the atmosphere is taken into account, but not with regard to specific components. Solar indoor simulators simulate only beam (direct) radiation as the total radiation. As a result, simulation of solar irradiance is inaccurate, as only direct solar radiation is represented in indoor tests. For this reason, mathematical models of solar energy collection and thermal conversion in a building envelope and its interior should be validated in natural conditions, typically in field laboratories.

In indoor laboratories, thermal properties of envelope elements, especially windows, but also walls, are tested. Tests are performed within airtight boxes, often called "hot" or "cold" depending on climate conditions under indoor solar simulation. Solar radiation is simulated by using light sources with a radiation spectrum similar to solar (individual components are not taken into account). Solar tracking systems simulating apparent movement of the sun on the sky are not always used. Influence of the sky expressed by its temperature changes in time is not taken into account.

Field tests of passive solar systems are usually performed with physical test modules of different sizes. The smallest ones, called test boxes, are considerably scaled down in reference to real objects; their individual dimensions normally do not exceed 1 m. Larger modules are called test rooms; their height is identical to that of the simulated rooms, while other dimensions are scaled down. The largest facilities are full-scale test buildings, which fully represent analyzed phenomena and processes.

Test boxes are primarily used for comparative testing of heat gains and losses, where convection is not taken into account. All investigated boxes have identical dimensions and are provided with good thermal insulation, except for the front, which holds various tested glazing types. In this case similarity to a real building is very vague: only the ratio between the glazed area and insulated area is kept at the same level as in a real building. Test boxes are used in both indoor and field tests.

Test rooms, externally similar to construction site containers, are usually configured in a permanent structural assembly, with exchangeable envelope elements [29]. Such rooms are typically built outdoors and used for comparative testing of different passive solar systems, both of indirect and direct gains type. Another application is verification of mathematical models describing thermal phenomena and solar energy collecting occurring in such systems. Ventilation often only concerns airflow by infiltration, which is assumed to be at low level, constant in time. However, nowadays forced ventilation is applied more often. Test rooms may be installed on moving platforms, which allow reasonably quick change of geographical orientation, i.e., change of insolation conditions. Usually experimental verification of mathematical models concerns individual components of a building, e.g., a window, transparent façade, opaque wall, etc.

In the past, solar test buildings were mostly uninhabited, located on the premises of research centers or universities, and used only for experimental research or validation of mathematical models describing phenomena occurring in those buildings. Nowadays, however, test buildings are demonstration facilities that are actually utilized for some practical purposes and act as "living laboratories." Parameters related to thermal comfort and process parameters of installed systems are monitored, recorded, and analyzed.

Contemporary theoretical and experimental research is particularly focused on studying and modeling dynamics of windows and glazed façades. Glazing is always an essential component of a passive solar system and allows utilization of available solar energy.

5.3.2 Selected Glazing Technologies

As already mentioned, the influence of solar radiation on the building envelope is particularly important in transparent structures—windows and other glazed envelope elements. There are many types of windows with different thermal and optical properties, and therefore different coefficients of solar radiation gains.

Essential components of a window are glass panes and frame. Currently in most European countries, double-glazed windows are used, although in passive houses (not passive solar), windows with triple glazing are installed. Old windows were typically double-glazed windows, with two separate panes. Nowadays compact windows, with at least two panes combined into one unit, are used. In the past a gap between panes was ventilated. In case of compact windows, the gap is never ventilated. In modern compact windows, gaps between panes are filled with a noble gas (usually argon, sometimes krypton). Until recently, when compact glazings were not yet in use, the gap between panes was filled with air. At present, glazing is formed by a single sheet of glass. However, in the past many small sheets separated by frame bars were used. In modern windows, mullions are sometimes used to give them an old-fashioned look. There is a spacer between glass panes. Different materials may be used for frames (e.g., wood, plastic, aluminum) and spacers (e.g., wood, aluminum, steel, sometimes with some content of insulation). The frame may be of full structure, usually made of wood, or of the chamber type, usually made of plastics with the chambers filled with air or insulation.

Optical phenomena related to transmission of solar radiation through a transparent medium are described in Section 5.3.3. Phenomena related to the temperature difference between the interior and outside (external surrounding) of a building concern heat transfer processes like thermal radiation, conduction, and convection. These phenomena can affect the energy balance of a window in various ways.

Total thermal resistance of glazing is the sum of elementary resistances, i.e., external thermal resistance (from the outside), thermal resistance of the glazing, and internal thermal resistance (from the inside). Thermal resistance of the glazing in a double-glazed window (consisting of outer pane—air gap—inner pane) depends on the intensity of phenomena in the glass (heat conduction, solar energy absorption) and in the gap (thermal radiation, convection, conduction). Typically resistance of a compact glazing unit is about 60% of the total value. This means that design and construction of a window has a decisive influence on heat transfer between the room and ambient air.

Research and development of new window technologies led to considerable reduction of heat transfer through glazing [30,31]. Most important solutions include a low-emission coating on panes and noble gas filling of gaps between panes. Heat transfer processes in the gas gap between glass panes in a standard window are in at least 60% of cases due to the thermal radiation, and in some 40% to natural convection. In order to reduce heat exchange through thermal radiation, low-emission coatings mentioned earlier of the inner glass pane surfaces are used (on the gap side). Convection in a gas gap is reduced by using a proper distance between the glass panes and filling the gas gap with a noble gas or, even better, evacuating it (vacuum gap).

Heat exchange with the external surroundings is mostly due to forced wind-driven convection. In a typical vertical window, this process is responsible for some 90% of total heat exchange, while only 10% is due to thermal radiation. Glazed surfaces protruding from the envelope (bay windows) are more affected by wind than typical glazing of windows installed in a wall structure. In buildings, heat exchange with the sky should also be considered. It is more intense in case of sloped surfaces than in vertical ones, as a view factor for such cases is higher (this factor for the sky decreases as the slope increases). Heat exchange by thermal radiation with the sky is the most intensive for horizontal surfaces, e.g., roofs or horizontal atrium glazing totally surrounded by the apparent sky, and it is particularly evident in winter. Heat transfer on the interior side is dominated by thermal radiation, with a smaller share of natural convection.

Optical properties of glass panes depend on the incident angle of solar radiation, which means they change with time, both in short daily cycles and long seasonal periods. For this reason, coatings ensuring certain angular selectivity are studied to enable getting specific transmittance in certain conditions (certain time of day and season).

In case of standard glass, transmittance of a single pane for solar radiation at a normal angle of incidence is 83.7%. Losses result from reflection on media boundaries (7.5%) and from absorption within the glass (8.8%). The solar absorption coefficient of typical (soda-lime) glass is 30/m. The purest form of glass is quartz; its absorption is practically zero. Absorption of solar radiation within glass is a result of impurities, primarily iron. In solar energy applications, glass with lowered iron content is typically used; its transmittance is 92%. Losses of 8% are due to reflection. The presence of impurities in glass does not affect refraction phenomena. One of the methods used to reduce reflectance (in order to increase transmittance) is polishing the glass surface. Unfortunately, material processed in this manner becomes sensitive to changing ambient conditions. Multi-interference coatings could be another option. This solution requires application of multiple layers of coatings with different transmittance for individual wavelengths of solar radiation. Unfortunately, it has not yet been introduced for practical applications. Another option is dielectric coating with a refractive index of 1.23 on both media boundaries. This would reduce reflection losses to 2% (1% at each boundary). However, few available materials really have such a low refractivity. Porous silicon dioxide SiO_2 is seen as a potential solution [30].

Besides improving the optical properties of glass, it is also important to improve the thermal insulating characteristics of windows. Considerable reduction of heat transfer through windows has already been achieved thanks to introduction of low-emissivity coatings. Such coatings are also characterized by high, although lower than in case of normal glass, solar radiation transmittance. Such coatings reduce solar gains by an average of 20%. Application of low-emissivity coatings increases thermal resistance of double-glazed windows more than twofold. In cold climates, low-emissivity coatings are applied on the internal surface (i.e., on the gap side) of the inner glass pane. In result, heat transfer through thermal radiation from the inner glass pane to the outer one is reduced. As a consequence, total heat losses from glazing decrease. In order to reduce the impact of the external environment during summer, i.e., to reduce excessive heat gains due to high ambient air temperature, low-emissivity coatings should be applied on the internal surface (i.e., on the gap side) of the outer glass pane. Thus it would be best to use a rotating window, which could be rotated by 180° depending on ambient conditions (primarily season—summer or winter). Using such a solution is relatively difficult for standard vertical large-sized windows, but it is considerably easier in the case of roof windows. It should be especially applied to windows exposed to solar radiation during the summer.

In triple-pane windows, a low-emissivity coating is also applied at the gas gap side, on the inner and outer panes. Coatings should not be applied to the inner side of the inner pane, because solar energy absorbed within the coating would increase the glass temperature, leading to excessive heating of the pane.

Utilization of low-emissivity coatings considerably reduces radiative heat exchange between panes. Filling the gap between panes with a noble gas reduces convective heat transfer (noble gas has fewer particles than air would have, so there are fewer heat carriers). The best way to reduce convection is a vacuum gap between the panes (no particles to carry heat), but this requires very good tightness of the glazing assembly and high mechanical strength of the panes [20]. In vacuum windows there may be considerable temperature differences between inner and outer panes; therefore the design must be able to deal with thermal expansions. The glazing assembly and entire window structure must be designed to withstand resulting stress.

If the gap between panes contains gas, convective heat transfer depends on the properties of that gas. Even if the convection is minimized, the heat conduction occurs in a gaseous medium. For that reason, the gas used to fill the gap should have as low a heat conductivity as possible. Filling the gap between panes with different gases does not affect optical parameters. Transmittance for solar radiation does not change, as the fillings, including air and vacuum, are transparent for solar radiation. However, as was already mentioned, gas filling influences the heat transfer in a gap. In an old standard double-glazed window with an air-filled gap, the heat transfer coefficient during winter is $2.86 \, W/(m^2K)$, and during summer $3.38 \, W/(m^2K)$. The solar heat gain coefficient $SHGC_{normal}$ (method for determining it is described in Chapter 6) is 0.76. In double glazing with a low-emissivity coating on the inner surface (from the gap side) of the inner pane ($\varepsilon = 0.08$) and an argon-filled gap, the heat transfer coefficient is $0.78 \, W/(m^2K)$ for winter and $0.89 \, W/(m^2K)$ for summer, and $SHGC_{normal}$ of 0.45 [30].

Another factor that influences natural convection inside the gap is distance between the panes. For small distances, between several millimeters and 1 cm, the convective heat exchange coefficient is high due to heat conduction within the gas. The convective heat transfer coefficient is directly proportional to the Nusselt number and thermal conductivity, and inversely proportional to the gap width. The convective heat transfer coefficient achieves its minimum at a certain spacing (distance) between panes. For air it is 12.7 mm, and this is a typical pane spacing in a standard window. For argon it is only slightly lower, around 12 mm, while for krypton and xenon it would be 7.5 and 5.3 mm, respectively [30]. At gaps of more than 10 mm, the convective heat transfer coefficient becomes practically constant.

Usually in energy analyses of buildings, heat transfer through windows is addressed in a simplified way (windows treated as one unit). One heat transfer coefficient is used for an entire window; it may be sourced from standards or catalogs, calculated from appropriate formulas with varying accuracy levels [32−37]. ASHRAE publications [37] provide equations or graphic relations that allow calculating average heat transfer coefficients separately for frames, pane edges, and the central part of the glazing, as well as total values. Those sources list various types of windows and relevant coefficient values. A simplified way of determining heat gains is also given; it takes into account energy absorption within glass and the total averaged solar heat gain coefficient for the frame and other opaque components of a window. Solar heat gain coefficients (SHGCs) are listed for certain latitudes, time of day (hour according to the solar time), month, and 16 main orientations. For detailed analyses of phenomena occurring in windows, a one-dimensional heat transfer model is proposed [16] for the central parts of panes (dimension along the window's thickness), while for the frame and pane edges at least two-dimensional models should be used.

Thermal and optical properties of glass, as well as application of advanced glazing technologies in buildings, are subjects of research and publications [16−20,22,30,37−41]. As already mentioned, materials ensuring variability of solar radiation transmissivity over time are under development. One possible solution is electrochromic glazing [18,22,42], mentioned earlier in the chapter. This type of glazing is listed among so-called "smart window" technologies, as it may modulate transmittance of the solar spectrum by modulating the short-wave absorptance. Typical electrochromic glazing is a five-layer structure made of the following layers: transparent conductor, ion storage film (or electrochromic film), ion conductor (electrolyte), electrochromic film, and transparent conductor. The first layer is transparent and conductive, doped on glass (usually as indium titanium oxide (ITO)). Next comes the electrochromic layer (EC), coated with a layer of ion-conducting electrolyte (solid or polymeric). The next layer is the ion-storage layer, and the last is another transparent conductive one.

Solutions with polymeric electrolytes require using two glass sheets, one with an ITO/EC layer (indium-titanium oxide/electrochromic) and another ITO/ion-storage. Polymeric electrolyte acts as laminate, holding together two coated glass panes. When voltage (typically low voltage of 3−5 V) is connected to two electrodes, it forces a flow of Li^+ ions or protons (depending on electrolyte type) from the electrolyte to the EC active layer. Electrons from the external circuit are also injected to this layer, to balance the charge. This changes electron density of the layer, and thus also color of the material. Reversing voltage polarity causes movement of ions and electrons in the opposite direction, thus returning the material to the original colorless condition and transmittance.

Application of automatic control over EC windows may lead to considerable reduction in building energy consumption. This applies especially to the summer season, when electrochromic windows may considerably reduce heat gains from solar radiation, and therefore decrease energy consumption for cooling and air conditioning. This is particularly important for public utility buildings and office buildings, which typically use large glazed areas to provide daylight access and also (or may be first of all) due to current architectural fashion trends where glazed office buildings are very popular. A new solution is combination of EC windows with vacuum window technology [42].

5.3.3 Transparent Insulation Technologies

In the 1960s, search for new solutions of passive solar systems that would simultaneously ensure thermal protection of a building led to commencing intensive research on materials transparent for solar radiation, but at the same time acting as thermal insulation. Development work came up with a structural and material solution known as a transparent insulation [43]. In fact, it had been "discovered" much earlier. In 1881, Edward Morse patented a solution for a building envelope's structure, which combined thermal insulating features with transparency for solar radiation (U.S. Patent 246626). Unfortunately this concept was ignored until 1964, when engineer Felix Trombe and architect Jaques Michel, authors of passive solar architecture fundamentals, started to promote it.

One of the earliest research and experimental projects focused on transparent insulation, which included construction of a major demonstration facility, took place at the *Fraunhofer Institute for Solar Energy Systems* in Freiburg, Germany [43–45]. In 1991, the first so-called "100% solar building" was built. It was a demonstration project, which presented for the first time the energy self-sufficient building. Its total energy demand was covered by different solar energy technologies. Building integrated solar thermal technologies, active and passive, including transparent insulation and PV technologies, were applied and integrated within the main southern façade structure. The building was also equipped with fuel cells running on hydrogen generated by hydrolysis of water powered by photovoltaic panels.

One of the consequences of using solar energy technologies in the first "100% solar building" was bringing attention to the concept of a curved southern façade in the shape of a flattened half-cylinder. This solution improves insolation conditions, increasing solar exposure of a building's envelope. Examples of such façades may be found in many European low-energy buildings. One of them has been already shown in Figure 3.2.

The shape of the southern façade of the Freiburg demonstration building was designed to use solar energy gains directly through regular windows and windows with transparent insulation. Solar energy was also used indirectly. Transparent insulation was used as a cladding on an opaque construction wall, which gives a solar collector—storage wall.

Windows with transparent insulations allow solar radiation to pass through them. They are so-called "daylight walls." They are a case of a direct solar gain system, but at the same time ensure considerable reduction of heat losses. The light transmission effect achieved with transparent insulation in daylight walls is practically equivalent to the effect of Japanese paper walls. Rooms are provided with daylight, but it is not possible to look through such a wall. Application of such partially transparent walls is also recommended on the northern side, if daylighting is required there (such a solution is more energy-efficient than regular windows). This solution also works well at roofs (e.g., at office buildings), where it plays the role of fixed skylight and reduces the demand for artificial lighting.

As was already mentioned, transparent insulation can be also used as a specific insulating material on external opaque walls. In this case it combines good transmittance for solar radiation and good thermal insulation properties [45]. Transparent insulation may be used in both newly designed low-energy buildings and in older structures with insufficient thermal comfort levels, which need refurbishment. Example of a refurbished building where transparent insulations were applied to external surfaces of opaque walls is shown in Figure 5.17.

The transparent insulation concept is schematically presented in Figure 5.18. Solar radiation passes through the insulation and is absorbed by a surface placed behind the insulation. Temperature of that surface increases.

FIGURE 5.17   Villa Tanheim, headquarters of the International Solar Energy Society (ISES), Freiburg, Germany. Transparent insulations on the western façade.

FIGURE 5.18 Concept of energy
flow through an external opaque
building wall with transparent
insulation.

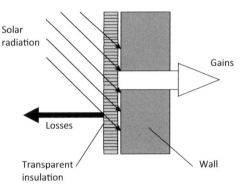

Absorbed energy is transferred toward interior of the building by conduction
within the wall protected by transparent insulation. Part of the absorbed energy
is irradiated toward transparent insulation, but thermal radiation cannot "get
out" to the external environment, because insulation is "opaque" for thermal
(long-wave infrared) radiation. This energy therefore is "trapped" within the
wall structure, and ultimately transferred to building's interior.

Transparent insulation technology of external surfaces on opaque walls
uses low-emissivity coatings (for thermal radiation), which considerably
reduce radiative heat losses. In order to avoid heat losses due to conduction,
space within the transparent material (e.g., polycarbonate) is filled with a
kind of thermal insulator (material of very low density and conductivity): air
or another gas. Theoretically, the thicker the air layer, the better the insulating effect gets. In fact, however, above a certain distance between two
solid surfaces, heat transfer by convection considerably increases, thus also
leading to increased convective heat losses to the environment. In order to
maintain insulating capabilities of air, but at the same time avoid convective
heat losses, the design should contain the air in relatively small volumes.
This is achieved either by using a honeycomb or capillary structure of
insulating material or by introducing low-conductivity materials into the
airspace, e.g., aerogels.

Transparent insulating materials may have the form of very thin tubes
perpendicular to the absorbing surface. Capillaries have small diameters of
1−3 mm. Another type of transparent materials are quasi-homogenous structures: aerogels or glass fibers. Silicon aerogels are most popular; they have
porous structure (open cells) with pore diameters of some 10 nm. Unfortunately aerogels have certain disadvantages. The most important is vulnera-
bility to water, which may destroy the gel's structure upon contact.
Honeycomb structures or other capillary materials used on daylight walls may
often cause excessive glare effect, undesirable in rooms continuously occupied

by people. Daylight walls made with aerogels are more "pleasing to the eye," but it has been discovered that after some time, when water or water vapor penetrates elements on windows within which they are installed, the aerogel structure gets damaged, and insulation "settles down."

To summarize, materials used to make transparent insulations are characterized by:

- High transmittance for solar radiation. This is achieved by using transparent materials, e.g., low-iron glass (so-called solar glass) and thin polycarbonate layers. Sometimes high transmittance only for the visible light is required, if daylighting is one of main functions to be played by insulation.
- Low transmittance for thermal radiation. This is achieved by using internal reflective coatings and low emissivity coatings for infrared thermal radiation.
- Low thermal conductivity. It is achieved thanks to using light materials (very low density) with high content of gases or vacuum.
- Limited convective heat exchange, thanks to using closed porous structure (gas spaces) to avoid gas particle movement.

This combination of properties allows constructing insulating elements with heat transfer coefficient U much lower than $1 \, W/(m^2K)$, but ensuring solar radiation transmittance of some 70% or even more (hence the name—transparent insulation). Transparent insulation acts in a way similar to a thermal diode, allowing heat transfer only in one direction—into the building (but not outside).

A dormitory of Strathclyde University in Glasgow, UK, was the first case of installing large areas of transparent insulation on a residential building. The dormitory named "Solar Residence" was commissioned in 1989. It could accommodate 376 students (94 four-person apartments). *Strathclyde Solar Residence* was built within an R&D (research and development) program sponsored by the European Commission. The dormitory was located on a terrain slightly sloped toward the south. Its structures were lined along the east–west axis. Design and construction of the building incorporated three fundamental and innovative (for those times) energy-saving technologies: transparent insulation, passive architectural solutions, and heat recovery from ventilation systems.

Design of the building maximized use of transparent insulations. Panels with such materials cover $1040 \, m^2$ of southern façades. Day working (living) rooms and bedrooms were placed on the southern side of the building. After the building had been completed, considerable overheating of those rooms was observed. For this reason, in 1992 southern windows were provided with shading-reflecting blinds.

It may be noted here that some of the solutions adopted in the Solar Residence nowadays are considered to be standard in low-energy buildings.

One of the examples is heat recovery from ventilation systems. Popularization of other solutions, like transparent insulation, have encountered a number of obstacles, primarily related to unpredicted excessive solar energy gains during summer and resulting room overheating. Because of that, certain modifications of proposed solutions have been made. Nevertheless, the overheating issue considerably restricted popularity of this technology for several years.

To sum up, key disadvantages of some transparent insulation technologies include glares, the possibility of room overheating during summer, and a blocked view to the outside. Due to overheating, special attention must be paid to ensuring good room shading during summer. An important advantage is good compatibility with other daylighting technologies and a system of active façades. Heat transfer coefficients of façades with transparent insulation are usually between 0.6 and 0.8 W/(m^{2}K), and their transmittance for solar radiation is around 0.8.

Recently a number of new transparent insulation technologies have been introduced to the market. They are integral elements of so-called multifunctional façade systems, aimed at controlling inflow of solar energy and light. They can be used as elements of daylighting systems and solar wall components [46,49].

5.3.4 Phase-Change Materials

The essential problem related to solar energy utilization is the periodic character of its availability and lack of coherence between the energy availability and heating demand. For that reason it is very important to ensure solar energy storage, both in the short and long term. Heat is generated due to photothermal conversion occurring within elements of the building's envelope. Then it is stored for various periods, depending on thermal capacity of a storage medium. The material's (medium's) ability to store the heat is described by its specific heat capacity. Specific heat capacity is expressed in kJ/(kg K), This value describes how much heat may be stored by 1 kg of a material at a temperature difference of 1 K. Except for specific heat capacity, capacity of a storage element depends on density and volume of material used to make it. In higher-latitude countries massive, large, and heavy (high-density) external (and often internal) walls are used to ensure high heat capacity. Such massive walls are used to keep heat inside a building as long as possible (to reduce heat losses).

Effective increase of a building's heat capacity may be achieved without increasing the mass and volume of walls and other construction elements, but even with reduction of these parameters. This can be done by use of so-called PCMs. The high heat capacity of such materials is a result of their capability to store large amount of energy at relatively constant temperature, at which they remain in liquid state. Then, when temperature decreases, the solidification process takes place and the latent heat is released out of such material. PCMs used in buildings are characterized by the ability to absorb heat in a liquid state at room temperature. When the temperature inside a room drops (e.g., at night), phase-change material undergoes the solidification process and the room is heated. Then during a day, when the temperature increases due to influence of solar radiation, the material melts again, going from solid to liquid. It absorbs excessive heat from the surroundings (a room). The amount of heat absorbed depends on the thermal capacity of the PCM. As a result, a room is cooled.

PCM application for heating and cooling of rooms was proposed already in 1970s. However, major progress in adoption of this technology has only been observed recently, along with development of technologies for production of durable construction materials with planned PCM content.

Utilization of PCMs in a building structure reduces consumption of conventional energy carriers, and even more importantly decreases the peak energy demand. By flattening the demand curve, so-called *peak shaving* is achieved [47−52]. PCM application improves energy balance of a building in a daily cycle. It allows solar energy to be stored in building components like walls, floors, and ceilings, or specially-designed accumulators during the day, and then to release the heat stored with a time shift, during the night. Table 5.1 presents key thermophysical parameters of materials used for energy storage in solar active and passive systems in buildings, mainly low-energy buildings. Table 5.2 presents the density and specific heat capacity of the main construction materials and one example of the PCM applied in buildings. In case of PCM the effective specific heat capacity refers to the phase-change heat. It is evident that the phase-change material has considerably higher heat capacity than other materials used in buildings.

Phase-change materials used in construction engineering are characterized by very high phase-change enthalpy for phase changes at room temperatures. They are able to absorb a very large heat flux at nearly constant temperature (temperature variation is very small). Utilization of PCMs within a building's structure is achieved by adding a determined dose of the PCM to construction materials: concrete, plaster, or hollow bricks. Examples of construction material created with PCM integrated into their structure are shown in Figure 5.19.

Numerous technologies of combining such materials with construction materials have been studied over the last decades [55−62]. The most important include:

● Direct incorporation (mixing) of pure PCM in construction material—e.g., cement, plaster—during the production process.
● Saturation of porous construction elements (bricks, concrete blocks, mineral granulate) with PCM by immersion in melted PCM.
● PCM encapsulation: production of PCM microgranulate in the form of spheres (balls) with a diameters of a few to a few hundred micrometers in polymer coating.

TABLE 5.1 Key Thermophysical Parameters of Materials Used for Heat Storage in Solar Active and Passive Systems in Buildings [53]

Storage Material	Temperature (Operating Range) (°C)	Density (kg/m³)	Specific Heat Capacity (J/(kg K))	Volumetric Density of Heat Capacity (kWh/(K m³))	Heat Conductivity (W/(mK))
Water	0–100	1000	4190	1.16	0.63 at 38 °C
Water (10 bar)	0–180	881	4190	1.03	–
Water + ethylene glycol (50/50%)	0–100	1075	3480	0.98	–
Therminol 66	−9 to 343	750	2100	0.44	0.106 at 343 °C
Salt, NaNO₃–KNO₅ (50/50% by weight)	220–540	1680	1560	0.72	0.61
Liquid sodium	100–760	750	1260	0.26	67.5
Aluminum	660 (melting point)	2700	920	0.69	200
Refractory clay	–	2100–2600	1000	0.65	1.0–1.5
Rock	–	1600	880	0.39	–

TABLE 5.2 Density and Heat Capacity of Major Construction Materials and an Example of PCM [53]

Material	Density (kg/m³)	Specific Heat Capacity (kJ/(kg K))	Specific Heat Capacity (MJ/(m³ K))
Concrete	2300	1.0	2.3
Steel	7800	0.47	3.67
Plasterboard	1400	0.84	1.18
Brick	1600	0.84	1.34
Wood	600	1.6	0.96
PCM[a]	870−1000	18.0	15.7−18

[a]PCM: mixture of fatty acids, melting point 22°C, effective specific heat capacity in temperature range 18−28°C [54].

FIGURE 5.19 Composite created of a plasterboard with incorporated PCM made for laboratory testing at the Institute of Heat Engineering, Warsaw University of Technology, Poland. *(Photo: R. Wnuk.)*

- Production of small polymer containers with a PCM, with a size of up to a few centimeters (spheres or cylinders filled with PCM, or flexible mats with groves filled with PCM).
- Production of special stable composite structures with high PCM content (up to 80%) with polymeric matrix, in most cases HDPE (high density polyethylene known as shape-stabilized PCMs).
- Production of laminated plates (most often of mineral wool) with a thin, internal layer of PCM.
- Production of light thermal mass panels with incorporated PCM (with PCM content usually about 40%).
- Incorporation of a small quantity of PCMs into indoor equipment, e.g., internal blinds or furniture, which then acts as solar storage−collectors.

Technologies in which PCM is in direct contact with the construction material should not be used due to incompatibility between materials, which could result with chemical corrosion or erosion.

Usually encapsulated PCMs are used. Spheres (balls) contain up to 80% of the pure PCM. Thanks to the small dimensions and polymer coating, mechanical properties of the created composite are not changed in time, i.e., degradation does not appear. During the melting process, PCM volume increases by 10–15%. Encapsulated PCM (micrometer-size PCM capsules) is then mixed with concrete or gypsum. At present the most popular product of encapsulated PCM available on the market is *Micronal* developed by the BASF company.

Another form of PCM applied quite often in buildings was already mentioned: small containers—spheres or tubes with a diameter of about 1–2 cm—placed in hollow spaces of bricks or embedded in concrete. Small containers with PCMs may be also placed in hollow spaces of buildings, e.g., above suspended ceilings, creating an effective free cooling solution, or in ventilation channels (ducts). They can be also placed below the floor, giving a floor heating system.

Light thermal mass panels with incorporated PCM have also become more popular. They can be put on internal and external walls. For example, DuPont panels are available on the market [54].

Thermal capacity of a PCM depends primarily on its latent heat (the phase-change heat), and for materials that may be used in buildings this is between 100 kJ/kg (ready product—microcapsules) and 250 kJ/kg (homogenous media). It needs to be noted that if PCM may undergo significant supercooling, then its utilization is not recommended. In such a case, conditions ensuring solidification might fail to develop and the material would not work as cyclic heat accumulator (to reduce this effect, substances acting as solidification nuclei may be added). Another important aspect is PCM stability at melting/solidification processes, as during regular operation the material is subjected to thousands of such changes.

Current research and development focuses on analyses of thermophysical properties of construction composites with PCMs such as thermal conductivity and capacity, for various content and forms of PCM. Composite samples of various PCM weight content are tested (15–40%). Other research concerns development of new methods for measuring effective thermal performance of structures containing PCM (e.g., Refs [58,62]).

PCMs may also be used within systems of photovoltaic cells integrated with buildings (on walls or roofs). Flat containers filled with PCM are attached to a photovoltaic (PV) panel backside. In such an arrangement, PCM acts as storage for part of the solar energy, which is later supplied to the building's interior as heat with a half-day delay (i.e., at night). At the same time, the PCM container acts as a cooler for PV panels, stabilizing their temperature and preventing loss of efficiency (PV cell efficiency drops when its temperature rises).

In conclusion, it needs to be underscored that much hope is put on the potential use of PCM in building structures. Key objectives are solar energy accumulation and considerable increase in building heat capacity with simultaneous reduction of weight and dimensions of external envelope elements [63].

REFERENCES

[1] The directive 2010/31/EU of the European Parliament and of the Council of 19 May, 2010 on the energy performance of buildings.

[2] Hidlgo A, Marandino A, Duvielguerbigny A, Riley B, Staniaszek D, Petroula D, et al. EU energy efficiency directive. Guidebook for good implementation. Draft. 30/1/13. In: Scheuer S, editor. The coalition for energy savings; 2013.

[3] Chwieduk D. Budownictwo niskoenergetyczne — Energia odnawialna, Budownictwo Ogólne. In: Klemm P, editor. Fizyka Budowli, t. 2; 2005. Arkady.

[4] Chwieduk D. Energetyka słoneczna budynku; 2011. Arkady.

[5] Grabarczyk S. Fizyka Budowli Komputerowe wspomaganie budownictwa energooszczednego. Warsaw: Oficyna Wydawnicza Politechniki Warszawskiej; 2005.

[6] Popiołek Z, editor. Energooszczedne kształtowanie środowiska wewnetrznego. Gliwice: Politechnika Śląska. Katedra Ogrzewnictwa, Wentylacji i Technik Odpylania; 2005.

[7] Anderson B. Solar energy: fundamentals in building design. Harrisville, New Hampshire: Total Environmental Action, Inc.; 1975.

[8] Athienitis AK, Santamouris M. Thermal analysis and design of passive solar buildings. James & James Ltd; 2002.

[9] Balcomb JD, Jones RW, McFarland RD, Wray WO. Passive solar heating analysis — a design manual. ASHRAE; 1984.

[10] Balcomb JD. Passive solar buildings. Cambridge, Massachusetts: The MIT Press; 1992.

[11] Solar Heating & Cooling Programme IEA. In: Hastings R, editor. Solar low energy houses of IEA Task 13. London: James & James; 1995.

[12] Mazria E. The passive solar energy book. Emaus: Rodale Press; 1979.

[13] Tombazis AN, Preuss SA. Design of passive solar buildings in urban areas. Sol Energy 2001;70(3):311−8. Elsevier Science Ltd., UK.

[14] Chwieduk D. Modelowanie i analiza pozyskiwania oraz konwersji termicznej energii promieniowania słonecznego w budynku. Warsaw: Prace IPPT; 11/2006.

[15] Kisilewicz T. Wpływ właściwości dynamicznych przegród budynku na zużycie energii i komfort cieplny, Czasopismo Techniczne Politechniki Krakowskiej, seria B, z. 14-B/2004.

[16] ASHRAE. Fundamentals, handbook, chapter fenestration. SI ed. Atlanta: American Society of Heating, Refrigerating and Air Conditioning Engineers; 1997.

[17] Hollands KGT, Hum JE, Wright JL. Analytical model for the thermal conductance of double-compound honeycomb transparent insulation, with validation. ISES; 2001 [Solar World Congress].

[18] Granqvist CG. Progress in electrochromics: tungsten oxide revisited. Electrochim Acta 1999;44:3005−15.

[19] Milburn D, Hollands K, Kehl O. On measurement techniques for spectral absorptance of glazing materials in the solar range. Sol Energy 1998;62(3):163−8.

[20] Eames PC, Zhao JF, Wang J, Fang Y. Advanced glazing systems. Pol Energ Słonecz 2006;1−2:27−30.

[21] Balcomb JD, Hedstrom JC, McFarland RD. Simulation analysis of passive solar heated buildings-preliminary results. Sol Energy 1977;18(3):277−82.

[22] Papaethimiou S, Leftheriotis G, Yianoulis P, Hyde TJ, Eames PC, Fang Y, et al. Development of electrochromic evacuated advanced glazing. Energy Build 2006;38:1455−67.

[23] Herzog T, editor. Solar energy in architecture and urban planning. Prestel; 1998.

[24] Bloem JJ. System identification applied to building performance data, PASLINK Conference, Joint Research Centre European Commission, Office for Official Publications of the European Communities.

[25] Eicker U. Solar technologies for buildings. West Sussex, UK: John Wiley& Sons; 2003.

[26] Smolec W. Fototermiczna konwersja energii słonecznej. Warsaw: Wydawnictwo Naukowe PWN; 2000.

[27] Zawidzki M. Eko-Osiedla jako rozproszone ośrodki rozwoju energetyki słonecznej w budownictwie. Pol Energ Słonecz; 2004:8−12. 1/2004, PTES-ISES, Warsaw.

[28] Kisilewicz T. Computer simulation in solar architecture design, vol. 3. Architectural Engineering and Design Management; 2007.

[29] www.paslink.org/dame/index.htm.

[30] Hollands KGT, Wright JL, Granqvist CG. In: Gordon J, editor. Glazing and coatings: solar energy the state of the art. UK: ISES Position Papers; 2001. pp. 29−107.

[31] Leftheriotis G, Yianoulis P. Glazing and coatings. Compr Renew Energy 2012;3:313−55. Elsevier.

[32] ISO 10077-1:2000(E). Thermal performance of windows, doors and shutters − calculation of thermal transmittance − part 1: simplified method; 2000.

[33] ISO 15099. Thermal performance of windows, doors and shading devices − detailed calculations; 2003.

[34] ISO 10077-2:2003(E). Thermal performance of windows, doors and shutters − calculation of thermal transmittance − part 2: numerical method for frames; 2003.

[35] ISO/DIS 10077-1 (Draft International Standard). Thermal performance of windows, doors and shutters − calculation of thermal transmittance − part 1: general; 2004.

[36] ASTM. Standard procedures for determining the steady state thermal transmittance of fenestration systems. ASTM Standard E 1423-91. In: 1994 annual book of ASTM standards 04.07. American Society of Testing and Materials; 1991. pp. 1160−5.

[37] ASHRAE. Standard method for determining and expressing the heat transfer and total optical properties of fenestration products. BSR/ASHRAE Standard 142P (Public Review Draft). Atlanta, Georgia: American Society of Heating, Refrigerating and Air-Conditioning Engineers (ASHRAE); 1996.

[38] Arasteh D. An analysis of edge heat transfer in residential windows. In: Proceedings of ASHRAE/DOE/BTECC conference, thermal performance of the exterior envelopes of buildings IV, Orlando, FL; 1989. pp. 376−87.

[39] Hutchins MG. Advanced glazing materials. In: Solar energy, vol. 62. School of Engineering, Oxford Brookes University; 1998. 3.

[40] Johnson ET. Low-e glazing design guide. Boston: Butterworth Architecture; 1991.

[41] Wigginton M. Glass in architecture. London: Phaidon Press Ltd; 1996.

[42] Fang Y, Eames PC, Norton B, Hyde T, Huang Y, Hewitt N. The thermal performance of an electrochromic vacuum glazing with selected low emittance coatings. Thin Solid Film 2008;516:1074−81. Elsevier.

[43] Jesch LF. Transparent insulation technology. ETSU − OPET for the Commission of the European Communities Directorate-General XVII for Energy, UK; 1993.

[44] Stahl W, Voss K, Goetzberger A. The self-sufficient solar house in Freiburg. Sol Energy 1994;52(1):111−25.

[45]   Wong IL, Eames PC, Perera RS. A review of transparent insulation systems and the evaluation of payback period for building applications. Sol Energy 2007;81(9):1058−71.

[46]   Daylight in buildings − a source book on daylighting systems and components. IEA; 2000.

[47]   Zhu N, Ma Z, Wang S. Dynamic characteristics and energy performance of buildings using phase change materials. Energy Convers Manag 2009;50(12):3169−81.

[48]   Kuznik F, Virgone J, Roux J-J. Energetic efficiency room wall containing PCM wallboard: full-scale experimental investigation. Energy Build 2008;40(2):148−56.

[49]   Darkwa K, O'Callaghan PW. Simulation of phase change drywalls in a passive solar building. Appl Therm Eng 2006;26:8−9. 853−8.

[50]   Kuznik F, Virgone J, Johannes K. Development and validation of a new TRNSYS type for the simulation of external building walls containing PCM. Energy Build 2010;42(7):1004−9.

[51]   Kuznik F, Virgone J, Johannes K, Roux J-J. A review on phase change materials integrated in building walls. Renew Sustain Energy Rev 2011;15(1):379−91.

[52]   Zhang Y, Zhou G, Lin K, Zhang Q, Di H. Application of latent heat thermal energy storage in buildings: state-of-the-art and outlook. Build Environ 2007;42:2197−209.

[53]   Martin CL, Goswami DY. Solar energy pocket reference. ISES; 2005.

[54]   http://energain.co.uk/Energain/en_GB/products/thermal_mass_panel.html; 2013.

[55]   Hed G. Service life estimations in the design of a PCM based night cooling system [Doctoral thesis]. Sweden: University of Gävle; 2005.

[56]   Zalba B, Martyn JM, Cabeza LF, Mehling H. Review on thermal energy storage with phase change: materials, heat transfer analysis and applications. Appl Therm Eng 2003;23(25): 251−83.

[57]   Mehling H, Cabeza LF. Heat and cold storage with PCM. An up to date introduction into basics and applications. Springer; 2008.

[58]   Jaworski M. Materiały zmiennofazowe (PCM) do zastosowań w budownictwie. In: Polska Energetyka Słoneczna. Warsaw: PTES-ISES; 2008. pp. 57−60. 1-4/2008.

[59]   Tyagi VV, Buddhi D. PCM thermal storage in buildings: a state of art. Renew Sustain Energy Rev 2007;11:1146−66.

[60]   Baetens R, Jelle BP, Gustavsen A. Phase change materials for building applications: a state-of-the-art review. Energy Build 2012;42(9):1361−8.

[61]   Soares N, Costa JJ, Gaspar AR, Santos P. Review of passive PCM latent heat thermal energy storage systems towards buildings' energy efficiency. Energy Build 2013;59:82−103.

[62]   Wnuk R, Jaworski M. Badania charakterystyk cieplnych elementów budowlanych akumulujacych cieplo, zawierajacych materiały PCM. Experimental study of thermal characteristics of thermal storage building materials with incorporated PCM. Pol Energ Sloneczna 2011;1:5−11 [In Polish].

[63]   Chwieduk D. Dynamics of external wall structures with a PCM (phase change materials) in high latitude countries. Energy 2013;59(2013):301−13.

Chapter 6

Energy Balance of a Building with Regard to Solar Radiation Exposure

6.1 FORMULATION OF ENERGY BALANCE OF ROOMS IN A BUILDING

6.1.1 Energy Balance of a Building

Any building is influenced by its external surroundings, especially climatic conditions, including solar radiation. This influence depends on a building design and construction. Building design should include architectural and structural solutions that increase solar radiation gains in winter and reduce in summer. This requires choosing the slope and orientation of envelope surfaces that receive high solar irradiation in winter and low in summer. This also requires using transparent elements of a building facade of high transmittance as well as opaque elements of high absorptance for solar radiation in winter. During the summer, elements of a building envelope should protect the room's interior from overheating. This also requires choosing the correct slope and orientation, but, in this case, low transmittance of transparent elements and low solar absorptance of opaque elements are needed. Presented requirements for winter and summer are contradictory. The simplest solution to solve the problem is shading. The influence of shading is variable over time. It would be recommended for the construction materials to have their thermal and optical parameters variable over time too, in short and longer term, especially with seasonal differences of summer and winter.

The energy balance of a building is influenced by heat (energy) flows through its envelope, air flows (as a result also heat flows) through infiltration and ventilation system, and internal heat sources. As a result of heat and air flow, and heat gains from internal heat sources, thermal conditions in a building are changing in time and internal air temperature and humidity are changing, too (humidity changes are neglected in the following discussion of energy balance of the air within the building). Thus, taking into account heat

Solar Energy in Buildings. http://dx.doi.org/10.1016/B978-0-12-410514-0.00006-2

flows into and out of the building interior, as well as internal heat sources, energy balance may be expressed by the equation:

$$\rho c V \frac{dT_{in}}{dt} = \sum \dot{Q}_{in}(t) - \sum \dot{Q}_{out}(t) + \sum \dot{Q}_{qv}(t) \qquad (6.1)$$

where:

$\sum \dot{Q}_{in}(t) =$ the sum of energy fluxes flowing into the interior through elements of a building envelope (W);

$\sum \dot{Q}_{out}(t) =$ the sum of heat fluxes flowing out of the interior through elements of a building envelope (W);

$\sum \dot{Q}_{qv} =$ the sum of internal heat sources in a building (room) (W);

$T_{in}(t) =$ the air temperature inside a building (room) (K);

$V =$ the air volume inside a building (m^3).

Equation (6.1) is an equation in general form with lumped parameters. This means that every component of the energy balance equation (i.e., every individual flow affecting the energy balance for the interior air), may be determined separately and calculated with different accuracy.

Building envelope consists of opaque and transparent elements. Opaque elements in the form of external walls create a structural envelope and their surfaces usually form a majority of the envelope surface. Other opaque envelope elements are formed by roof surfaces and elements buried in the ground, such as foundations and the basement floor. Transparent envelope elements include windows and other glazed surfaces (e.g., facades).

If the whole interior of a building is heated and the air temperature in all rooms is the same and constant in time, then such a building may be considered as having one uniform temperature zone. As a result, heat transfers between individual rooms may be neglected, and the energy balance may be formulated for the building as a whole (however, the thermal capacity of internal walls and other partitions shall be taken into account). Determining heating loads in different rooms of a building and dimensioning room-specific heating systems (e.g., size of radiators) is a separate task. It may be noted here, that due to development of modern low-energy buildings an idea of applying several temperature zones in a building, mainly depending on the room functions, has been introduced.

Heat fluxes, through different envelope elements characterized by specific thermal resistance coefficients, are exchanged between the building's interior of a certain air temperature and its exterior of ambient air, the sky temperature, and ground temperature variable in time.

Internal heat sources are essentially people and electrically powered or gas fired appliances, devices, and equipment (if any, e.g., gas cookers). Room heating systems (e.g., radiators, floor heating systems) may be either considered as heat flows supplied to the interior or as internal heat sources. Solar energy gains are usually described in a simplified manner, assuming that they depend on window's surface area, their location (inclination and orientation), and shading (e.g., given in percentage). Variability of solar radiation over time is usually reflected in a simplified manner by using averaged values: daily, monthly, or even seasonal. This significantly simplifies a mathematical model, but it does not reflect the actual dynamics of a building and its surroundings. Shading is usually taken into account by introducing fixed, averaged (monthly, seasonal) coefficients.

This publication focuses on the influence of solar radiation on the energy balance of a building. For this reason, even simplified energy balance will include solar radiation flux. Taking into account internal heat gains Q_{in}, solar radiation gains Q_{sol}, heat flow supplied by heating system Q_h, and the total heat transfer coefficient U_t for all the envelope elements and their surfaces, in case of a building losing heat, the Eqn (6.1) may be transformed into:

$$(\rho c V)\frac{dT_{in}}{dt} = \dot{Q}_{in}(t) + \dot{Q}_h(t) + \dot{Q}_{sol}(t) - U_t(t)A_t(T_{in}(t) - T_a(t)) - \dot{Q}_{ven}(t)$$

$$(6.2)$$

Equation (6.2) is formulated assuming, that the external envelope elements (opaque and transparent) are lumped together and the total heat loss coefficient is averaged (weighted), depending on specific characteristics of specific elements and their surfaces. Heat losses via ventilation are included as a total, but they may also be expressed in a similar way to the heat losses via the envelope (i.e., as a function of temperature difference between exterior and interior), and the averaged ventilation heat loss coefficient.

6.1.2 Heat Transfer through a Wall in a Steady-State

In simplified analyses, heat flows through a building envelope are considered in a steady state. Typically, the description of related thermal phenomena is based on the thermal resistance method, which may be applied with various levels of detail [1−3]. Generally, a heat flow through an opaque wall is described by a total heat transfer coefficient for that wall. It holistically addresses phenomena in a wall and in its internal and external surroundings. This coefficient is inversion of the total thermal resistance. Heat transfer phenomena in a wall and in its surroundings may be considered separately by using elementary resistances for each medium (external surroundings, internal surroundings, and the wall itself). In case of an opaque multilayer external wall, specific elementary resistances may be determined for each layer. Windows are usually treated as a whole, without addressing their individual components (i.e., frames and panes). The total heat transfer coefficient for a whole window is used (i.e., inversion of its total thermal resistance).

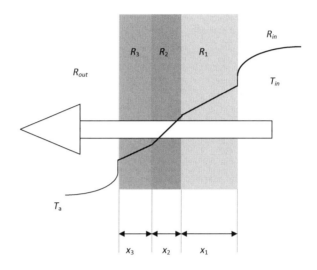

FIGURE 6.1 Steady state heat flow through an opaque wall in a one-dimensional model.

Heat flow through an opaque wall in a steady state and one-dimensional model, based on thermal resistance method, is schematically shown in Figure 6.1. Air temperature inside the room is higher than outside, and the heat flows outward (heat loss). The magnitude of this loss depends on the ambient and internal conditions, and on the structure of a wall, described by external resistance R_{se}, internal resistance R_{sin}, and wall resistance R_{wall}, respectively. In case of a multilayer wall, its total resistance depends on resistance values of each layer. Usually, the resistance of each layer, which is a solid body, is described as thermal conductivity resistance.

Coefficient of heat transfer through an opaque wall in the discussed case (Figure 6.1) is described by the equation:

$$U = \frac{1}{R_{se} + R_1 + R_2 + R_3 + R_{si}} = \frac{1}{\frac{1}{h_{se}} + \frac{x_1}{\lambda_1} + \frac{x_2}{\lambda_2} + \frac{x_3}{\lambda_3} + \frac{1}{h_{si}}} \qquad (6.3)$$

Thermal resistances on the external and internal sides are the combination of elementary resistances of radiative and convective heat transfer. Usually, typical construction materials are characterized by thermal radiation emissivity around 0.9, whereas in case of aluminum plating (roof elements) emissivity is 0.2. In case of polished aluminum, the emissivity drops to around 0.05. This shows how surface material change may affect radiative characteristics, and therefore also radiative heat transfer. Apart from radiative performance of a surface, radiative heat transfer is influenced by temperature of surfaces exchanging heat and their mutual position (configuration—view factor). Obviously, the impact of temperature is most significant (heat flow depends on the fourth power of temperature). Average coefficients of radiative heat exchange between the surroundings and traditional wall surfaces are around 4−6 W/(m²K).

In rooms with gravity ventilation, heat is transferred through natural convection. Depending on the heat flow direction, convective heat transfer coefficients are different. In case of walls, heat transfer is horizontal, and convective heat transfer coefficient is, by average, around 3.0 W/(m²K). For the ceilings, with the upward heat flow, the convective heat transfer coefficient is 4.3 W/(m²K), and for the floors and downward flow it is 1.5 W/(m²K) [4].

From the external side of a building, usually there is wind-driven forced convection and the convective heat transfer coefficients depend on the wind speed, as presented in Chapter 4.4. This may be approximately described by Eqns (4.73)−(4.63″). Also relations given by published standards may be used. Standard EN ISO 6946:2007 [5] specifies a method for calculating thermal resistance and heat transfer coefficients for construction components and other elements of a building, except for doors and windows, and through the ground. For cases when data needed for calculating individual thermal resistances is not available, the averaged standard heat transfer coefficients can be used. They jointly describe both radiative and convective heat transfer. These values apply to steady state heat transfer on internal and external surfaces, depending on the heat flow direction, and are listed in Table 6.1 [5].

As already mentioned, influence of surroundings and heat transfer through a wall may be described by elementary thermal resistances. Graphic interpretation of applying thermal resistance methods for describing heat transfer through a wall is shown in Figure 6.2. Total thermal resistance of the analyzed wall (as in Figure 6.2.) may be expressed by relation:

$$R_t = R_{se} + R_{wall} + R_{si} = R_{12} + R_{23} + R_{34} \qquad (6.4)$$

Figure 6.2 shows a two-layer wall with individual layers characterized by resistances of R_2 and R_3, accordingly, which only depend on thermal

TABLE 6.1 Standard Heat Transfer Resistance Values on Internal and External Surfaces, Depending on the Heat Transfer Direction [5]

Heat Transfer	Upward	Horizontal	Downward
Surface Heat Transfer Resistance R (W/(m²K))			
On internal surface R_{si}	0.10	0.13	0.17
On external surface R_{se}	0.04	0.04	0.04

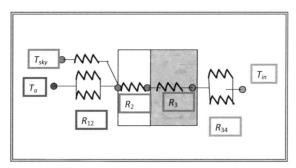

FIGURE 6.2 Graphic interpretation of heat transfer phenomena in a wall and its surroundings, described using the thermal resistance method.

conductivity of the respective layer and its thickness. Those resistances are connected in series, and upon adding up yield the total thermal resistance of the wall R_{23}, which may be expressed as:

$$R_{wall} = R_{23} = R_2 + R_3 \tag{6.4a}$$

Influence of the external surroundings on any external element of a building envelope, both opaque (wall, roof) and transparent (window), is a result of the same thermal phenomena: convection and radiation. Heat is exchanged by thermal radiation with the sky (of T_{sky} temperature), as described by radiative heat transfer resistance $R_{1'2}$, and in parallel (at the same time) with direct surroundings (of a temperature T_a), as described by radiative heat resistance $R_{1''2}$. At the same time (in parallel), there is a convective heat exchange with direct external surroundings (of T_a temperature); this is described by the convective heat transfer resistance $R_{12'}$. All the above-mentioned processes of heat exchange with direct and far (sky) surroundings take place simultaneously and are mutually independent (i.e., they are parallel). The total resistance of external heat transfer R_{12} takes into account all individual resistances just mentioned and can be expressed as:

$$R_{se} = R_{12} = \frac{1}{R_{1'2}} + \frac{1}{R_{1''2}} + \frac{1}{R_{12'}} = \frac{R_{1'2} + R_{1''2} + R_{12''}}{R_{1'2}\,R_{1''2}\,R_{12'}} \tag{6.4b}$$

Internal resistance R_{34} consists of two resistances for convective and radiative heat transfer, which are parallel to each other, and may be expressed as:

$$R_{si} = R_{34} = \frac{1}{R_{3'4}} + \frac{1}{R_{34'}} = \frac{R_{3'4} + R_{34'}}{R_{3'4}\,R_{34'}} \tag{6.4c}$$

Finally, total thermal resistance of the wall under consideration is a combination of individual elementary resistances. Heat transfer between external surroundings and the wall, through the wall, and between the wall and internal surroundings takes place in series. Therefore, total thermal resistance is a sum of: total external resistance, total resistance of the wall itself, and total internal resistance. This is expressed in general terms by Eqn (6.4) and in detail, in reference to the phenomena shown in Figure 6.2 and described by Eqns (6.4a)–(6.4c), by the following relation:

$$R_t = R_{se} + R_{wall} + R_{si} = R_{12} + R_{23} + R_{34}$$
$$= \frac{R_{1'2} + R_{1''2} + R_{12''}}{R_{1'2}\, R_{1''2}\, R_{12'}} + (R_2 + R_3) + \frac{R_{3'4} + R_{34'}}{R_{3'4}\, R_{34'}} \quad (6.5)$$

Then, in case of a transparent envelope element (i.e., a window), the total thermal resistance in a simplified analysis can be described by a relation analogical to Eqn (6.5); the only difference being the resistance R_{23}, which, in this case, becomes a total resistance of a window as one unit (i.e., jointly for the frame, panes, and gas gap).

It needs to be noted that in case of foundations and the basement floor in the ground, heat transfer to the external surroundings is expressed with reference to the ground. Heat transfer between elements settled in the ground (foundations and the basement floor) and the ground is driven by heat conduction. Two conductive resistances R_{12} and R_{23}, describe heat conduction in the ground and through elements settled in the ground, accordingly. Those conductive resistances are connected in series (one by one). Then heat is transferred between a floor and internal surroundings (a room) through convection and thermal radiation, which are parallel to each other. Graphic interpretation of heat transfer from a room to the ground and relevant thermal resistances are schematically presented in Figure 6.3. For this case, Eqn (6.4) transforms into:

$$R_t = R_{se} + R_{base} + R_{si} = R_{12} + R_{23} + \left(\frac{R_{3'4} + R_{34'}}{R_{3'4}\, R_{34'}} \right) \quad (6.6)$$

In this particular case, symbols stand for:

R_{12} = thermal resistance of the ground surrounded elements settled in the ground;

R_{23} = thermal resistance of the elements settled in the ground, foundation, and basement floor;

$R_{3'4}$ = thermal resistance between the room floor and internal surroundings by convection;

$R_{34'}$ = thermal resistance between the room floor and internal surroundings by thermal radiation.

If a building has a basement and this basement is heated, then heat is transferred to the ground through the floor (as shown in Figure 6.3) and through the side walls of the basement, which may also, partially or fully, act

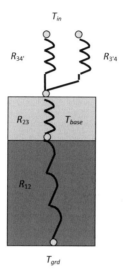

FIGURE 6.3   Graphic interpretation of heat transfer phenomena in an element settled in the ground and surrounding ground described using the thermal resistance method.

as foundations. In the latter case, the R_{12} resistance must be a combination of the horizontal heat transfer through side walls (analogically to Figure 6.2) and vertical heat transfer through a floor (as shown in Figure 6.3). If the basement is not heated, then its internal temperature is variable and such a case requires a more complex description, similar to the heat transfer through a non-heated attic. Heat exchanged with the ground can also be determined in a way given by standards on the thermal characteristics of buildings, especially referred to heat transfer via the ground (EN ISO 13370 [6]).

Heat transfer through the roof depends on its design and shape. In case of a horizontal flat roof with a simple multilayer structure, without an air gap, heat flow through the roof into external and internal surroundings may be considered in a similar way as for the external wall, but of different direction (i.e., vertical one). If a building has a complex roof structure, the standard-derived method of upper and lower limit [5] may be used. If a building has an attic, it requires addressing heat flow direction with regard to the roof slope angles. If the attic is a living space and heated, the heat transfer model through opaque walls of the attic and roof is formulated analogically to the one presented in Figure 6.1, taking into account that an inclined roof structure is usually more complex than of a typical vertical wall. The attic may also be unheated (like the basement mentioned before) and then internal air temperature is variable in time. The energy balance of air in the attic can be described in a general way by Eqn (6.1). Temperature T_{in} at the left side of Eqn (6.1) is now temperature

T_{att} of the internal air and the heat losses are through the roof to the surroundings of temperature T_a. A more detailed analysis also addresses the influence of farther surroundings (i.e., the sky with the temperature T_{sky}). Heat gains are due to the vertical heat flow from rooms below the attic of temperature T_{room}.

For averaged calculations (e.g., for monthly averages), it can be assumed that individual thermal resistances are constant in time, surface areas of all elements of a building envelope are known, and internal air temperature in a building and ambient air temperature are given (the sky impact is neglected). The energy balance can be written in simplified form of Eqn (6.7). This equation enables relatively simple calculation of the air temperature of the attic, if this temperature is constant in considered period (e.g., in a month). Then the simplified energy balance for the air in the attic, which does not take into account heat capacity of a building and heat exchange with the sky ($T_{room} > T_{att} > T_a$) takes a form:

$$\dot{Q}_{in,build}(t_{month}) = \dot{Q}_{out,a}(t_{month})$$
$$\frac{A_{in}(T_{room} - T_{att})}{R_{in}} = \frac{A_r(T_{att} - T_a)}{R_{att}} \tag{6.7}$$

In this case, resistance R_{in} is a combination of elementary resistances resulting from heat transfer from a room through the ceiling to the air in the attic. Resistance R_{att} is a combination of elementary resistances describing heat transfer between the attic air and the ambient air through the roof. Equation (6.7) can be presented in a simpler form introducing constant coefficients a and b and showing how the unknown averaged attic temperature can be determined for a given period of time (e.g., a month).

$$\frac{A_{in}}{R_{in}}T_{room} - \frac{A_{in}}{R_{in}}T_{att} = \frac{A_r}{R_{att}}T_{att} - \frac{A_r}{R_{att}}T_a \Rightarrow aT_{room} + bT_a = aT_{att} + bT_{att}$$
$$\Rightarrow T_{att} = \frac{aT_{room} + bT_a}{a+b}, \quad \text{where } a = \frac{A_{in}}{R_{in}} \quad b = \frac{A_{in}}{R_{in}}$$

(6.7a)

If the heat exchange with the sky is taken into account, it is necessary to include view factors for the radiative heat transfer with near surroundings (F_{r-a}) and with far surroundings (the sky, F_{r-sky}), where $F_{r-a} + F_{r-sky} = 1$.

It is also possible to write the energy balance for air in the attic taking into account that attic temperature is variable in time. The energy balance (Eqn (6.1)) can be changed from differential to difference form taking into account changes of the attic temperature in time, when ambient air and room air temperatures are also variable in time, but they are known. Coefficients a and b

are constant as before (Eqn (6.7a)). Now the considered energy balance for air in the attic takes a form:

$$\left(\rho c_p V\right)\frac{dT_{att}}{dt} = \dot{Q}_{in,build}(t) - \dot{Q}_{out,a}(t) \Rightarrow C_p \frac{T_{att,i} - T_{att,i-1}}{\Delta t}$$

$$= \frac{A_{in}\left(T_{room,i} - T_{att,i}\right)}{R_{in}} - \frac{A_r\left(T_{att,i} - T_{a,i}\right)}{R_{att}}$$

$$\Rightarrow T_{att,i} = T_{att,i-1} + \frac{C_p}{\Delta t}\frac{A_{in}}{R_{in}}T_{room,i} - \frac{C_p}{\Delta t}\frac{A_{in}}{R_{in}}T_{att,i} - \frac{C_p}{\Delta t}\frac{A_r}{R_{att}}T_{att,i}$$

$$+ \frac{C_p}{\Delta t}\frac{A_r}{R_{att}}T_{a,i} \Rightarrow T_{att,i} = \frac{T_{att,i-1} + CaT_{room,i} + CbT_{a,i}}{1 + Ca + Cb}$$

$$\text{where } C = \frac{C_p}{\Delta t} \tag{6.8}$$

Attic air temperature may be calculated with Eqn (6.8), if the ambient air temperature, the room temperature, and the attic temperature at the defined time step before (e.g., an hour) are known (with some simplification it can be assumed that the attic temperature in the initial time is equal to averaged T_{att} from Eqn (6.7a)). The analogy between Eqns (6.7) and (6.8) can be noticed. It can be seen that Eqn (6.8) in a simplified way takes into account dynamics of attics (i.e., changes in time of its thermal state and surroundings), and includes its thermal capacity. The thermal resistance method usually considers steady state conditions and does not take into account heat capacities of walls. Thermal resistance of a building envelope is significant when the heat fluxes are calculated using averaged ambient and envelope thermal conditions. Each envelope element should be highly insulating, and therefore considerably reduce heat losses and gains, depending on climatic conditions. Insulating capabilities of an envelope element are decided by its thickness, which should be possibly high, and its thermal conductivity, which should be possibly low.

Heat (or cold) stored in a building should stay inside as long as possible. It may be "contained" in a building thanks to the high thermal capacity of envelope elements. Internal and external walls should be characterized by high thermal capacity (i.e., ability to store large amounts of heat at possibly small temperature differences). This, in consequence, means high thermal mass. Thermal capacity of an envelope element depends on its volume, density, and specific heat, which should all possibly be high. Heat capacity is therefore defined as:

$$C_p = V\rho c_p \text{ (J/K)}$$

Heat that is contained in a building because of the operation of heating systems and influence of solar radiation, especially through transparent envelope elements, is primarily accumulated in massive walls and other elements of a building interior, and affects the energy balance of a building.

It is possible to take into account the heat capacity of a building when using the thermal resistance method, although it will only be an approximation. We may consider the heat storage of the external surface area of A with a medium of mass m, temperature T_1, and specific heat capacity c. The storage is placed in a space with a temperature T_0 lower than T_1, therefore, there are heat losses from the storage to the surroundings. If the total thermal resistance is designated as R_{1-0}, then the energy balance of the storage may be expressed as:

$$mc\frac{d}{dt}(T_1 - T_0) = \frac{A(T_1 - T_0)}{R_{1-0}} \tag{6.9}$$

Analyzing Eqn (6.9) analogy between this equation and Eqn (6.8) can be seen. Both equations refer to temperature of the storage medium and in a similar simplified way takes into account the thermal capacity of storage medium, and, because of that, dynamics of thermal phenomena. Using the thermal resistance method, the energy balance, expressed by Eqn (6.9), may also be expressed graphically, as illustrated in Figure 6.4.

6.1.3 Some Simplified Forms of Heat Balance of a Building

Equation (6.10) is another simplified energy balance equation for air within a building. It is formulated in a way reflecting (in a simplified way) the influence of solar radiation on the energy balance and the role of heat capacity of the walls. Internal heat generation caused by electrical appliances, people, and ventilation losses are not taken into account.

$$\begin{aligned}(\rho c V)\frac{dT_{in}}{dt} &= U_t A_t(T_{in}(t) - T_a(t)) - \dot{Q}_{sol}(t) - \dot{Q}_h(t) \\ &= \frac{A_t(T_{in}(t) - T_a(t))}{R_t} - (\tau\alpha)A_g G_s(t) - \dot{Q}_h(t)\end{aligned} \tag{6.10}$$

The left side of the Eqn (6.10) refers to the variability of air temperature in the room caused by changing thermal conditions in the external and

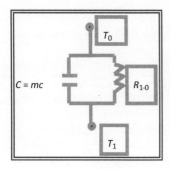

FIGURE 6.4 Heat flow closed loop taking into account heat storage, based on analogy to the electrical current flow.

internal surroundings. On the right side, there are: gains because of solar radiation—the second term; heat supplied to the room by heating system (operation of radiators)—third term; the first term represents heat losses to the environment because of the influence of external surroundings (of temperature T_a), reduced thanks to the total thermal resistance ($R_t = 1/U_t$) of the envelope with a total surface area of A_t—third term. If it is assumed that solar radiation gains are sufficient to ensure required air temperature inside the room, then the operation of radiators (i.e., supply of heat Q_h) is no longer needed. Mathematically, this means that $\dot{Q}_h = 0$, and Eqn (6.10) transforms into:

$$(\rho cV)\frac{dT_{in}}{dt} = \frac{A_t(T_{in}(t) - T_a(t))}{R_t} - ((\tau\alpha)A_{win}G_s(t)) \tag{6.11}$$

Equation (6.11) may be further simplified. It may be assumed that the room is insulated so well that heat transfer through opaque external walls is negligibly small. As a result, heat is lost only through windows (i.e., total thermal resistance is the thermal resistance of windows ($R_t = R_{win}$)). If it is assumed that the temperature within the room is maintained at a constant level thanks to solar irradiation, then Eqn (6.11) is simplified to:

$$(\tau\alpha)A_{win}G_s(t) = \frac{A_{win}(T_{in}(t) - T_a(t))}{R_{win}} \tag{6.12}$$

Using the simplified form of the energy balance in Eqn (6.12), the solar irradiance that is sufficient to maintain constant temperature in a room for certain assumed optical parameters of a window (glass pane transmittance) and thermal parameters of a window (holistically described by its thermal resistance), such as a window, and solar absorptance of internal walls. For example, simplified calculations may be performed for the glass pane transmittance $\tau = 0.8$ and internal wall solar absorptance $\alpha = 0.6$ and heat transfer coefficient for the windows $U_{win} = 1.4$ W/(m^{2}K) (inverse of thermal resistance). At the temperature difference between the ambient air and interior equal to 20 K (ambient temperature 0 °C and interior 20 °C), the irradiance required to maintain constant temperature within the room would be only 60 W/m^2. In case of the older type of windows with higher heat transfer coefficients ($U_{win} = 3$ W/(m^{2}K)), the required irradiance would be twice as high and the approach would be 125 W/m^2. Those calculations are of course only rough estimations. However, even such a simplified model allows drawing certain conclusions, quite important for an energy balance of a modern building: the higher heat resistance of the envelope causes more pronounced influence of solar radiation on the energy balance, which is especially evident in case of large glazed areas.

Also, other simplifications of the energy balance Eqn (6.11) may be introduced to carry out simplified calculations and draw certain conclusions. The conclusions are very rough approximations in quantitative terms, but, nevertheless, are quite important in qualitative meaning. After sunset, there are no more solar radiation gains. If the heating system is not operating, then the thermal state of a building will depend on the heat capacity of a building and insulating properties of the envelope. If there are no gains from solar radiation and no heating system is in use ($G_s = Q_h = 0$), then Eqn (6.11) transforms into:

$$(\rho c V)\frac{dT_{in}}{dt} = \frac{A_{win}(T_{in}(t) - T_a(t))}{R_{win}}$$ (6.13)

It may be assumed that the internal air temperature is equal to the temperature of internal walls, which act as heat accumulators, and heat capacity $C = V\rho c_p$ describes the capacity of those walls. After those simplifications, Eqn (6.13) may be expressed as:

$$\frac{dT_{in}}{dt} = \frac{A_{win}(T_{in}(t) - T_a(t))}{CR_{win}}$$ (6.14)

The solution of Eqn (6.14) is an equation describing the internal temperature T_{in} at the time t_k (i.e., after the time period $t = t_k - t_0$), as a function of the ambient air temperature T_a, also at the time t_k, and of the initial conditions described by the temperature difference $T_{in} - T_a$ at the initial moment t_0, and time constant $\tau = rC$ of the building (where r (K/W) is elementary resistance and C (J/K) is thermal capacity). Thus, the time constant of a building is a product of total elementary thermal resistance of an envelope and the total heat capacity of a building (room). It reflects the comprehensive thermal quality of the building. The time constant grows along with the heat capacity and the total thermal resistance of the building envelope. It should be stated that the time constant is a measure of energy intensity of a building. This energy intensity decreases when the time constant grows. The solution of Eqn (6.14) may be expressed as:

$$T_{in}(t_k) = T_a(t_k) + (T_{in} - T_a)|_{t=0}\, \exp\left(\frac{-t}{rC}\right)$$ (6.15)

Using Eqn (6.15), it is possible to carry out estimated calculations for various opaque walls (of different mass and made of different materials) and various transparent envelope elements—windows (single-pane, double-pane, etc.). This allows evaluating the importance of selecting construction materials for reducing the energy demand for heating. The time constant describes the thermal mass of a building and its insulation capability to be independent from the impact of external surroundings and to contain (store) heat inside.

Equation (6.15) allows carrying out simplified calculations of internal air temperature after a certain time. For example, it can be assumed that the analyzed room has massive accumulating internal walls (and their temperature is equal to the internal air temperature T_{in}). As in the previous example, the opaque external envelope elements (i.e., walls, ceiling, and floor) are assumed to be insulated so well that there is no heat flow through those elements. Heat is transferred only through a window with the heat transfer coefficient of $U = 1.4$ W/(m^{2}K) (as before, typical standard window, with argon in a gas gap, and low emissivity coating) and a surface area of 4 m^2. The Sun sets at 4 p.m. when the internal air temperature is $T_{in} = 20$ °C. Ambient air temperature is assumed to be constant and equal to $T_a = 0$ °C and no heating system is in use (i.e., $\dot{Q}_h = 0$). The thermal state of the room (building) after a certain time results from the initial thermal conditions described by temperature $T_{in}((t_0)$ at 4 p.m.) $= 20$ °C, as well as the thermal mass of the building and the thermal insulating properties of the building (in this case, the window only). A massive internal wall has a surface area of 4 m^2, thickness of 0.25 m, and is made of concrete with a density of $\rho = 2000$ kg/m^3 and specific heat capacity $c = 0.8$ kJ/(kgK). The time constant for the analyzed room is 80 h. Using Eqn (6.15), the internal temperature on the next day at 8 a.m. can be calculated. This temperature is above 16 °C. If the heat capacity is changed by reducing the thickness of the massive wall to 0.1 m, while retaining all other parameters, then the internal air temperature in the morning drops below 12 °C. For this altered case, the time constant is less than 32 h. If, except for decreasing heat capacity, we also reduce the thermal insulation of the envelope by increasing the heat transfer coefficient to, for example $U = 2.6$ W/(m^{2}K) (standard value for double-pane window with an air gap), then the air temperature in the morning drops below 8 °C and the time constant is only slightly above 17 h. This example illustrates the role of heat capacity and thermal insulation capabilities of the envelope and their significance for the energy balance of a building. It also shows how important the time constant is for evaluation of a building's thermal quality. Although the time constant is usually calculated in the process of creating the energy characteristic of a building, its impact is often not well reflected and presented. For example, the standard calculation method of the thermal capacity of a building in Poland is based on the assumption that always only a 10 cm thickness of external, internal walls, and other partitions is taken into account. This does not allow showing the real thermal capacity of a building and causes big errors in evaluating its thermal mass and time constant.

It should be noted that structural internal walls and ceilings do have heat storage capability. If their heat capacity is high and the envelope is highly insulating, even in case of big temperature differences between the ambient air and the interior of the building, the thermal state of the building expressed by the internal air temperature practically would not have been changed, if ventilation and window losses had not existed. The thermal state of a building is a result of earlier thermal conditions of the external and internal surroundings, including solar radiation influence, as well as the ability to accumulate previously available heat, and contain it thanks to the high capacity of the walls and good insulating properties of the envelope.

The discussion presented in this section is valid for a steady or quasi-steady state. In case of a steady state, all parameters are constant. The quasi-steady state for a building and its surroundings is understood as a series of consecutive steady states, described by heat transfer coefficients or thermal resistances constant in time of every steady state in the series. Changing conditions of the internal and external surroundings are kept at constant levels for given time periods (time steps). Those conditions may be expressed by air temperature and solar irradiance that are constant during every time step. When one steady state is finished then a new one described by a new set of parameters starts (for the same time step).

Assuming steady state conditions of heat exchange with the internal and external surroundings by applying constant thermal resistances for the entire year is a major simplification and does not permit modeling actual dynamics of phenomena occurring within a building and its surroundings. It is so, because thermal resistances depend on convective and radiative heat exchange coefficients, which, in reality, depend on temperatures of media between which heat is transferred. In order to obtain a realistic description of the actual phenomena, the variability of both temperatures and heat transfer coefficients needs to be taken into account. In case of walls, the variability of temperatures in their structure and of related parameters should be addressed. A detailed study should take into account the fact that heat conductivity depends on the temperature of the body in question. This dependency is particularly pronounced in case of humid media. It is also considerably altered at phase changes (e.g., upon solidification conductivity increases greatly).

6.2 DYNAMICS OF PROCESSES IN OPAQUE WALLS AND SURROUNDINGS, INCLUDING THE IMPACT OF SOLAR RADIATION

6.2.1 Extension of General Form of the Energy Balance Equation

As already mentioned, Eqn (6.1) in its general form and with lumped parameters, is solved by finding every component of the energy balance separately, with a freely chosen level of detail. The sum of energy flows from internal heat sources in the analyzed room comes from sources like electrical appliances, light sources, and human beings (presence of a single person in a room is equivalent to a source of heat with an approximate output of 50–100 W, depending mainly on that person's activity). In some cases, heat gains provided by appliances may be high and may considerably affect the energy balance of a room. In extreme cases, there is no need to use room heating and even constant cooling might be required (e.g., server rooms). Figure 6.5 shows the essential energy flows coming into and out of a building, which together constitute its energy balance.

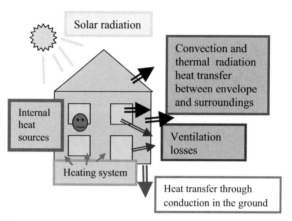

FIGURE 6.5 Energy exchange between a building and its surroundings.

If we assume that there are no internal heat sources in a building or room, then:

$$\sum \dot{Q}_{qv}(t) = 0 \qquad (6.16)$$

If we assume that all the rooms within a building have the same temperature T_{in}, which is constant in time and space (which makes the discussion and comparative analysis of energy balances for different rooms easier), then the left side of Eqn (6.1) is zero i.e.,

$$\left(Vc_p\rho\right)\frac{dT_{in}}{dt} = 0 \qquad (6.17)$$

At the same time this means that a heating/cooling system's operation follows variable ambient conditions to maintain constant temperature level. Referring to the energy balance in Eqn (6.1), it needs to be noted that energy fluxes flowing into or out of a building may have different directions, depending on ambient conditions and its main parameters: solar radiation and ambient air temperature. Solar radiation (if solar radiation is present) is always an energy flux directed into the building coming through a transparent envelope (windows, glazed facades). Heat flows transferred through both opaque and transparent envelope elements may have different directions, depending on the ambient conditions described by the ambient air temperature, and thermal state of the building related to heat capacity of walls, and the insulating properties of the envelope. Heat flow related to room ventilation can also have different directions (heating or cooling), depending on ambient air temperature and indoor air temperature. If there are unheated rooms in a building, then heat may be exchanged between rooms depending on their thermal state. Heat also flows into rooms because of heating system operation or flows out of them when a cooling (air conditioning) system is in use.

Because of the fact that envelope elements (i.e., external walls and windows) have different thermal resistances and heat capacities, it may happen that heat flows through individual components will have opposite directions. Usually, the heat flows through those components are analyzed separately.

Solar radiation influences a building interior in a direct way only via transparent envelope elements. Indirect influence takes place on the entire envelope and depends on solar radiation absorptance of surfaces (coatings) of envelope elements, and on their structure, especially thermal insulation properties, and location (orientation, inclination).

Demand for heating or cooling for ventilation purposes may be calculated in a simplified manner, as a function of averaged standardized volumetric flow V_{ven} of the ventilation air exchanged by a given room. This flow depends on the function of the room, which decides air exchange intensity a [7], room volume V, and type of ventilation system. The heat demand for ventilation purposes may be assumed or calculated using standards or a simplified equation:

$$\sum \dot{Q}_{ven}(t) = a\left(V_{ven}\rho c_p\right)\left(T_{in}(t) - T_a(t)\right) \tag{6.18}$$

If we take into account the presence of equipment for heat recuperation from ventilation systems with certain efficiency η, and the interior temperature T_{in} constant over time and space, then Eqn (6.18) may be expressed as:

$$\sum \dot{Q}_{ven,r}(t) = (1 - \eta) \sum \dot{Q}_{ven}(t) = (1 - \eta)\, a\left(V_{ven}\rho c_p\right)\left(T_{in} - T_a(t)\right) \tag{6.18a}$$

Assuming constant room interior temperature means that constant temperature conditions are maintained in any room. As already mentioned, this may be achieved by using appropriate control systems for heating/cooling (air conditioning) systems, which maintain preset internal air temperature. Based on the presented assumptions, including those described by Eqns (6.16) and (6.17), the energy balance Eqn (6.1) for the case of heat losses from a room may be expressed as:

$$\sum \dot{Q}_h(t) = \sum_j \dot{Q}_{j,wall}(t) + \sum_i \dot{Q}_{i,win}(t) + \dot{Q}_{ven}(t) - \sum_i \dot{Q}_{i,sol}(t) - \sum_n \dot{Q}_{n,qv}(t) \tag{6.19}$$

In case there are only heat gains, the energy balance equation takes the following form:

$$\sum \dot{Q}_c(t) = \sum_j \dot{Q}_{j,wall}(t) + \sum_i \dot{Q}_{i,win}(t) + \dot{Q}_{ven}(t) + \sum_i \dot{Q}_{sol}(t) + \sum_n \dot{Q}_{n,qv}(t) \tag{6.20}$$

As mentioned before, there may be a situation when heat flow directions are more diverse (For example, there are heat losses through windows), and heat gains through opaque envelope elements gained thanks to their heat storage capacity (influence of previous thermal conditions), despite lower ambient temperatures. Depending on the situation, individual components of the energy balance equation may be positive or negative.

6.2.2 Dynamics of Heat Transfer through Opaque Envelope Elements

The analysis may have different levels of detail, depending on its purpose. In this publication, the discussion focuses on the influence of solar radiation on a building, and therefore this particular issue is analyzed in detail, while other phenomena are addressed in a simplified manner. This section discusses the heat transfer through opaque elements of the envelope, while neglecting air and moisture flow (infiltration) through such elements. It is assumed that the heat flow within the opaque element is only by conduction. Therefore, the temperature field within such an opaque element is described with the conductivity equation:

$$\left(\rho c_p\right)_g \frac{\partial T}{\partial t} = \lambda \nabla^2 T + q_v \tag{6.21}$$

Most standard opaque elements are single- or multilayer solids (although there are also slotted walls, walls with ventilated air spaces, etc., but they will not be discussed here). It may be assumed that each layer of the element is homogenous and isotropic. As a result, the heat conductivity of such a layer may be assumed to be constant within that layer. Additionally, it may be assumed that conductivity is constant over time (i.e., it does not depend on temperature). Analyzed structure may be considered as a one-dimensional body, where the envelope element is seen as an infinite plate with a thickness of δ (coordinate x goes through the thickness), without internal heat sources. With these assumptions, governing temperature field Eqn (6.21) simplifies and in case of a single-layer wall (or for each layer of a multilayer wall structure) takes form:

$$\frac{\partial T_{pi}(x,t)}{\partial t} = a_{pi} \frac{\partial^2 T_{pi}(x,t)}{\partial x_{pi}^2} \quad \left(\rho_{pi} c_{pi}\right) \frac{\partial T_{pi}(x,t)}{\partial t} = \lambda_{pi} \frac{\partial^2 T_{pi}(x,t)}{\partial x_{pi}^2} \tag{6.22}$$

The a_{pi} symbol represents thermal diffusivity for the i-th layer of the wall structure, where $a_{pi} = \lambda_{pi}/(\rho_{pi} c_{pi})$. Here, we may refer to the description of the heat conduction within an elementary volume, as this description well reflects the dynamics of this phenomenon and the very concept of heat capacity. According to Fourier's law, elementary heat flow $q_{x,in}$ into an elementary volume of the wall in a one-dimensional model may be expressed as:

$$\dot{q}_{x,in} = -\lambda \frac{\partial T_i}{\partial x}$$

Then an elementary heat flow $\dot{q}_{x,out}$ of the same elementary volume with a length of dx is:

$$\dot{q}_{x,out} = \dot{q}_{x,in} + \frac{\partial \dot{q}_{x,in}}{\partial x} dx$$

The second addend of this sum takes into account heat flux contained within the elementary volume because of the phenomena that took place earlier. Absence of this flux would lead to a simple relation $q_{x,in} = q_{x,out}$ (i.e., it would not take into account the "heat transfer history"—dynamics of phenomena in time).

Equation (6.22) may be solved, if the initial condition and the heat exchange conditions on boundary surfaces of the i-th layer of a wall are known. It may be noted here that the description of heat transfer processes on boundary surfaces of an opaque envelope element, from the internal and external sides, is analogous to the case of a transparent element. Thus, the mathematical model is also identical.

6.2.3 Heat Exchange with Surroundings through Building Envelope in Unsteady State

Elements of a building envelope are in contact with the interior of a temperature T_{in} and with the building's direct surroundings of a temperature T_a. The building is also influenced by farther surroundings—the sky of a temperature of T_{sky}. Figure 6.6 schematically shows individual temperatures and location of boundary points (nodes) in a wall, for which boundary conditions are formulated for solving the unsteady heat transfer through a double layer wall in a one-dimensional model.

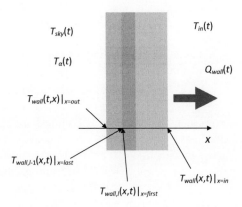

FIGURE 6.6   Wall under consideration with boundary point (nodes) locations.

At the internal boundary surface (i.e., on the building's interior side), the flux transferred through a wall by heat conduction at the wall side is equal to the heat flux transferred by natural convection and thermal radiation at the room side:

$$-\lambda_p \frac{\partial T_p(x,t)}{\partial x_p}\bigg|_{in} = h_{in}(t)\left[T_p(x,t)\big|_{in} - T_{in}(t)\right] \qquad (6.23)$$

In case of natural convection in rooms, empirical relations for vertical surfaces may be used; such equations have been proposed by various researchers (e.g., Awbi [8], Churchill and Chu [9]). Heat flow by natural convection in case of sloped surfaces, heated from the bottom or from the top, has been studied, for example, by Fugi and Imura [10]. Simplified relations for sloped surfaces exchanging heat by natural convection have also been proposed by Holman [10]. Those relations allow finding convective heat exchange coefficient depending on the Ra number and the type of surface that exchanges heat with surrounding air. Modeling the heat exchange phenomena in rooms was also studied by Curcij and Gross [11]. In the empirical relations they proposed, Nu and Ra numbers are based on characteristic dimension defined as the height of an analyzed element. The Ra number and individual air parameters are determined for a boundary air layer near the internal side of the wall, for the averaged boundary layer temperature. The averaged boundary layer temperature may be calculated with equation:

$$\overline{T}_{pi,l}(x,t) = 0.75 \cdot T_{in}(t) + 0.25 \cdot T_{pi}(x,t) \qquad (6.24)$$

It may be assumed [3], that properties of air of the boundary layer change linearly with the temperature of this boundary layer in the following way:

$$\text{Air parameter }(t) = a_{parameter|T_{min}} + b_{parameter|T_{max}} \cdot \overline{T}_{pi,l}(x,t) \qquad (6.25)$$

Coefficients a and b of the linear Eqn (6.25) are equal to the air parameter (property) under consideration at expected extreme (minimum and maximum) temperatures (here 250 and 350 K).

For example, the dynamic viscosity of the air can be determined from Eqn (6.25), that now takes a form:

$$\mu(t) = \mu_{|T_{min}} + \mu_{|T_{max}} \cdot \overline{T}_{pi,l}(x,t)$$

Convective heat transfer coefficient is determined as a function of the Nusselt number, heat conductivity (in this case, conductivity λ_l of the boundary layer), and the characteristic dimension L, in this case, thickness of the boundary layer. Therefore, the convective heat transfer coefficient is expressed as:

$$h_{con,in}(t) = Nu(t)\frac{\lambda_l}{L} \qquad (6.26)$$

The Nusselt number Nu is usually described as a function of the Rayleigh number Ra, $Nu = f(Ra)$, and the Rayleigh number is calculated as:

$$Ra(t) = \frac{g\beta\rho^2 c_p L^3}{\lambda\mu}\,\Delta T(t) \tag{6.27}$$

Relation between Nu and Ra depends on the inclination of studied surface and heat flow direction. For horizontal surfaces and surfaces with small inclination β, where $0° \leq \beta < 15°$ (nearly horizontal), when the heat flows from the building to the surroundings (heat losses), Curcij and Goss [11] recommend using the following relation:

$$Nu = 0.13\,(Ra_H)^{1/3} \tag{6.28}$$

For surfaces inclined at larger angles, in range $15° \leq \beta \leq 90°$, when heat flows from the interior to the exterior (heat losses), the relation between the Nusselt and Rayleigh numbers is:

$$Nu = 0.56\,(Ra_H \cdot \sin\beta)^{1/4} \quad \text{for } Ra_H \leq Ra_{cv} \tag{6.29}$$

$$Nu = 0.13\left(Ra_H{}^{1/3} - Ra_{cv}{}^{1/3}\right) + 0.56(Ra_{cv} \cdot \sin\beta)^{1/4} \quad \text{for } Ra_H > Ra_{cv} \tag{6.30}$$

$$Ra_{cv} = 2.5 \cdot 10^5 \left(\frac{e^{0.72\beta}}{\sin\beta}\right)^{1/5} \tag{6.31}$$

In case of horizontal surfaces when heat flows from the outside to the interior (heat gains), it is recommended [12] to use equation:

$$Nu = 0.58(Ra_H)^{1/5} \quad \text{for } Ra_H \leq 10^{11} \tag{6.32}$$

For surfaces inclined when heat flows from the outside into the building (heat gains), it is recommended to use the equation identical to Eqn (6.29), but with a different Rayleigh number range:

$$Nu = 0.56\,(Ra_H \cdot \sin\beta)^{1/4} \quad \text{for } 10^5 \leq Ra_H \sin\beta \leq 10^{11} \tag{6.33}$$

Radiative heat transfer takes place in every room between the internal surface of an envelope element and the room interior. The interior is considered to be a black body totally surrounding the surface of the envelope element. Thus, the radiative heat transfer coefficient for the analyzed surface may be calculated as:

$$h_{i,r,in}(x,t) = \varepsilon_{i,o}\sigma\left(T_{i,in}(x,t)^2 + T_{in}(t)^2\right)\left(T_{i,in}(x,t) + T_{in}(t)\right) \tag{6.34}$$

Index i in Eqn (6.34) refers to the relevant envelope element. The temperature of the surfaces of those elements is variable over time and may also vary on the surface (depending on the dimension of the model used).

The external surface of the building envelope is in contact with external surroundings. The description of the heat transfer on the external envelope surface is more complex than the one for the internal surface. The envelope exchanges heat with near (direct) surroundings with a temperature T_a by thermal radiation and by convection, usually of a forced type, driven by wind. At the same time, there is also heat exchange by thermal radiation with the farther surroundings (i.e., the sky of a temperature T_{sky}). Finally, there is also influence of solar radiation with irradiance G_s.

Heat generated by photothermal conversion is received by the boundary surface of the envelope element and is partly lost to the surroundings by convection and thermal radiation, and partly transferred through the envelope, mainly by conduction (depending on the structure, also convection and radiation may be taken into account (e.g., in case of slotted or porous walls) but such situations will not be discussed in this study). The boundary condition for the external surface of an envelope element that is in contact with external surroundings may be expressed as [12]:

$$-\lambda_p \frac{\partial T_p(x, t)}{\partial x_p}\bigg|_{x=out} = h_{out}(t)\left[T_a(t) - T_p(x, t)\big|_{x=out}\right] - \dot{q}_{sky}(t) + \dot{q}_{sol}(t) \quad (6.35)$$

Equation (6.35) is valid for a situation when the ambient temperature is higher than the wall surface temperature, but the sky temperature is lower (which is usually the case). Heat exchanged with the sky and the influence of solar radiation are taken into account. Heat exchange with close surroundings with a temperature of T_a is included in Eqn (6.35) as a total, including both radiation and convection with a single coefficient h_{out}.

In case of convective heat exchange with external surroundings, wind-driven forced convection is analyzed; natural convection is a much rarer case. In section 4.4 of Chapter 4, the interaction between a solar energy receiver and its surroundings has been described, including relations used to determine forced and natural convection, which also refer to a building. They are expressed by Eqns (4.73)–(4.77b). In simplified energy analyses of buildings, constant co-efficients of heat exchange by forced convection may be used as follows [13]:

- for winter: $h_{con} = 20$ W/(m²K);
- for summer: $h_{con} = 8$ W/(m²K).

It seems reasonable to introduce another constant convective heat transfer coefficient for transitional spring and autumn seasons. It would be on inter-mediate level [12]:

- transitional period: $h_{con} = 12$ W/(m²K).

Heat exchange by radiation between the outer surface of a building en-velope and direct external surroundings (ground and neighboring objects) is analyzed by considering surroundings as a black body totally surrounding the building under consideration. Heat exchanged with near surroundings is described by Eqn (4.79). Surfaces of the envelope elements are characterized by specific emissivity values, and may have different temperatures variable over time (t) and location on the surface (x, y). The coefficient of radiative heat exchange with near surroundings may be calculated with the following equation:

$$h_{i,r,a}(x, t) = \varepsilon_{i,o}\sigma \left(T_{i,a}(x, t)^2 + T_a(t)^2\right)\left(T_{i,a}(x, t) + T_a(t)\right)F_g \qquad (6.36)$$

In this equation, index i refers to a specific envelope element. Heat transfer coefficient as well as surface and ambient temperatures are variable over time, or over time and space (depending on number of coordinates (dimensions) of the analyzed model). Equation (6.36) includes a view factor F_g (like in Eqn (4.79)). It describes "how well" the surface "sees" its direct surrounding. Complement of the view factor (to 100%) is the view factor for farther surroundings (i.e., the sky). The view factor F_g may be expressed in a way analogical to the correction factor for reflected radiation in the isotropic diffuse radiation model defined by Eqn (2.33). In this case, angle β stands for the slope of the analyzed surface.

Equations that describe heat transfer with near surroundings include ambient air temperature T_a. This temperature may be determined upon actual data, averages provided by meteorological stations, or upon weather databases maintained by various institutions. Also, the approximations of real weather data may be used. In this section, the author's own method for approximating daily ambient temperature variability is presented.

The variability of ambient air temperature in time may be identified upon the average daily temperature values $T_{day}(t_m)$ and average daily temperature amplitudes $A_{day}(t_m)$ specified for every month t_m of a year. It may be assumed that the ambient air temperature T_a changes during an average day of each month of a year in asymmetrically sinusoidal way. According to weather observations [14,15] (for higher latitude countries), it has been assumed that the maximum temperature during a day is observed around 2:30 p.m., while the daily minimum occurs just before sunrise. Thus, the sinusoidal function should assume its extremes for those moments. In order to obtain the appropriate function of temperature variability described by an asymmetrical sinusoid, special correcting factor $F(t_h, t_m)$ is introduced for the m-th month of a year, as an hourly variable. Ultimately, the equation describing variability of ambient air temperature T_a in time of the averaged day for specific months of a year (t_m) for any hour (t_h) of this day has a following form:

$$T_a(t_h, t_m) = T_{day}(t_m) + \left[A_{day}(t_m)\cdot\sin(\pi\cdot(F(t_h, t_m)))\right] \qquad (6.37)$$

The correcting factor is defined with three different functions, depending on the time of day:

$$F(t_h, t_m) = \frac{0.5 + t_h - t_{\max-1d}(t_m)}{(t_{rise}(t_m) - 1) - t_{\max-1d}(t_m)} \quad \text{for } t_h < t_{rise}(t_m) \qquad (6.38)$$

$$F(t_h, t_m) = \frac{-0.5 + t_h - (t_{rise}(t_m) - 1)}{t_{max} - (t_{rise}(t_m) - 1)} \quad \text{for } t_{rise}(t_m) \leq t_h < t_{max} \quad (6.39)$$

$$F(t_h, t_m) = \frac{0.5 + t_h - t_{max}}{t_{min+1d}(t_m) - t_{max}} \quad \text{for } t_h \geq t_{max} \quad (6.40)$$

Calculation of the ambient air temperature correcting factor $F(t_h, t_m)$ involves the sunrise time $t_{rise}(t_m)$, which depends on solar day length $T_d(t_m)$ obtained with Eqn (2.21). Time of sunrise is calculated with Eqn (2.42). Just before sunrise, the ambient air temperature reaches its daily minimum. As it has been already mentioned, at the time equal to $t_{max} = 14.5$ (i.e., 2:30 p.m.) there is the maximum of ambient temperature. Calculation of the temperature correcting factor also involves time (hour) of the maximum daily temperature on the previous day, designated t_{max-1d} and equal to $t_{max-1d} = t_{max} - 24$, as well as time (hour) of the minimum daily temperature on the following day, designated t_{min+1d} and equal to $t_{min+1d} = (t_{rise} - 1) + 24$.

Heat exchanged with farther surroundings (i.e., with the sky), is described by Eqn (4.78). The radiative heat exchange coefficient between the surface of a building envelope and the sky may be calculated with the following equation:

$$h_{i,r,sky}(x, t) = \varepsilon_{i,o}\sigma\left(T_{i,a}(x, t)^2 + T_{sky}(t)^2\right)\left(T_{i,a}(x, t) + T_{sky}(t)\right)F_s \quad (6.41)$$

The view factor F_s in this case is analogous to the correction factor for diffuse radiation in the isotropic model of diffuse radiation described by Eqn (2.32).

As already mentioned, publications on solar energy [4] provide averaged coefficients of radiative heat transfer with external surroundings treated jointly (i.e., direct surroundings and the sky). Averaged coefficients refer to cloudless sky and are approximately between 5 and 6 W/(m²K).

The calculation of heat exchanged by radiation with the sky (and also of radiative heat transfer coefficient described by Eqn (6.41)) requires knowledge of the sky temperature. The method for determining this temperature is presented in Section 1.5 and described by Eqns (1.6)–(1.15).

The solar radiation flux incident on the external surface of opaque envelope elements is expressed as a function of solar irradiance on the surface under consideration and solar absorptance of this surface $\alpha_{s,wall}$. Solar irradiance $G_s(t, \beta, \gamma)$ is calculated for the specific time taking into account inclination β and azimuth angle γ of the surface. The hourly solar irradiation of the surface may be determined with Eqn (2.42) for the isotropic model or with Eqn (2.62) for the anisotropic model. The absorbed solar energy flux may be described as:

$$\dot{q}_s(t) = G_s(t, \beta, \gamma) \cdot \alpha_{s,wall} \quad (6.42)$$

In case of a multilayer external wall, apart from boundary conditions on its outer and inner surfaces, also the boundary conditions for the boundary contact surface between individual contacting layers need to be specified. As already mentioned, it may be assumed that every layer is isotropic and homogeneous, and the heat transfer is carried by conduction (with both layers being solid). It may be assumed that the boundary condition is described by equal temperatures at the contact surface and equal heat flow through this contact boundary surface. This specific boundary condition in a one-dimensional model may be expressed as:

$$T_{p,l-1}(x,t)\big|_{x_{br}} = T_{p,l}(x,t)\big|_{x_{br}}$$

$$-\lambda_{p,l-1} \frac{\partial T_{p,l-1}(x)}{\partial x_{br}}\bigg|_{X_{br}} = -\lambda_{p,l} \frac{\partial T_{p,l}(x)}{\partial x_{br}}\bigg|_{X_{br}} \tag{6.43}$$

Index l refers to the analyzed layer, and $l-1$ to the previous one. Governing equations of the temperature field in an envelope body and boundary conditions on its surfaces (from the outside and inside) allow finding the temperature in any point (node) in that body including its boundary surfaces. Thus, the variability of temperature on the inner wall surface over time $T_{p,in}(t)$ may be determined (in one-dimensional model $T_p(x,t)_{x=in}$). Then, having calculated the heat exchange coefficient h_{in} for the internal surface (for convective heat transfer, e.g., from Eqn (6.26) and for radiative heat transfer from Eqn (6.34)), and knowing the surface area of the envelope element $A_{wall,in}$, we may calculate heat flow between the wall surface and room interior with a temperature of $T_{in}(t)$. In case of heat gains (surface temperature higher than air temperature), this flow would be equal to:

$$\dot{Q}_{wall}(t) = h_{in}(t)\left(T_p(x,t)\big|_{x=in} - T_{in}(t)\right)A_{wall,in} \tag{6.44}$$

The heat flow described by Eqn (6.44) is one component of the energy balance equation given by Eqn (6.1), or Eqn (6.19) or Eqn (6.20).

6.2.4 Selected Examples of Unsteady Heat Transfer through Opaque Envelope Elements

This section presents selected results of numerical simulations of heat flow through an opaque multilayer building envelope based on a mathematical model presented in the previous sections of this chapter. It was assumed that the temperature in heated or cooled rooms should be maintained at the constant level of 22 °C, and this is reached with the operation of automatically controlled heating/cooling system. The value of 22 °C has been selected because of one type of the considered walls (light thermal mass panel with a phase-changing material (PCM)) shows up its storage ability at this temperature (phase change temperature for selected material is 21.7 °C). The analysis has included a few structures of the external envelope, reflecting solutions currently used or potentially available for the use in higher latitude countries, but not only. Typical construction materials and innovative structural solutions

(PCM, as mentioned above) have been taken into account. Simulations have been performed for four selected vertical elements (walls) and two horizontal ones (roofs). According to the aim of this publication, particular attention has been paid to the influence of solar radiation. Surface coatings of different solar absorptance have been analyzed.

Essential thermal properties of analyzed elements and their dimensions are listed in Tables 6.2—6.6. The following figures present graphic interpretation of selected results of numerical simulation of wall and roof dynamics and their surroundings over time for two selected months of a year: January and July. Those months represent opposite weather conditions in a year (i.e., winter and summer). For the sake of comparison, it has been assumed that all external walls have the identical thickness of 45 cm, with the exception of the unconventional PCM wall. Smaller thickness of the wall does not affect the thermal comfort (temperature) within the building, as it had been identified during simulations for an equivalent to a standard structure (20 cm—thermal insulation + 25 cm—structural material with a high thermal mass). Assumed structures are purely hypothetical and do not reflect any real structure of elements used in buildings in practice (plasterwork and other structural details are omitted). However, individual assumed structures do reflect the thermal behavior of selected structure types. Single- and double-layer elements with good or poor insulating properties have been considered to present influence of the environment in case of insufficient insulation. Four vertical walls facing

TABLE 6.2 Main Thermo-Physical Parameters of the Double Layer Wall 1 and 2 [16]

Layer from Outside	Material	Thickness δ (m)	Specific Heat c (kJ/kgK)	Density ρ (kg/m^3)	Thermal Conductivity λ (W/(m K))
1	Mineral wool	0.20	0.70	24	0.038
2	Concrete	0.25	0.88	2000	0.76

TABLE 6.3 Main Thermo-Physical Parameters of the One Layer Wall 3 [16]

Layer from Outside	Material	Thickness δ (m)	Specific Heat c (kJ/kgK)	Density ρ (kg/m^3)	Thermal Conductivity λ (W/m K)
1	Concrete	0.44	0.88	2000	0.76

TABLE 6.4 Main Thermo-Physical Parameters of the Double Layers Wall 4 with a Thermal Mass Panel with PCM Content (60% Paraffin Wax) [17]

Layer from Outside	Material	Thickness δ (m)	Specific Heat c (kJ/kgK)	Density ρ (kg/m³)	Thermal Conductivity λ (W/m K)	Latent Heat Q_{PC} (kJ/kg)	Melting Temperature T_{melt} (°C)
1	Mineral wool	0.25	0.70	24	0.038	–	–
2	PCM panel	0.015	2.2_{solid}	856	$0.18_{solid}/0.14_{liq}$	70	21.7

PCM, phase-change material.

TABLE 6.5 Main Thermo-Physical Parameters of the Two-Layer Flat Roof 5 [16]

Layer from Outside	Material	Thickness δ (m)	Specific Heat c (kJ/kgK)	Density ρ (kg/m³)	Thermal Conductivity λ (W/m K)
1	Mineral wool	0.25	0.70	24	0.038
2	Concrete	0.20	0.88	2000	0.76

TABLE 6.6 Main Thermo-Physical Parameters of the Single-Layer Flat Roof [16]

Layer from Outside	Material	Thickness δ (m)	Specific Heat c (kJ/kgK)	Density ρ (kg/m³)	Thermal Conductivity λ (W/m K)
1	Concrete	0.45	0.88	2000	0.76

south have been analyzed to ensure the best annual exposure to solar radiation. Three of them are insulated, and one is not. The latter one is a single-layer wall. Two double-layer walls have insulation at the external side. They have identical structure and dimensions, but different solar radiation absorptance of the external surface. The third insulated wall has a special structure. It contains a light mass thermal panel with PCM at the internal side. Also, two horizontal flat roofs have been analyzed: one of them was insulated, and the other was not. Their dimensions and thicknesses are identical (which is, of course, a purely theoretical assumption made for the sake of comparison). Roof surface coatings are assumed to be black, as is often the case at higher latitudes. It has been assumed that mineral wool would be used as thermal insulation and concrete as the structural thermal capacity material. Simulations for such conceptual structures show certain characteristic features of heat transfer processes in envelope elements and their surroundings, as well as the influence of solar energy.

Thus, the following envelope elements have been analyzed:

- Wall 1, insulated, double layer, with external insulating layer and black coating, and thermal mass internal layer;
- Wall 2, insulated, double layer, with external insulating layer and white coating, and thermal mass internal layer;
- Wall 3, uninsulated thermal mass wall with white external coating;

- Wall 4, double layer, with external insulating layer and white coating and thermal mass internal layer in the form of a PCM panel;
- Roof 5, insulated double layer, with external insulating layer and black coating, and thermal mass internal layer;
- Roof 6, uninsulated roof, with black external coating.

In the case of walls 1 and 2, two types of external coating have been assumed:

- black with solar radiation absorptance $\alpha_{sol} = 0.9$ and thermal radiation emissivity $\varepsilon_{therm} = 0.8$;
- white with solar radiation absorptance $\alpha_{sol} = 0.1$ and thermal radiation emissivity $\varepsilon_{therm} = 0.8$.

Apart from wall 2, also walls 3 and 4 have white external coating. While, apart from wall 1, roofs 5 and 6 are black. In all cases, the internal surface is white with a thermal radiation emissivity $\varepsilon_{therm} = 0.8$.

Figures 6.7 and 6.8 show the assumed daily distribution of the averaged ambient air temperature and the sky temperature for January and July, respectively. Then, 36 further Figures 6.9–6.45 present three diagrams for each of six analyzed wall and roof structures for two selected averaged days of January and July. These diagrams show:

- temperature distribution across wall thickness at six different times of day with 4-h intervals, starting at 4 a.m.;
- distribution of heat exchanged by the wall external surface with near surroundings and far surroundings (sky) by convection and radiation for consecutive hours of a day, solar irradiance incident on a wall is also presented;
- distribution of heat flux density exchanged between the internal surface of the wall and internal cavity (room), taking into account individual components (i.e., heat exchanged by convection and radiation), for consecutive hours of a day.

This section describes the essential observations made during analysis of obtained results shown in the figures just presented. In case of a wall with a white surface of the insulating layer, the external surface temperature generally follows the ambient air temperature variations, both in January and in July. In case of a wall with a black external coating of the insulating layer, the temperature of the external surface rises considerably during the day, exceeding the ambient air temperature, because of the absorption of solar radiation. Absorbed radiation gets converted into heat, and then is unable to easily penetrate insulation, thus staying at the surface of the wall.

The direction of the heat flow exchanged between the wall and surroundings, and its intensity, depend on the surface temperature and ambient temperature, because convective and radiative heat exchange are functions of temperature difference between the surface and ambience (all elements in the surroundings are assumed to have a temperature equal to the ambient

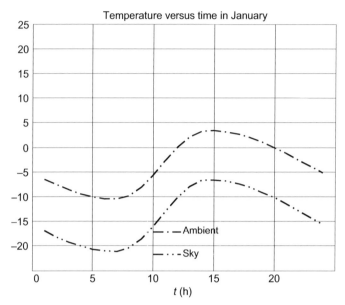

FIGURE 6.7 Distribution of ambient and sky temperature versus time (hours of an averaged day) in January.

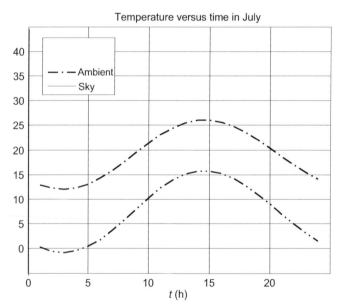

FIGURE 6.8 Distribution of ambient and sky temperature versus time (hours of an averaged day) in July.

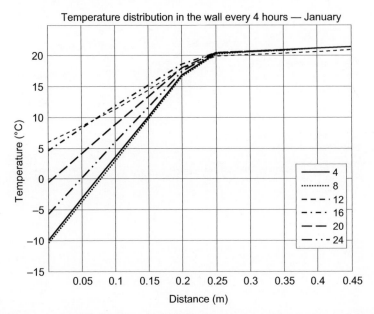

FIGURE 6.9   Distribution of external wall temperature through its thickness versus time (hours of an averaged day) in January; wall 1.

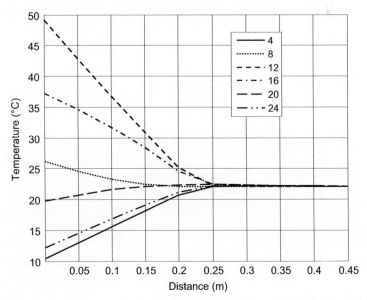

FIGURE 6.10   Distribution of external wall temperature through its thickness versus time (hours of an averaged day) in July; wall 1.

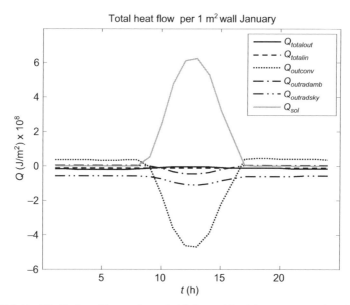

FIGURE 6.11 Distribution of heat exchanged with the outside of the room versus time (hours of an averaged day) in January; wall 1.

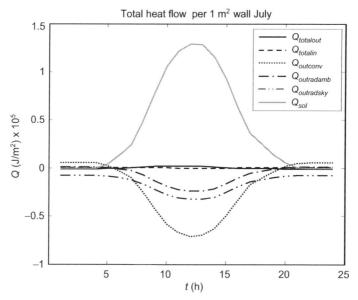

FIGURE 6.12 Distribution of heat exchanged with the outside of the room versus time (hours of an averaged day) in July; wall 1.

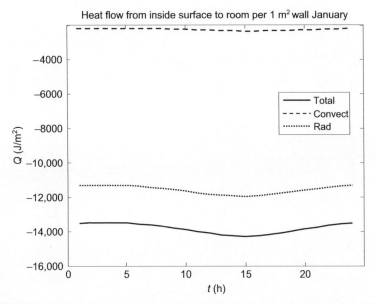

FIGURE 6.13   Distribution of heat exchanged with inside cavity (room) versus time (hours of an averaged day) in January; wall 1.

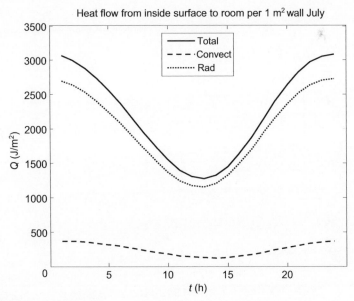

FIGURE 6.14   Distribution of heat exchanged with inside cavity (room) versus time (hours of an averaged day) in July; wall 1.

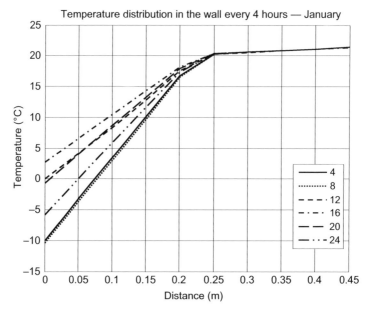

FIGURE 6.15   Distribution of external wall temperature through its thickness versus time (hours of an averaged day) in January; wall 2.

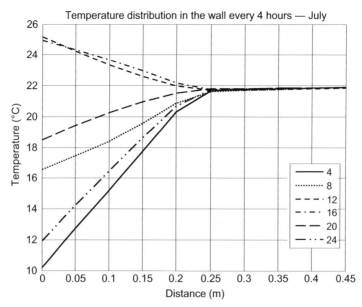

FIGURE 6.16   Distribution of external wall temperature through its thickness versus time (hours of an averaged day) in July; wall 2.

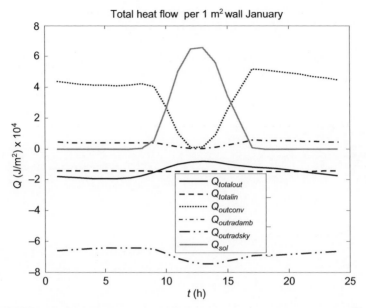

FIGURE 6.17 Distribution of heat exchanged with the outside of the room versus time (hours of an averaged day) in January; wall 2.

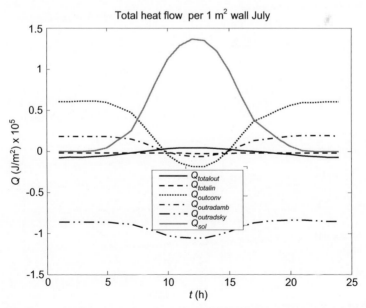

FIGURE 6.18 Distribution of heat exchanged with the outside of the room versus time (hours of an averaged day) in July; wall 2.

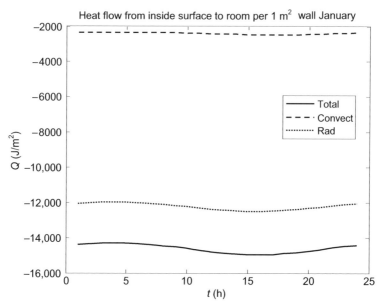

FIGURE 6.19 Distribution of heat exchanged with inside cavity (room) versus time (hours of an averaged day) in January; wall 2.

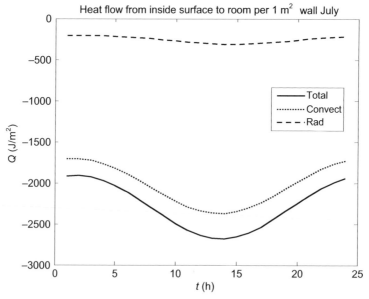

FIGURE 6.20 Distribution of heat exchanged with inside cavity (room) versus time (hours of an averaged day) in July; wall 2.

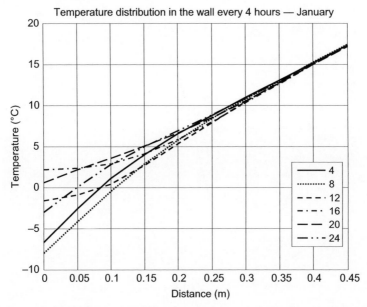

FIGURE 6.21 Distribution of external wall temperature through its thickness versus time (hours of an averaged day) in January; wall 3.

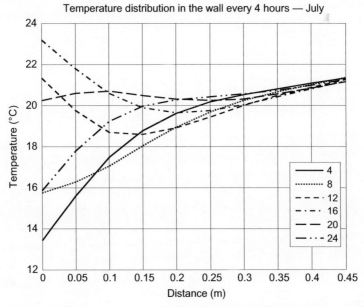

FIGURE 6.22 Distribution of external wall temperature through its thickness versus time (hours of an averaged day) in July; wall 3.

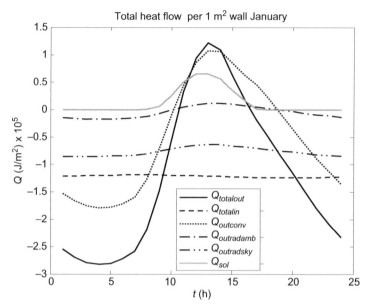

FIGURE 6.23 Distribution of heat exchanged with the outside of the room versus time (hours of an averaged day) in January; wall 3.

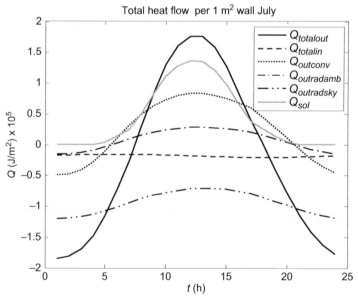

FIGURE 6.24 Distribution of heat exchanged with the outside of the room versus time (hours of an averaged day) in July; wall 3.

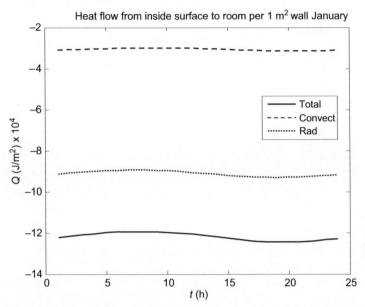

FIGURE 6.25   Distribution of heat exchanged with inside cavity (room) versus time (hours of an averaged day) in January; wall 3.

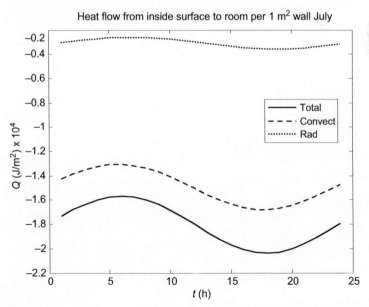

FIGURE 6.26   Distribution of heat exchanged with inside cavity (room) versus time (hours of an averaged day) in July; wall 3.

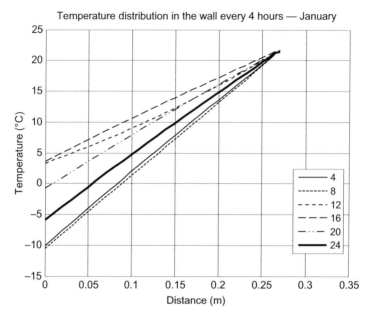

FIGURE 6.27 Distribution of external wall temperature through its thickness versus time (hours of an averaged day) in January; wall 4.

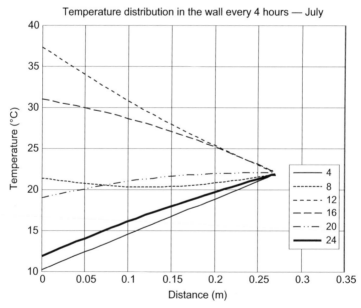

FIGURE 6.28 Distribution of external wall temperature through its thickness versus time (hours of an averaged day) in July; wall 4.

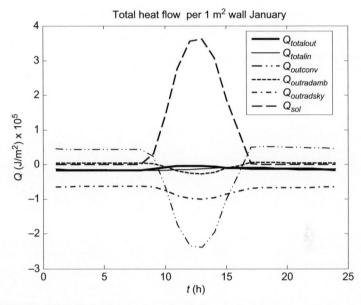

FIGURE 6.29 Distribution of heat exchanged with the outside of the room versus time (hours of an averaged day) in January; wall 4.

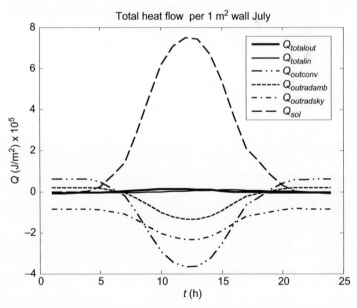

FIGURE 6.30 Distribution of heat exchanged with the outside of the room versus time (hours of an averaged day) in July; wall 4.

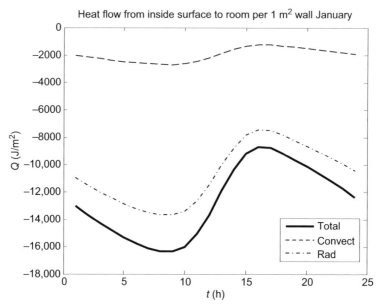

FIGURE 6.31   Distribution of heat exchanged with inside cavity (room) versus time (hours of an averaged day) in January; wall 4.

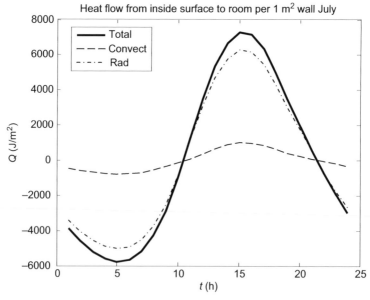

FIGURE 6.32   Distribution of heat exchanged with inside cavity (room) versus time (hours of an averaged day) in July; wall 4.

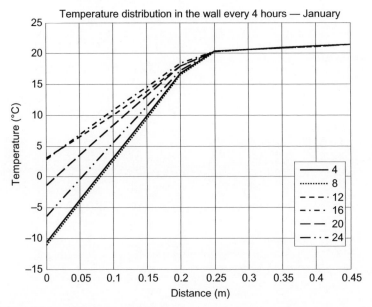

FIGURE 6.33   Distribution of external wall temperature through its thickness versus time (hours of an averaged day) in January; wall 5.

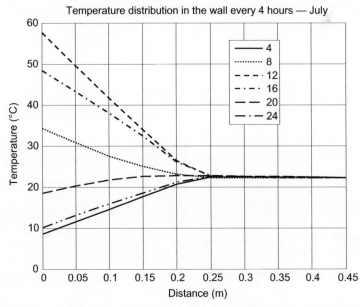

FIGURE 6.34   Distribution of external wall temperature through its thickness versus time (hours of an averaged day) in July; wall 5.

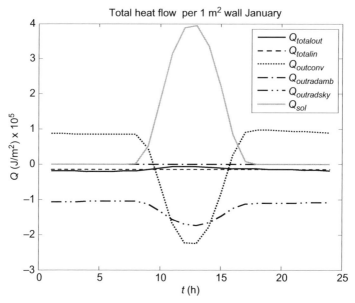

FIGURE 6.35 Distribution of heat exchanged with the outside of the room versus time (hours of an averaged day) in January; wall 5.

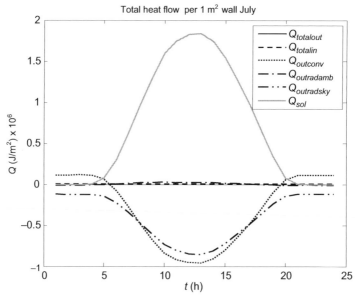

FIGURE 6.36 Distribution of heat exchanged with the outside of the room versus time (hours of an averaged day) in July; wall 5.

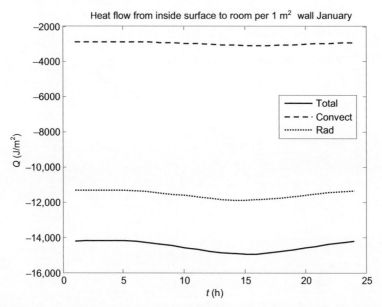

FIGURE 6.37 Distribution of heat exchanged with inside cavity (room) versus time (hours of an averaged day) in January; wall 5.

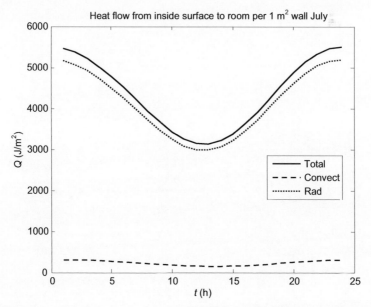

FIGURE 6.38 Distribution of heat exchanged with inside cavity (room) versus time (hours of an averaged day) in July; wall 5.

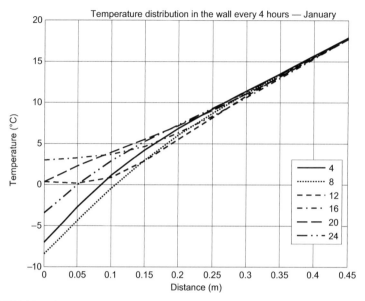

FIGURE 6.39   Distribution of external wall temperature through its thickness versus time (hours of an averaged day) in January; wall 6.

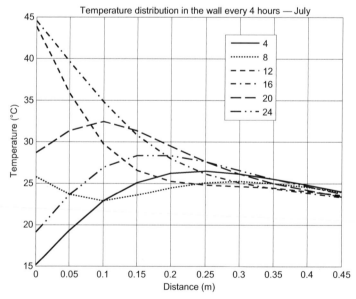

FIGURE 6.40   Distribution of external wall temperature through its thickness versus time (hours of an averaged day) in July; wall 6.

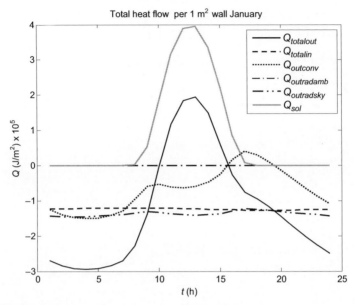

FIGURE 6.41 Distribution of heat exchanged with the outside of the room versus time (hours of an averaged day) in January; wall 6.

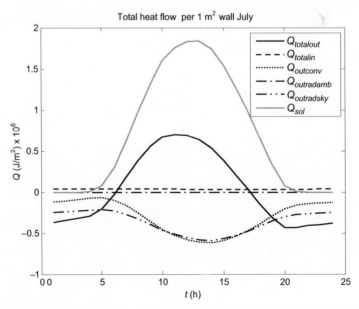

FIGURE 6.42 Distribution of heat exchanged with the outside of the room versus time (hours of an averaged day) in July; wall 6.

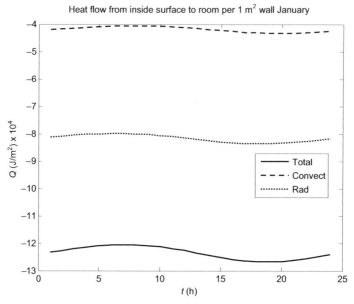

FIGURE 6.43 Distribution of heat exchanged with inside cavity (room) versus time (hours of an averaged day) in January; wall 6.

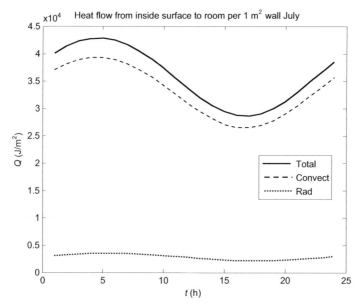

FIGURE 6.44 Distribution of heat exchanged with inside cavity (room) versus time (hours of an averaged day) in July; wall 6.

temperature). Usually, the most important heat flow is the wind-driven convective heat transfer. In case of high temperature of a wall surface (e.g., with black coating) there are big temperature differences. As a result, the convective heat transfer between the wall surface and surroundings is high. However, as already mentioned, in case of using insulation as an external layer, this phenomenon is limited to the surface and does not lead to major heat losses from the entire wall.

Using insulation considerably reduces heat losses from a building envelope. It may be noted that, thanks to the use of insulation as an external layer (elements 1, 2, 4, and 5, Figures 6.9, 6.10, 6.15, 6.16, 6.27, 6.28, 6.33, and 6.34, respectively) influence of external surroundings is stopped. The ambient impact is significantly limited, and it can be noticed in the insulating layer, while in the structural layer it is nearly negligible. If an element is not insulated (structures 3 and 5, Figures 6.21, 6.22, 6.39, and 6.40, respectively), influence of weather conditions is evident across the entire structure.

In case of structures insulated from outside (walls, roofs) temperature at the boundary between the insulation and thermal mass layer (e.g., concrete or PCM panel), stabilizes at the level close to the room temperature. In case of uninsulated structures, in winter (January) the temperature of the internal surface is below the required room temperature during the entire day (a difference of several degrees, changing depending on the time of day), which indicates major heat losses from the interior.

During summer (July), the internal surface temperature of an uninsulated flat roof is always higher than the required room temperature, which indicates heat gains through such a roof. In case of an uninsulated vertical wall of big thermal capacity (see Table 6.3) with a white coating of the external surface, the temperature of the internal wall surface is around 21 °C, which theoretically means minor heat losses. However, such a situation is caused by the averaged distribution of the ambient air temperature in July, with a maximum of 26 °C only and a minimum of 12 °C (see Figure 6.8). The averaged ambient air temperature for July is 19 °C. It is lower than the assumed indoor temperature of 22 °C, because of the melting point of the PCM panel. A horizontal uninsulated roof with a black external surface is especially exposed to solar radiation, which, during summer, is particularly intense for horizontal (or slightly sloped) surfaces. Because of that, there is a possibility of room overheating (ceiling surface temperature is higher than assumed internal air temperature). Vertical surfaces are much less exposed to solar radiation during summer. In the considered case, also the solar radiation absorptance of the investigated white walls is much lower. Influence of solar radiation on the building's interior is significant in case of a horizontal black uninsulated roof structure. If an element is insulated, its color is not that significant. For example, in case of an insulated horizontal roof with a black external surface, the temperature of this surface during summer may approach 60 °C, but, thanks to the insulation, this has practically no effect on the air temperature inside the building. This shows the importance of the material of which the external layer is made: essential properties are its density, thermal conductivity, and thickness. In case of uninsulated envelope elements (walls, roofs), most important parameters are: orientation, slope, and coating (color) related to solar radiation absorptance. When analyzing obtained results, it can be underlined, that in warmer climates when the temperature fluctuations are around that in July, as presented in Figure 6.8, the external walls with high thermal mass and white coating do not need thermal insulation. However, the insulation should always be used on roofs to protect them against excessive solar gains. Only white coatings should be used.

If there is insulation as an external layer, and a surface (coating) with low solar radiation absorptance (e.g., with white plasterwork), then the temperature of the external surface is very close to the ambient air temperature, both in summer and winter. As a result, the heat exchange between this surface and the near surroundings is greatly reduced (Figure 6.15—January and Figure 6.16—July). If the external insulating layer is covered with black paint (e.g., on plasterwork) of high solar radiation absorptance (Figure 6.9—January and Figure 6.10—July), then the temperature of this surface is close to that of ambient air only at night. During the day, the surface temperature gets much higher than that of ambient air (in January approaching 7 °C, and in July around 50 °C), and, as a result, there is a heat flow from the surface to the surroundings. Intensive absorption of solar radiation absorption causes the surface temperature to rise quickly, as insulation limits heat flow across its thickness (inward) and the temperature increase is only at the surface. As a result, it can be said that information (data) only on the external surface temperature is not enough to evaluate the total heat transfer through the external wall and to estimate the air temperature inside the building.

If the wall surface temperature is lower than the temperature of the air in contact with that surface, then the heat flow driven by this temperature difference (i.e., convection and radiative exchange with near surrounding) is directed toward the structure. It needs to be noted that, in any case, there is heat flow from the external surface of a building envelope (especially roofs) toward the sky, as the sky temperature is always considerably lower than the temperature of the surface. Heat exchange between the external surface of the walls, roofs, and external surroundings for individually considered structures is presented in Figures 6.11, 6.12, 6.17, 6.18, 6.23, 6.24, 6.29, 6.30, 6.35, 6.36, 6.41, and 6.42.

Heat flow because of forced convection between the external surroundings and the surfaces of a building envelope is directly dependent on the temperatures of the media involved. If a roof is horizontal, and there are no higher buildings that would obscure it nearby, then such a roof can only "see" the sky and radiative heat transfer is only to the sky (this is shown in Figures 6.35, 6.36, 6.41, and 6.42). This heat transfer becomes very significant in the entire energy balance of a building, especially for the rooms located at the top story. The "night cooling" of a building is especially severe in winter, and is particularly pronounced in case of horizontal roofs. A vertical wall only "sees" the sky in approximately 50%, whereas the remaining 50% is a "view" on neighboring objects and the ground, and this reduces the cooling effect of the sky. Cooling during winter is of course an undesirable effect, but, in summer, it can be beneficial by reducing overheating. It needs to be reminded here that, in case of insulated structures, heat exchange between the envelope surfaces and external surroundings is not essential for actual heat losses or gains of the interior. If an element is uninsulated, then the heat exchange from the external surface does influence heat transfer into the room or out of it, and then the influence of the sky is also more pronounced.

In this part, analysis of the heat losses or gains is based on heat transfer inside a building, because thermal conditions of the rooms surrounded by various types of considered walls and roofs are better described by the variability of heat flows between the interior and surfaces of these structures. Figures 6.13, 6.14, 6.19, 6.20, 6.25, 6.26, 6.31, 6.32, 6.37, 6.38, 6.43, and 6.44 show the distribution of total heat flows and its components for consecutive hours of a day, in January and July, respectively, exchanged between the internal wall and ceiling surfaces and the interior of a room. Heat exchange is driven by radiation and natural convection, and the radiative heat transfer is a dominating component.

In case of double-layer walls with external insulation, in January, heat losses are minor. Those losses are observed during the entire day and are slightly lower in case of a black external surface (Figure 6.13, where they oscillate around $14 \, kJ/(m^2h)$) than in case of a wall with a white coating (Figure 6.19, around $15 \, kJ/(m^2h)$). In case of a wall with a PCM panel, heat losses in January are even lower, around $12 \, kJ/(m^2h)$ (Figure 6.31). Differences in heat losses through insulated vertical walls, including the wall with a PCM panel, are insignificant. However, it could be advisable to use black coating for winter and white for summer. This gives an idea of building "skin" (cover) changing in time, especially between summer and winter. Of course, as it can be expected, in case of a single-layer uninsulated wall (Figure 6.25), the heat losses are much more evident and they are about 10 times higher than in case of insulated walls (around $120 \, kJ/(m^2h)$). It can be concluded that even high thermal mass (without insulation) does not help to keep internal temperature at the required (thermal comfort) level in winter weather conditions.

In case of an insulated horizontal roof in January (Figure 6.37), daily heat loss distribution is similar to that of a vertical insulated wall, with the magnitude only slightly higher (around $14.5 \, kJ/(m^2h)$). But in case of an uninsulated roof (Figure 6.43), just like in case of an uninsulated wall, heat losses are much larger (around $120 \, kJ/(m^2h)$ in this case note that the investigated uninsulated wall is white, while the uninsulated roof is black).

In July, in case of a double-layer wall with external insulation and black coating (Figure 6.13), negligibly small heat gains are observed throughout the day (3 kJ/(m²h) at night, 1.5 kJ/(m²h) at noon). For the same wall structure but with white coating (Figure 6.19), minor heat loss is observed (1.9 kJ/(m²h) at night, up to 2.7 kJ/(m²h) during the day), and it can be also neglected. As it was mentioned before, these theoretical heat losses are caused by ambient air temperature distribution in July. Temperature at night is low with a minimum of 12 °C, while the required indoor temperature is 22 °C. In case of a PCM wall (Figure 6.32), there are slight losses during the nights (up to 5.8 kJ/(m²h) in analyzed case) and minor gains during the days (up to 7.5 kJ/(m²h); they can be also neglected). For uninsulated wall 3 of high thermal capacity (Figure 6.26), there are small heat losses (oscillating around 18 kJ/(m²h)) because of the assumed ambient air distribution. In case of a horizontal roof, heat gains are continuously observed in July. They are minor in case of an insulated roof (Figure 6.38), with values higher at night (maximally 5.5 kJ/(m²h)) than during the day (maximally 3.5 kJ/(m²h)), and much more significant in case of an uninsulated roof (Figure 6.44), also higher at night (up to 42 kJ/(m²h)) than during the day (about 30 kJ/(m²h)). The thermal capacity of the wall or roof structure gives the effect of "delay in time" influence of the surroundings.

Summarizing, as it might have been expected, uninsulated structures are clearly sensitive to ambient conditions, especially in winter. In case of horizontal roofs, there are heat gains during summer, especially in case of an uninsulated roof structure, which is a result of solar radiation. However, the overheating effect is reduced because of the night cooling. In case of vertical walls, the lowest heat losses during winter are observed for insulated (double-layer) structures, first for a wall with a PCM panel, then for a wall painted black, and then for the white one. However, the differences are very small. Then, in summer, an insulated wall with a white surface practically fully isolates the room from the ambient conditions (theoretical heat losses are very small and can be neglected), whereas in case of a wall painted black, very slight gains are observed. For a wall with PCM panel, the gains and losses observed are minor. As a result, the analyzed insulated structures may be recommended for application in higher latitude countries. Admittedly, it would be reasonable to use a wall with a changing outer appearance "skin" (black in winter, white in summer).

The discussion concerning the application of PCM thermal panels in buildings in high-latitude countries may be found in publication by Chwieduk [18]. The results of simulation studies (also for high-latitude countries) presented in this chapter show that, in winter, the heating effect of a room (building) because of the PCM storage ability is particularly evident. In summer, the cooling effect is not so pronounced, because the ambient air temperature is not very high (see Figure 6.18). Only for 6 h per day the ambient temperature increases a few degrees (maximum 4°) above the melting point of the considered PCM.

It needs to be stressed that the proposed innovative PCM wall structure combines two essential energy-related features: good insulating quality and high thermal capacity. Proposed wall structure ensures good performance at small dimensions (thickness of 26 cm equivalent to a traditional wall of at least 45 cm) and low weight (PCM panel of 1 cm thickness, density 856 kg/m^3) when compared to a traditional construction material (25 cm of concrete or brick with a density of some 2000 kg/m^3). The proposed solution makes the wall structure light and reduces the footprint of a structural element, thus extending the living or utility area within the building, or reducing the external size of a building, giving more free space outside (land can be used for other purposes).

In conclusion, it may also be said that an uninsulated envelope elements should not be used in higher latitude countries. Also, horizontal roofs should be avoided because of major heat losses during winter, including losses to the sky, and high heat gains in summer resulting from incident solar radiation.

6.3 DYNAMICS OF ENERGY FLOW THROUGH TRANSPARENT ELEMENTS OF A BUILDING ENVELOPE

6.3.1 General Discussion

Solar radiation directly influences the interior of a building through transparent elements of its envelope (i.e., through windows and other glazed facade elements). The energy balance of the air in a building (room) has been presented in form of Eqn (6.19) (in case of heat losses) or Eqn (6.20) (in case of heat gains) for internal air temperature constant over time and space. Those balances include two components describing the energy flow through windows. The first one concerns heat flow $Q_{win}(t)$ with the direction depending on the temperature gradient, the other is the solar energy flow $Q_{sol}(t)$ depending on the irradiance G_s variable over time and position of the exposed surface (described by angles β and γ), and it is always directed into a building.

Transmission of solar radiation as an electromagnetic wave through a transparent element is related to optical phenomena. Thanks to certain transmittance of the glazing, most radiation reaches the interior directly, while some part undergoes photothermal conversion because of absorption within the glazing, and some very small part is reflected from the glazing surface. At the same time, the heat transfer phenomena driven by the temperature difference existing between the building's interior and external surroundings occur within a transparent element. The heat exchange with internal and external surroundings is analogical to the case of an opaque wall, as described in detail in Section 6.2. In this chapter discussion is focused on the heat transfer phenomena in glazed panes and between them, in a gas gap. Modern windows usually feature two glass panes (double glazed windows).

Heat transfer between the glass panes in a gas gap in a standard window is at least 60% because of radiative heat exchange and in some 40% to natural convection [4]. Intensive heat transfer occurs at glass pane edges and is mainly driven by design solutions and materials used in a window structure. In case of a modern double glazed window with noble gas in the gap between panes, heat transfer through glazing is less intense than through the frame. This means that enlarging the surface of the frame or using decorative elements (e.g., muntins) intensifies heat exchange between the interior and external surroundings, so heat transfer coefficient increases.

In a detailed discussion of transfer of heat and solar radiation through a window, its structure is divided into three essential components: frame, glass pane edges, and the central part of glazing. Bearing in mind those components, the total energy flow transferred through a window into the room (energy gains) may be expressed as:

$$
\sum_i \left(\dot{Q}_{i,win}(t) + \dot{Q}_{sol}(t) \right) = \dot{Q}_{Cgl}(x,t) \big|_{x=in} + \int_x \dot{Q}_{edge}(x,y,t) \big|_{in} \, dx
$$

$$
+ \int_y \dot{Q}_{edge}(x,y,t) \big|_{in} \, dy + \int_x \dot{Q}_{fr}(x,y,t) \big|_{in} \, dx \quad (6.45)
$$

$$
+ \int_y \dot{Q}_{fr}(x,y,t) \big|_{in} \, dy + G_s(t)\tau A_{Cgl+edge}
$$

Energy flow through a window into a room may be described with various levels of detail. Usually, the central part of glazing is considered as a whole in a one-dimensional simplified model. The pane edges and frame are analyzed in a more detailed way in a two or three-dimensional model [19−21].

The problem of energy flow through a window is a complex one. Energy balance for a window and its solution method may be characterized by various accuracy levels. As noted before, in case of energy flow through a window, we are usually dealing with two different energy flows, with identical or opposite directions. During the day, between sunrise and sunset, when solar irradiance G_s and the solar heat gain factor (SHGF) [4,22], are bigger than zero, then solar energy enters the room. Then, if solar irradiance is zero (i.e., between sunset and sunrise) or SHGF is equal to zero, then there is no solar energy flow. Regardless of the solar energy flow, there is continuous heat flow through a window between the interior with a temperature of T_{in} and external surroundings with a temperature of T_a caused by difference of those temperatures. In order to find a net energy flow through a window, it is possible to formulate separate energy balances for the transparent central part of glazing, transparent pane edges, and for the opaque frame, taking into account mutual interactions between those elements. Both glazing and the frame have complex structures and mathematical descriptions of thermal phenomena in those elements and in their surroundings, which is also complex.

The problem of the energy flow through a window may be analyzed by assuming that a window consists of a few essential components (i.e., two panes at a certain distance from each other), a gas gap between those panes, and a frame ensuring rigidity and stability of a window structure. Presence of the frame influences heat transfer processes within the glazed part of the window, especially in the area of the pane edges. Previous research [4,19,21,23], has proved that the mutual interaction between the frame and glazing is particularly important at the pane edge area within 0.063 m from the frame, regardless of the pane (window) dimensions. The distance of 0.063 m from the frame has been adopted by the American Society of Heating, Refrigerating, and Air Conditioning Engineers (ASHRAE) as a standard dimension unambiguously defining the zone of the frame influence on glazing to be used in window thermal analyses. This distance has also been adopted by the standard ISO 15099 [13] concerning detailed calculations used to investigate thermal properties of windows, doors, and shading devices, and by the standard EN ISO 10077-1:2002 [24].

6.3.2 Fundamentals of Energy Transfer through Glazing

A window consists of two glass panes installed at a certain distance from each other. Panes may be made of the same material and have the same thickness. The gap between them contains homogeneous transparent gas medium (e.g., air, argon). Because of the large size of glass panes in reference to their thickness, a one-dimensional heat transfer model across the thickness may be used. Under this assumption, the central part of each pane is represented by a single node located centrally. This node has physical properties of the pane: density, specific heat, thermal conductivity, and temperature of the inner and outer panes, respectively. The pane temperature depends on conditions of its surroundings at a specific time and the influence of the thermal state of the frame. Discussion of energy transfer through glazing (two compact panes) may be carried out in reference to five main temperatures: the air inside the room T_{in}, ambient air temperature T_a, the sky temperature T_{sky}, inner glass pane temperature $T_{gl,in}$, and outer glass pane temperature $T_{gl,out}$. The sky temperature T_{sky} is variable over time and is expressed as a function of the ambient air temperature: $T_{sky} = f(T_a(t))$. It may be described by Eqns (1.7)−(1.12).

Solar radiation of irradiance G_{sol} passing through the panes is partially absorbed within them, causing their temperature to rise. However, most of solar radiation is transferred directly to the building's interior.

Discussion of the heat transfer through a window along its thickness may be based on the thermal resistance method [4,24,25]. In this method, temperatures under consideration (i.e., T_a, T_{in}, $T_{gl,in}$, $T_{gl,out}$, T_{sky}) are nodes at the heat transfer path, across the thickness of a window. Description of a heat transfer through a window based on the thermal resistance method is shown in a general way in Figure 6.45 and in more detail (i.e., taking into account individual heat transfer processes: radiation and convection, but without conduction within the panes) in Figure 6.46. Heat fluxes $\dot{q}_{out}$, $\dot{q}_{in}$, and $\dot{q}_{gl}$ are equivalents of electric current. Solar irradiance absorbed in the outer glass pane $S_{gl,out}$ and in the inner one $S_{gl,in}$ are equivalent to the internal current sources.

In order to determine solar irradiance absorbed in the outer pane $S_{gl,out}$ and in the inner pane $S_{gl,in}$ (both with identical structure and dimensions), it is necessary to take into account absorptance of those panes and their mutual position (order). Equations (4.57) and (4.58) describe solar radiation absorptance of a single pane. Knowing that $\alpha = 1 - \tau - \rho$, and knowing the values of single pane's transmittance and reflectance, we may express irradiance absorbed at a certain time in each of two planes $S_{gl,out}$ (W/m^2) and $S_{gl,in}$ (W/m^2) as:

$$S_{gl,in}(t) = (1 - \rho - \tau)\frac{\tau}{1 - \rho^2}G_{sol}(t) = a_{gl,in} \cdot G_{sol}(t) \qquad (6.46)$$

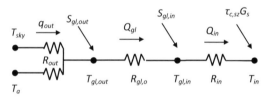

FIGURE 6.45 Application of the thermal resistance method for modeling heat transfer through a window across its thickness.

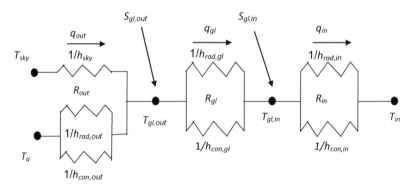

FIGURE 6.46 Application of the thermal resistance method for modeling heat transfer through a window, taking into account resistance components.

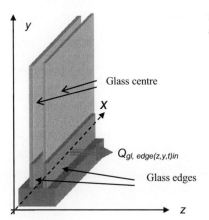

FIGURE 6.47 Window and its edges with assumed coordinates.

$$S_{gl,out}(t) = (1 - \rho - \tau)\left(1 + \tau\frac{\rho}{1 - \rho^2}\right)G_{sol}(t) = a_{gl,out}G_{sol}(t) \qquad (6.47)$$

Apart from solar energy absorbed by individual panes, solar energy passes directly into the room and transmitted irradiance $\dot{q}_{sol}$ is calculated as:

$$\dot{q}_{sol}(t) = \tau_{t,gl}G_{sol}(t) \qquad (6.48)$$

The method for determining total solar radiation transmittance $\tau_{t,gl}$ for a glass pane has been described in Section 4.2.3. In case of two identical panes, the value may be calculated using Eqn (4.50).

Application of the thermal resistance method, as shown for the analyzed case in Figures 6.45 and 6.46, allows formulating an energy balance for analyzed glazing. This balance consists of heat fluxes passed through the glazing at a specific time t: flow $q_{in}(t)$ flowing into the room interior, $q_{gl}(t)$ transferred between the panes, $q_{out}(t)$ flowing from the outside (case under consideration is shown in Figures 6.45 and 6.47 with heat flow inwards; i.e., heat gains). Calculation of the flow $q_{out}(t)$ is complex, as it refers to heat coming from two different sources with different temperatures: direct external surroundings (ambient air, temperature $T_a(t)$) and farther external surroundings, the sky (temperature $T_{sky}(t)$). For that reason, it is often assumed that the heat exchange with external surroundings refers only to the direct surroundings with a temperature of $T_a(t)$. In such a simplified model, using the thermal resistance method for the case of heat flow from the outside into the room, and taking into account the total glass area, individual heat flows are calculated with the following relations:

$$\dot{Q}_{in}(t) = \frac{A_t\left(T_{gl,in}(t) - T_{in}(t)\right)}{R_{in}(t)} \qquad (6.49)$$

$$\dot{Q}_{gl}(t) = \frac{A_t\left(T_{gl,out}(t) - T_{gl,in}(t)\right)}{R_{gl}(t)} \tag{6.50}$$

$$\dot{Q}_{out}(t) = \frac{A_t\left(T_a(t) - T_{gl,out}(t)\right)}{R_{out}(t)} \tag{6.51}$$

If no solar radiation is present, and the heat capacity of glass panes is neglected, all flows are equal:

$$\dot{Q}_{out}(t) = \dot{Q}_{gl}(t) = \dot{Q}_{in}(t) \tag{6.52}$$

Heat flow exchanged between the interior and exterior may be expressed in reference to relevant temperatures and the total glazing thermal resistance R_{tot}, that is a combination of all resistance components. This flow may be expressed as:

$$\dot{Q}_{gl,tot}(t) = \frac{A_t(T_a(t) - T_{in}(t))}{R_{gl,tot}} \tag{6.53}$$

while also the following is true:

$$\dot{Q}_{win}(t) = \dot{Q}_{out}(t) = \dot{Q}_{gl}(t) = \dot{Q}_{in}(t) \tag{6.54}$$

If solar radiation is being absorbed in panes, then heat flows are not equal, as in Eqns (6.52) and (6.54). Instead, the following energy balance equations referring to specific panes at the time t may be written:

$$\dot{Q}_{out}(t) + S_{gl,out}(t)A_t = \dot{Q}_{gl}(t) \tag{6.55}$$

$$\dot{Q}_{gl}(t) + S_{gl,in}(t)A_t = \dot{Q}_{in}(t) \tag{6.56}$$

Equations (6.55) and (6.56) allow finding the heat flow coming into the room, while taking into account the absorbed solar radiation. This flow is:

$$\dot{Q}_{in}(t) = \dot{Q}_{out}(t) + S_{gl,out}(t)A_t + S_{gl,in}(t)A_t \tag{6.57}$$

Taking into account all components of heat transfer between the external surroundings and the outer pane (i.e., convection and radiation), and assuming that heat is exchanged only with the direct surroundings with a temperature of T_a, Eqn (6.51) may be transformed into:

$$\dot{Q}_{out}(t) = \frac{A_t\left(T_a(t) - T_{gl,out}(t)\right)}{R_{out}(t)}$$

$$= A_t\left(T_a(t) - T_{gl,out}(t)\right)\left[h_{con,a}(t) + \frac{\varepsilon_o\sigma\left(T_a(t)^4 - T_{gl,out}(t)^4\right)}{\left(T_a(t) - T_{gl,out}(t)\right)}\right] \tag{6.58}$$

The heat transfer through the gas gap between two glass panes by convection and radiation, expressed by Eqn (6.50), now transforms into:

$$\dot{Q}_{gl}(t) = \frac{A_t \left(T_{gl,out}(t) - T_{gl,in}(t) \right)}{R_{gl}(t)} = A_t \left(T_{gl,out}(t) - T_{gl,in}(t) \right)$$

$$\left[h_{con,gl}(t) + \frac{\sigma}{\frac{1}{\varepsilon_o} + \frac{1}{\varepsilon_o} - 1} \frac{\left(T_{gl,out}(t)^4 - T_{gl,in}(t)^4 \right)}{\left(T_{gl,out}(t) - T_{gl,in}(t) \right)} \right] \qquad (6.59)$$

Finally, to describe the heat transfer between the internal surroundings (room) and the inner glass pane by convection and thermal radiation, Eqn (6.49) is expressed as:

$$\dot{Q}_{in}(t) = \frac{A_t \left(T_{gl,in}(t) - T_{in}(t) \right)}{R_w(t)}$$

$$= A_t \left(T_{gl,in}(t) - T_{in}(t) \right) \left[h_{con,in}(t) + \frac{\varepsilon_o \sigma \left(T_{gl,in}(t)^4 - T_{in}(t)^4 \right)}{\left(T_{gl,in}(t) - T_{in} \right)} \right] \qquad (6.60)$$

Thus, a set of eight simultaneous equations (i.e., Eqns (6.49)–(6.51), (6.55), (6.56), and (6.58)–(6.60)) with eight variables $T_{gl,in}(t)$, $T_{gl,out}(t)$, $Q_{gl}(t)$, $Q_{in}(t)$, $Q_{out}(t)$, $R_{gl}(t)$, $R_{out}(t)$, and $R_{in}(t)$ may be solved through iteration for consecutive time steps, if ambient air temperature and interior temperature are known. Solving these equations is complex, despite introduced simplifications. The method of solution may be further simplified with the introduction of certain new assumptions, and especially by using simple formulas for heat exchange coefficients [4,24], or even constant values of those coefficients (e.g., as given in published standards [13]).

In detailed studies, individual phenomena are described with complex equations for convective heat transfer expressed as functions of variable temperatures of heat exchanging media. In such a situation, very often the solution of a problem of heat transfer through a window requires computer simulations. Apart from Eqns (6.49)–(6.60) referring to the heat flow through glazing, also the direct flow of solar energy, described by Eqn (6.48), needs to be included in the energy balance of the analyzed room.

Coefficients of convective and radiative heat transfer on boundary surfaces on both sides (internal and external surroundings) are discussed in Section 6.2.3. They refer to both opaque and transparent elements. In case of glazing, additionally, the phenomena of the gas gap between the panes need to be analyzed, as they have major impact on thermal performance of the window itself.

6.3.3 Heat Transfer Coefficients of a Gas Gap between Panes

Heat transfer by convection can be described using empirical relations based on dimensionless numbers. Convective (natural) heat transfer coefficients for horizontal and vertical surfaces of building envelope can be found in the literature [26]. For convective heat transfer in a gas (air) gap, relations related to different inclinations of the gap may be found in text books on heat transfer [9]. The convective heat transfer coefficient for the gap is described by a function of a Nusselt number, given by Eqn (6.26). Characteristic dimension L is in this case the gap width $d_{gl,cav}$.

In solar energy analyses, in case of solar receiver surfaces (e.g., solar collectors) inclined at an angle β (to the horizontal plane) $\leq 60°$, the Nusselt number is described by the Hollands empirical relation [4,27], as a function of the Rayleigh number (expressed by Eqn (6.27)) and angle β:

$$
Nu = 1 + 1.44\left(1 - \frac{1708}{Ra\,\cos\beta}\right)^{+}\left(1 - \frac{1708\,\sin(1.8\beta)^{1.6}}{Ra\,\cos\beta}\right)
$$
$$
+ \left[\left(\frac{Ra\,\cos\beta}{5830}\right)^{\frac{1}{3}} - 1\right]^{+}
$$

(6.61)

Index $+$ in the equation means that only positive values of expressions within brackets are to be taken into account; if a value is negative, then zero is used in its stead.

Equation (6.61) is true, if a gap height is at least 20 times higher than its width (i.e., $Y_{gap}/d_{gap} \geq 20$). Equation (6.61) may also be applied for roof windows (at a slope $\leq 60°$). It needs to be noted that the Eqn (6.61) is valid for the heat flow from the interior to the exterior (i.e., if the outer pane is colder than the inner one, heat losses). For a reverse situation (i.e., the heat flow from the outer glass pane toward the inner one, heat gains), the value of β in Eqn (6.61) should be replaced with $(180° - \beta)$.

The reference temperature is assumed at the level of temperature of the boundary surfaces of a gap, in this case, understood as the central part of the outer and inner glass panes. Individual properties of the gas contained in a gap are therefore specified for the average temperature of a gas in the gap $\overline{T}_{gap}(t)$, which is understood as the mathematical average of temperatures of the central parts of both panes at a given time. This may be expressed as:

$$
\overline{T}_{gap}(t) = \frac{1}{2}\left(T_{gl,outC}(t) + T_{gl,inC}(t)\right)
$$

It is also assumed that individual properties of the gas (e.g., air) change as linear functions of temperature, as it has been explained and described by Eqn (6.25).

For a vertical air gap, an empirical relation proposed by Shewen [28] may be used:

$$Nu = \left[1 + \left(\frac{0,0665Ra^{1/3}}{1 + (9000/Ra)^{1/4}}\right)^2\right]^{1/2} \tag{6.62}$$

Using this relation, Wright [29] developed its graphical form for dimensionless numbers $Nu = f(Ra)$ for gas gaps with height-to-width ratio larger than 20. Then he created the following relation for vertical gaps in windows [30]:

$$Nu_{1,90} = 0.06738\,Ra^{1/3} \qquad 5 \times 10^4 < Ra \tag{6.63}$$

$$Nu_{1,90} = 0.02815\,Ra^{0.4134} \qquad 10^4 < Ra < 5 \times 10^4 \tag{6.64}$$

$$Nu_{1,90} = 1 + 1.75967 \times 10^{10}\,Ra^{2.2985} \qquad Ra \le 10^4 \tag{6.65}$$

$$Nu_{2,90} = 0.242\left(\frac{Ra}{\left(\frac{H}{L}\right)}\right)^{0.272} \tag{6.66}$$

Ultimately, the value of the Nusselt number is calculated as:

$$Nu_{90} = \left(Nu_{1,90},\ Nu_{2,90}\right)_{max} \tag{6.67}$$

Convective heat transfer coefficient for a gas gap is determined from Eqn (6.26) using the Nu number calculated from Eqn (6.67).

In case of a gap inclined at an angle β where $60° \le \beta \le 90°$, for heat flow from the interior to the exterior (inner pane temperature higher than that of the outer pane), El Sherbiny empirical relation [31] showing Nusselt number as a function of auxiliary factor G may be used. The value G is defined as:

$$G = \frac{0.5}{\left[1 + \left(\frac{Ra}{3160}\right)^{20.6}\right]^{0.1}} \tag{6.68}$$

For inclination of $60°$, Nu is calculated as:

$$Nu_{1,60} = \left[1 + \left(\frac{0.093 \cdot Ra^{0.314}}{1 + G}\right)^7\right]^{1/7} \tag{6.69}$$

$$Nu_{2,60} = \left(0.104 + \frac{0.175}{\frac{H}{L}}\right)Ra^{0.283} \tag{6.70}$$

where $H = Y_{gap}$ i $L = d_{gap}$.

For a gap inclined at $60°$, the Nusselt number is equal to the higher of the results yielded by Eqns (6.69) and (6.70) (i.e., analogically to Eqn (6.67)):

$$Nu_{60} = \left(Nu_{1,60}, Nu_{2,60}\right)_{max} \tag{6.71}$$

Convective heat transfer coefficient for a gap inclined at $\beta = 60°$ (for $L_{gap} = d_{gap}$) is calculated from Eqn (6.26) using the Nu number from Eqn (6.71).

For gaps inclined at β such that $60° < \beta < 90°$, El Sherbiny recommends using direct linear interpolation of convective heat transfer coefficient for a vertical gap and gap inclined at $60°$ in a following form:

$$h_{con,\beta} = h_{con,90} + \left(h_{con,60} - h_{con,90}\right) \cdot \left(\frac{90 - \beta}{30}\right) \tag{6.72}$$

This relation is valid for: $10^2 < Ra < 2 \times 10^7$ and $5 < (H/L) < 100$.

If the heat flow is directed outward (inner pane temperature is higher than that of the outer pane; i.e., there are heat losses), then for a gap inclined at $\beta < 90°$, the Arnold relation [32] may be used:

$$Nu_\beta = 1 + (Nu_{90} - 1) \sin \beta \tag{6.73}$$

In calculations of natural convection, depending on the window inclination, and on the heat flow direction, relations described by Eqns (6.61) and (6.63)–(6.73) may be used. There is also heat exchange by thermal radiation in a gap between glass panes. If the panes do not have low-emissivity coatings, then the share of thermal radiation in the total heat transfer through glazing is higher than that of convection. Radiative heat exchange coefficient is mathematically described by the second addend of the sum in Eqn (6.59) (on the right side of the equation). Panes may be considered as two identical parallel gray bodies with given thermal emissivity.

6.3.4 Description of Complex Phenomena of Energy Flow through Glazing

Just like in the case of simplified problems presented in Section 6.3.2, heat flows through a glass with a surface area A_t at the time t considered. In case of heat gains, those flows are: heat flux $Q_{in}(t)$ flowing into the room, heat flux $Q_{gl}(t)$ flowing between the glass panes, and heat flux $Q_{out}(t)$ incoming from the outside. Usually (like in case of the analysis of opaque elements: wall, roof), determining the $Q_{out}(t)$ is most complex, as this flow refers to the heat inflow from two different sources of different temperatures: direct surroundings with temperature of $T_a(t)$ and the sky with temperature of $T_{sky}(t)$. Heat flows incoming from the external surroundings and related coefficients of radiative and convective heat transfer may be calculated in the same way as in case of an opaque wall. Boundary condition for the outer boundary surface, which is in contact with the external surroundings is analogical to Eqn (6.35), and it refers to the outer glass pane and frame. In case of glazing, additionally absorbed solar radiation $S_{gl,out}$ described by Eqn (6.47) must be taken into account. Thus, the boundary condition takes the following general form:

$$-\lambda_{gl}\frac{\partial T_{gl,out}(x,t)}{\partial x_{gl,out}}\bigg|_{out} = h_{out}(t)\big[T_a(t) - T_{gl}(x,t)\big|_{out}\big] + \dot{q}_{sky}(t) + \dot{q}_{sol}(t) + S_{gl,out}(t)$$

$$(6.74)$$

Heat transfer from the edges of the inner glass pane to the room is described by a boundary condition that results from the heat transfer via the natural convection and thermal radiation. Thus, the boundary condition equation is analogous to Eqn (6.23). Just like in case of an opaque wall it is expressed in reference to the air boundary layer adjacent to the window on the room side, using averaged boundary layer temperature and relevant properties. Additionally, absorbed solar radiation $S_{gl,in}$, described by Eqn (6.46) is taken into account, and the boundary condition transforms into:

$$-\lambda_{gl}\frac{\partial T_{gl,in}}{\partial x_{gl,in}}\bigg|_{in} + S_{gl,in}(t) = h_{in}\big[T_{gl,in}(t)\big|_{in} - T_{in}(t)\big] \qquad (6.75)$$

Heat transfer between the glass panes in a gas gap through natural convection depends on the gap inclination and heat transfer direction and is described by the appropriate empirical relation, just discussed in Section 6.3.3. There is also thermal radiation between two panes considered as gray bodies. Heat flow exchanged between the panes (in case of heat gains) may be expressed as:

$$\dot{Q}_{gl}(t) = A_t(h_{con}(t) + h_r(t))(T_{gl,out}(t) - T_{gl,in}(t)) \qquad (6.76)$$

Heat transfer coefficient for a gap is expressed in a standard way as:

$$h_{gl}(t) = h_{con,gl}(t) + \frac{\sigma}{\frac{1}{\varepsilon_o} + \frac{1}{\varepsilon_o} - 1}\frac{\left(T_{gl,out}(t)^4 - T_{gl,in}(t)^4\right)}{\left(T_{gl,out}(t) - T_{gl,in}(t)\right)} \qquad (6.77)$$

Numerical simulation is needed to solve the presented complex problem of energy transfer through glass panes. Usually, analyses carried out according to the presented assumptions consider the heat transfer through the central part of glazing as one-dimensional, across window thickness. Heat transfer through glass edges is analyzed in at least a two-dimensional model (i.e., apart from the heat flow through glazing; window thickness) also heat flow along the glass surface caused by the frame impact is taken into account.

Energy flow into the room consists of a heat flow from the inner glass pane with a surface area of A_t and energy flow of solar radiation of irradiance G_{sol}

reduced by the given transmissivity of two glass panes, expressed by coefficient $\tau_{t,gl}$ (τ from Eqn (4.48) or for two identical panes τ_t from Eqn (4.51)), which may be expressed as:

$$\sum_i A_{t,i}\dot{q}_{tot}(t) = \sum_i \left(A_{t,i}\dot{q}_{in,i}(t) + A_{t,i}G_{sol,i}(t)\tau_{t,gl}(t)\right) \qquad (6.78)$$

Equation (6.78) has a general form and addresses both the heat transfer and solar radiation transmission through all glass panes, taking into account their position (inclination and geographic orientation) and relevant transmittance of glass panes.

Problems related to the heat transfer through glazing edges have been analyzed in a quasi-three-dimensional model [3,33]. The following coordinates are assumed: z across the window (its thickness), y along window height/width. The assumed coordinate system is shown in Figure 6.47.

If a frame consists of four box-shaped elements (which hold the glass panes) of the same thickness and width (i.e., where only lengths of vertical and horizontal elements are different), then the entire frame may be mathematically considered as a single element. A similar assumption may be made for the glass edge. In this situation, the frame and glass edge may be analyzed directly in a two-dimensional system, with the z coordinate across their thickness and y coordinate along the width of the frame or glass edge. As already mentioned, discussion concerning the z direction in case of glazing may be simplified, as it is possible to consider only selected fixed nodes along this coordinate (four nodes of the computational grid, representing external surroundings, outer pane, inner pane, and external surroundings). The third dimension is also taken into account, albeit in a specific manner. It is assumed that two-dimensional heat flow (across the window and along the width of the frame and glass edges) also flows along certain perimeters of the frame and the glass edges. It needs to be noted that the length of the perimeter is variable. The largest is the outermost perimeter, which for the frame corresponds to the line of contact with the wall, while for the glass pane edge it stands for the boundary line between the glass edge and frame. The smallest is the internal perimeter: in the case of the frame it is its boundary line between the frame and glass edge, and in case of the glass edge it is a theoretical boundary line between the edge and the central part of the glazing. For selected elements (frame or glass edge), a number of further intermediate perimeters may be defined (each of different length), and their number depends on the intended calculation accuracy (i.e., step of spatial grid related to accuracy of the computational model).

The boundary between the glass edge and central part of the glazing is, according to ASHRAE standards, a line at a distance of 0.063 m from the frame (as has been already mentioned). The temperature of the central part of the individual panes is always equal to the temperature of the glass edge at the boundary between the edge and central part of each pane. Presence of the frame causes heat flows by conduction from the glass toward the frame or in the opposite direction, depending on the temperature gradient. Conduction also occurs across the pane's thickness (i.e., along the coordinate z; this conduction across the thickness often is neglected as insignificant because of the low thickness of a pane in relation to its height/width). The temperature field within a glass edge is described by the heat conduction equation, which may be solved when an initial condition and appropriate boundary conditions are mathematically described.

When formulating boundary conditions, it may be assumed that the heat transfer at the frame-glass edge boundary takes place by conduction. Heat transfer at that boundary is described by the equity of the temperature and the equity of the heat flows exchanged by conduction at the contact surfaces of both media. Then, at the boundary between the glass edge and the central part of the pane, the heat conduction disappears. The temperature at the boundary of the pane edge (0.063 m from the frame) is equal to the temperature of the central part of that pane, which is identical for the whole central part.

Heat exchange with external and internal surroundings by convection and radiation is taken into account when formulating boundary conditions for the z coordinate, both in case of the central pane part and for the glass edges.

The problem of unsteady heat transfer within the glass edge may be solved by various numerical simulation techniques. One of the possibilities is a relatively simple method of finite volume, which well reflects the character of the actual phenomena. Its application yields a distribution of the temperature field within the glass edge and allows finding heat flow transferred through the glass edge and central part of a pane into or from the room interior, depending on the temperature gradient. The solution of the problem of energy transfer through glazing based on computer simulation and finite volume method may be found in the literature [3,33,34].

When the temperature fields within glass panes are known, it is possible to determine the heat flow exchanged between the inner pane and the internal surroundings. Calculation of the heat flow between the inner pane and the room is performed by adding up the flow of heat exchanged with the room through the pane edge and its central part. For any time t this flow may be expressed as:

$$\sum_i \left(\dot{Q}_{i,gl.in}(t) + \dot{Q}_{sol}(t) \right) = \dot{Q}_{Cgl}(x,t)\Big|_{x=in} + \int_x \dot{Q}_{edge}(x,y,t)\Big|_{in} dx$$
$$+ \int_y \dot{Q}_{edge}(x,y,t)\Big|_{in} dy + G_s(t)\tau A_{Cgl+edge} \qquad (6.79)$$

A window also includes an opaque frame, which plays a significant role in its energy balance.

6.3.5 Heat Flow through the Frame

A frame is usually placed between an opaque wall (roof slope) and transparent glazing (a double glazed compact system). The frame is also in contact with the external and internal surroundings (i.e., ambient air and air in a building). Just like in the case of the glass edge, heat transfer may be analyzed in a quasi-three-dimensional model, as already partially explained in the previous section.

If a thermal analysis is focused on selecting frame structure and material, then it needs to be carried out in detail, in a three-dimensional model. In less detailed analyses, when the entire heat balance of a certain room or building is a subject, like in the presented studies, certain simplifications may be introduced. It may be assumed that the frame is made of one material (e.g., wood) and is considered as homogenous and isotropic solid body. When considering the frame design, it may be taken into account that glass panes are partially embedded in it, thus distorting the structure and requiring an appropriately higher number of boundary conditions. Heat transfer within the frame is carried out by conduction, also in the place where the frame and glass are in contact. The temperature field within the frame is described by the unsteady conduction equation. The equation is solved when the initial and boundary conditions are known. Detailed expressions of boundary conditions may be found in publications [3,33,34]. Only selected boundary equations are presented in this section.

At the boundary surfaces, parallel to the glass panes, at the outside and inside, the boundary conditions are as follows:

$$-\lambda_{fr}\frac{\partial T_{fr}(y,z,t)}{\partial z}\bigg|_{z=z_{out}} = h_{z,out}(y,t)\left[T_a(t) - T_{fr}(y,z,t)\big|_{z=z_{out}}\right]$$
$$+ \dot{q}_{sky,fr}(t)\big|_{z=z_{out}} + \dot{q}_{s,r}(t)\big|_{z=z_{out}} \tag{6.80}$$

$$-\lambda_{fr}\frac{\partial T_{fr}(y,z,t)}{\partial z}\bigg|_{z=z_{in}} = h_{z,in}(y,t)\left[T_{fr}(y,z,t)\big|_{z=z_{out}} - T_{in}(t)\right] \tag{6.81}$$

Heat transfer coefficients h_{out} and h_{in} include both convective and radiative heat transfer between the two media with relevant temperatures. Additionally, in case of the outer face of the frame, boundary condition Eqn (6.80) includes the influence of the sky and solar radiation. Apart from the surfaces of the frame parallel to the glazing, there are also parts of the frame face perpendicular to the glazing, which are in contact with the external or internal surroundings. Defining the boundary conditions for those surfaces on the external side needs to take into account the fact that solar exposure of such surfaces to solar radiation is not the same, and also they "see" the sky to a different extent. This means that the solar radiation incident on a frame faces perpendicular to the glass has different irradiance, as different sections of such frame faces have different azimuth angles (different orientation in reference to cardinal directions) and different slope. Thus, in case of vertical windows, two frame faces (perpendicular to the glazing) are horizontal and two are vertical. In case of roof windows, two frame faces are vertical and two are inclined at an angle equal to the window tilt $\pm 90°$. Moreover, configuration factors in the equations describing the heat transfer by thermal radiation with the direct external surroundings and the sky also differ. Detailed discussion of this problem may be found in publication [3].

The window frame is placed in a wall. In case of a wall that is very well insulated, it may be assumed that the wall-frame boundary surface is adiabatic. Then, the following boundary condition can be applied for this surface:

$$\left.\frac{\partial T_{fr}(y, z, t)}{\partial y}\right|_{y=y_{\min}} = 0 \tag{6.82}$$

In case of a frame boundary surface on the glazing side, there is a contact surface with a spacer (of compact glazing unit) made, for example aluminum or wood, or insulated. The spacer is also in contact with the gas gap between the glass panes. In case of well insulated spacer, it may be assumed that there is no heat transferred through the spacer (the contact surface is adiabatic). Then the boundary condition is identical to Eqn (6.82), with relevant coordinates (if the spacer is not insulated, the model must be more complex). Boundary conditions between the outer glass pane and the frame and between the inner glass pane and the frame have been described in the previous section focused on heat transfer through the pane edges. The structure of a frame can be much more complicated when it is made of different materials (e.g., plastics, insulation, metal), and of different shape, and it contains an air an cavity. This case is not considered here.

Having formulated the governing equation of unsteady conduction heat transfer in a frame in a two-dimensional model with initial and boundary conditions, the problem of heat transfer can be solved by applying numerical simulation method. Then, knowing the temperature field in the frame, including the boundary surfaces, the heat flow into the room can be determined from equation:

$$\sum_{i}^{n} \dot{Q}_{fr,i} = \sum_{i}^{n} h_{fr,i} A_{fr,i} \left[T_{fr,i}(y, z, t)\big|_{in_i} - T_{in}(t) \right] \tag{6.83}$$

Symbol i stands for i-th of n frame surfaces exposed to the influence of inner surroundings (interior of the modeled room). $A_{fr,i}$ is the area of the i-th surface, across which the heat flow is transferred to the room from the external surroundings. Equation (6.79) may also be expressed as:

$$\sum_{i}^{n} \left(\dot{Q}_{fr,i}(t) \right) = \int_x \dot{Q}_{fr}(x, y, t)\bigg|_{in} dx + \int_y \dot{Q}_{fr}(x, y, t)\bigg|_{in} dy \tag{6.84}$$

Then, knowing the heat transfer across the frame, as well as heat flows across the glass edges and central parts of glass panes, the total heat flow across the window can be determined.

6.3.6 Some Examples of Unsteady Energy Flow through Windows

This chapter presents selected results of simulations of heat transfer through a window. Calculations have been performed using the mathematical model of phenomena occurring within windows and their surroundings, as discussed in the previous subsections of the Section 6.3. Results of performed calculations highlight the role of solar radiation in total energy transfer across windows. A window is an element of a building envelope with the strongest relation to the external surroundings; in particular, it is most affected by solar energy. Because of the improvement of thermal insulation properties of opaque envelope elements, heat transfer across those elements is now greatly reduced, as described in the previous Section 6.2. But the presence of a window often causes sudden major changes in room temperature, as demonstrated by graphic interpretation of simulation results shown in this section.

Energy flow across a window, into the room, or out of the room, is a result of the influence of external surroundings described by its essential parameters variable over time, such as: ambient air temperature, the sky temperature, and solar irradiance. Variability of energy flow across transparent glazing has a different character than in case of opaque frame. This section presents selected results of simulations performed for some selected rooms. Other results and detailed discussions may be found in publications [3,33,34].

Figures 6.48—6.55 present the solar irradiance incident on four vertical surfaces of standard orientations: southern, eastern, western, and northern, and four inclined surfaces with the same orientations. These figures are provided to show variability of the averaged daily solar irradiance incident on discussed surfaces in all months of a year and enable comparison with distribution of energy flow across an irradiated window of the analyzed position into a room at the same time.

Figures 6.56—6.63 show changes of energy flow $Q = f(t_h)$ through a window with an elementary surface (1 m × 1 m) as a whole (i.e., including heat transfer through the central part of the glazing, through the glass edges, and through the frame). Direct solar energy transfer through glazing is also presented. Distributions refer to average days of all months of a year, for vertical windows (Figures 6.56—6.59) and for windows inclined at 45° (Figures 6.60—6.63). Energy flow direction may be directed into the room or out of the room. Hourly energy flows are presented. As it has been already mentioned, energy flow should not be presented by a continuous function, but a discrete one and should be presented, for example, as a column diagram. However, because of the high number of values (hourly sums), graphic presentation in the form of columns or bars at a single diagram and with limited figure dimensions is not possible.

A characteristic feature of energy flow across glazing is its strong dependence on the solar radiation incident on analyzed glazing at a certain time.

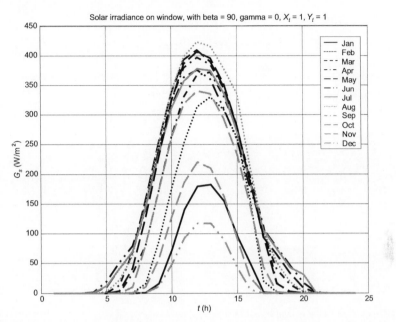

FIGURE 6.48   Hourly distribution of solar irradiance incident on vertical south window surface.

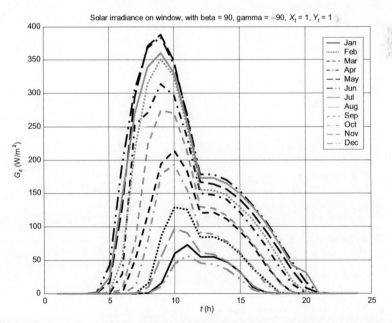

FIGURE 6.49   Hourly distribution of solar irradiance incident on vertical east window surface.

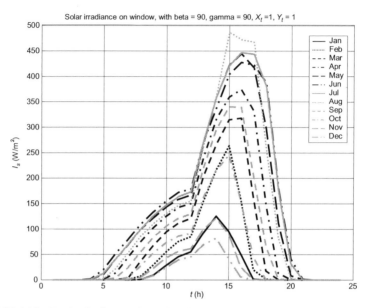

FIGURE 6.50 Hourly distribution of solar irradiance incident on vertical west window surface.

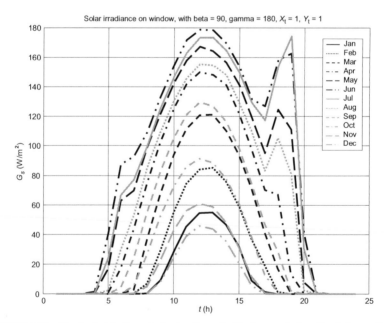

FIGURE 6.51 Hourly distribution of solar irradiance incident on vertical north window surface.

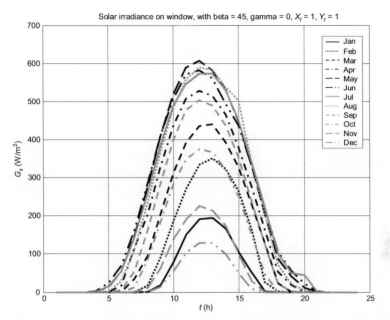

FIGURE 6.52   Hourly distribution of solar irradiance incident on inclined (45°) south window surface.

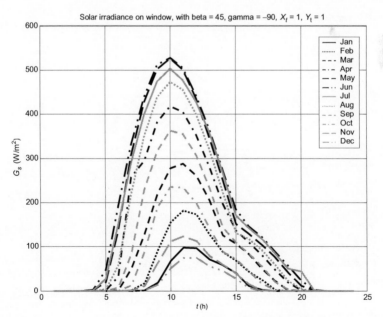

FIGURE 6.53   Hourly distribution of solar irradiance incident on inclined (45°) east window surface.

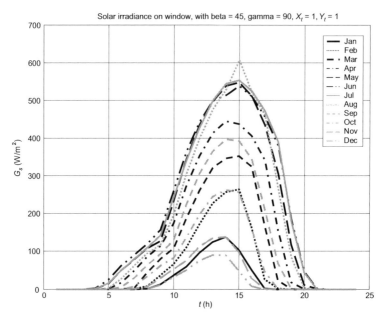

FIGURE 6.54 Hourly distribution of solar irradiance incident on inclined (45°) west window surface.

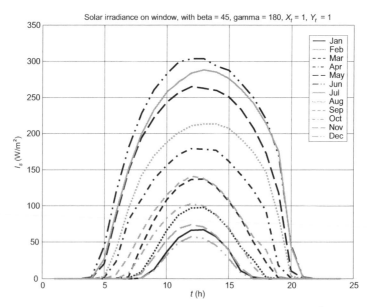

FIGURE 6.55 Hourly distribution of solar irradiance incident on inclined (45°) north window surface.

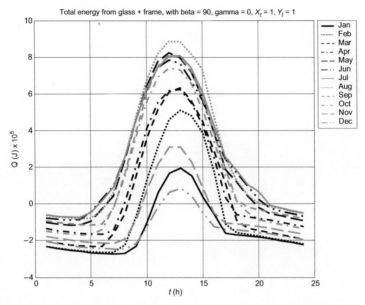

FIGURE 6.56   Daily distribution of energy flow through a south vertical window for all months of the averaged year.

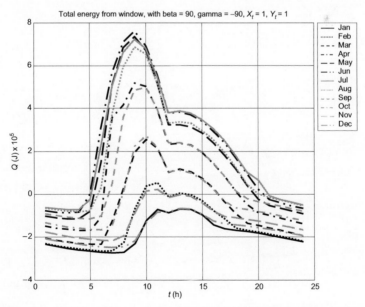

FIGURE 6.57   Daily distribution of energy flow through an east vertical window for all months of the averaged year.

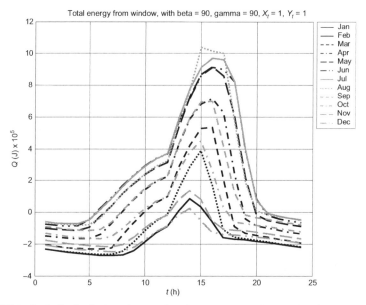

FIGURE 6.58 Daily distribution of energy flow through a west vertical window for all months of the averaged year.

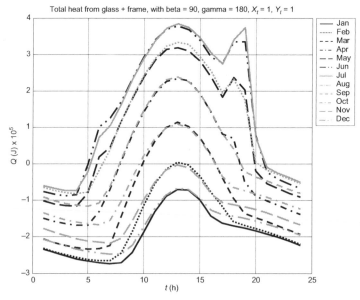

FIGURE 6.59 Daily distribution of energy flow through a north vertical window for all months of the averaged year.

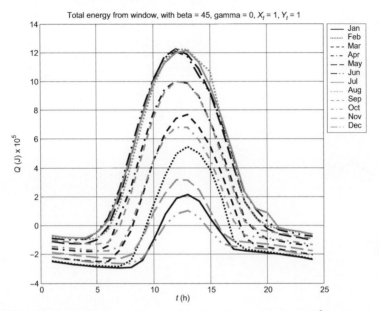

FIGURE 6.60 Daily distribution of energy flow through a south inclined ($45°$) window for all months of the averaged year.

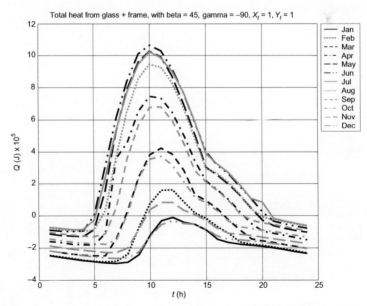

FIGURE 6.61 Daily distribution of energy flow through an east inclined ($45°$) window for all months of the averaged year.

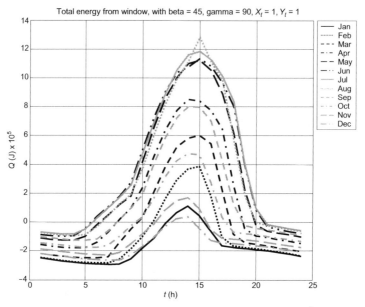

FIGURE 6.62 Daily distribution of energy flow through a west inclined $(45°)$ window for all months of the averaged year.

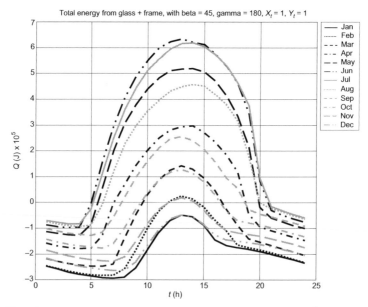

FIGURE 6.63 Daily distribution of energy flow through a north inclined $(45°)$ window for all months of the averaged year.

The highest energy flows from the external surroundings into the room are observed during summer and late spring. In case of vertical glazing, the highest energy flow is observed for west- and south-oriented windows, followed by east-oriented ones. Peak energy gains for different orientations are observed at different times of the day and year, strictly connected with this orientation. Decreasing surface inclination (i.e., using roof windows), leads to increased energy gains of the interior during days (because of the increased solar radiation exposure). In case of roof windows, the highest annual energy gains are observed for southern orientation.

When comparing Figures 6.56—6.63 it is possible to note the fact, which has already been highlighted in previous chapters, that energy flow across inclined windows is higher than that through vertical ones. The character of daily solar irradiation distribution for considered windows, as shown in Figures 6.48—6.55, is directly reflected by daily distribution of energy flow from the external surroundings into a room under consideration. During the daytime, when solar radiation is present, fluctuation of energy flow curves reflect the solar irradiance variability. When there is no solar radiation (i.e., at night) then the heat transfer results from the ambient temperature distribution.

Analysis of Figures 6.56—6.63 also allows noticing that during the nights, in case of inclined windows, heat losses are slightly bigger than for vertical ones. This phenomenon is partially caused by more intensive radiative heat exchange with the sky (with a temperature lower than the ambient air temperature). This is the night cooling effect already mentioned before.

During winter days, at high solar irradiance, heat flows into the room through southern windows may be observed for a few hours. This is also true for western windows, but only during afternoons and for a much shorter time. The character of heat transfer variability during nighttime is smoother than during daylight hours, and is related to the ambient air temperature fluctuation. Obviously, the lower the window heat resistance, the more pronounced the variability. In case of a sudden temperature drop at night, heat transfer would be, of course, more intense. Heat losses reach their maximum just before sunrise (according to the assumed ambient air temperature distribution).

Results obtained have been used to analyze temperature distributions on the inner and outer window surfaces and on main window elements (i.e., frame, glass edges, and the central part of glass panes). Graphical interpretation of obtained results may be found in the literature [3,33]. This study contains only essential conclusions from the analysis of simulation results for two extreme (climate-wise) months (i.e., January and July).

As already mentioned, during winter, the solar irradiation is low, and the parameters that have the largest impact on thermal phenomena occurring in windows are ambient air temperature and the sky temperature. The temperature of the outer surfaces of the frame follows the ambient air temperature variations. Also, distribution of temperature in the outer glazing (both glass edge and central part) is similar, but higher than that of the ambient air. Window attenuates influence of external surroundings. Compact double pane glazing has better insulating features than the frame. Temperature differences between individual window elements on their internal side are higher than on the external side. Temperature of the inner pane edges is noticeably influenced by the phenomena taking place in the internal surroundings. This influence is transferred through the frame, which acts as a kind of a thermal bridge. During winter, solar irradiation of a building envelope is low. As a result, not much solar radiation is absorbed in glass and temperature growth caused by solar radiation is not significant. Most of the solar radiation is transferred directly into the interior, thanks to the high transmittance of the glazing.

The outer glass pane temperature may considerably vary during short periods in case of major and sudden irradiation changes. Outer pane temperature variability is "transferred" to the inner pane with certain delay. The outer pane is quite fast to "react" to the changes in external conditions: ambient air temperature and solar irradiance. "Reaction" of the inner pane is attenuated thanks to the thermal resistance of the compact system: outer pane—inner pane, related to the window design and structure.

During summer, the frame temperature on its internal side oscillates quite closely around the temperature of the air inside the room. The inner glass pane temperature in its central part is usually lower than the ambient air temperature during the days and higher than the ambient air temperature during the nights. Glass edge temperature is close to the temperature of the glass center part; during nights it is slightly higher, during days slightly lower.

Calculations have been performed for different window surfaces. It turns out that increasing the window surface area does not influence the temperature distribution in window components. What does affect the temperature distribution is the window inclination. In summer, with the decrease of inclination, daily temperature fluctuations get smoother as the surfaces get exposed to direct solar radiation for a longer time. Glazing temperature is clearly increasing because of the absorption of solar radiation caused by its better availability. In case of inclined roof windows, in summer, the highest temperature is observed on southern glass panes (in case of vertical windows the highest temperature has been seen for western orientation). Inner pane temperature is, for most of the time, higher than the temperature of the interior, which indicates increased tendency for overheating of rooms with roof windows. Apart from solar energy being absorbed in glazing, most of the solar energy enters a room directly, and this considerably intensifies overheating. For this reason, the glass temperature does not fully reflect the amount of energy transferred into a room during the daytime. However, in summer, fluctuations of pane temperatures are much more dynamic than in winter. This is caused by higher solar irradiation, and, as a result, higher solar energy absorption in the glass. Absorbed energy is converted into heat and causes an increase of glass temperature. In winter, the solar irradiation is very small and the increase of pane temperature practically cannot be seen.

6.4 ANALYSIS OF SIMULATION RESULTS OF ENERGY BALANCES OF SOME ROOMS OF A BUILDING

6.4.1 The Method Applied to Solve Problems

A building is influenced by multiple factors from its external and internal surroundings. The following discussion of an energy balance concerns a structure built according to the current energy efficiency standards, following relevant legal regulations. Opaque elements of its envelope are characterized by high thermal insulation quality and good thermal capacity (thermal capacity of inner partitions is also high). Forced ventilation system with heat recovery is used. Heat demand caused by heat transfer through an opaque envelope is greatly reduced. Windows are of a standard type and good quality, with insulating elements in the gas gap between panes (within spacer). The following discussion is based on the assumption that the indoor temperature is maintained on a constant level over an entire year thanks to an automatic control system, which manages the operation of the heating and cooling systems.

The building under consideration is located in Warsaw, Poland. Climate conditions for this location have already been presented in detail. Hourly global radiation sums are given in Chapter 3.2, the method for determining daily distribution of ambient air temperature (averaged for every month of a year) are described in Section 6.2.3 (Eqns (6.37)−(6.40)). Average monthly ambient air temperatures are presented in Table 6.7 (averaged monthly sums of global diffuse radiation are presented in Table 3.1 in Chapter 3).

Solar irradiation for differently located envelope surfaces has been calculated using an anisotropic model, as described in Chapter 3 (the model itself is discussed in Section 2.4.3). The mathematical model for the energy balance of air in a building in general form is presented in Section 6.2.1. Heat transferred through multilayer opaque envelope elements and exchanged with surroundings is described in Sections 6.2.2 and 6.2.3. The mathematical model of energy transfer through windows is presented in Section 6.3.

Numerical simulations of dynamics of a building and its surroundings have been used to elaborate a number of patterns of heating/cooling demand and

TABLE 6.7 Average Climatic Conditions in Warsaw: Mean Monthly Ambient Air Temperature

Month	I	II	III	IV	V	VI	VII	VIII	IX	X	XI	XII
Temperature (°C)	−3.5	−2.6	1.2	7.8	13.8	17.3	19.1	18.2	13.9	8.1	3.0	−0.6

other components of energy balance of a building. Calculations have been performed for some selected examples of room locations and structure of envelope elements. Obtained results have been analyzed, and selected cases are shown graphically in the following section of this chapter. Monthly heating and cooling demand and other energy balance components throughout a year for different room locations and different sizes of windows are presented in the tables. Simulation studies of room dynamics have been based on purely hypothetical cases. However, analysis of obtained results allows drawing interesting general conclusions of qualitative and quantitative character concerning the influence of external surroundings, especially solar energy, on a building. It may be noted that even if the geographic location of a building is precisely defined (latitude 51° N), the essential principles of solar energy influence on a building are more universal. Therefore, also the discussion and drawn conclusions should be seen as much more general.

6.4.2 Heating and Cooling Demand of Selected Rooms

It was assumed that the analyzed building has three-layer external walls (brickwork—insulation—brickwork) and standard windows. Living rooms can be located on any story or in the attic. They are oriented toward four cardinal directions. Roof slopes are inclined at 45°. Key thermophysical parameters of assumed three-layer walls are presented in Table 6.8. Taking into account the average heat transfer resistances on external and internal surfaces, as well as data from Table 6.8, it is possible to determine an averaged heat transfer coefficient. It is equal to 0.26 W/(m²K). Essential thermal and optical parameters of windows are shown in Table 6.9. Windows are of a standard design with a heat transfer coefficient around 1.4 W/(m²K). Results are presented graphically for windows with relatively large dimensions (2 m × 2 m = 4 m²; i.e., 25% of the floor area of a heated room). Tables 6.10−6.25 present results for both large windows, as mentioned above, and for small windows (1 m × 1 m; i.e., slightly above 6% of the floor area of the heated room, exactly 1/16).

TABLE 6.8 Main Thermo-Physical Parameters of the Considered Three-Layer External Wall [16]

Layer from Outside	Material	Thickness δ (m)	Specific Heat c (kJ/kgK)	Density ρ (kg/m³)	Thermal Conductivity λ (W/m K)
1	Brick	0.12	0.84	1600	0.69
2	Mineral wool	0.20	0.70	24	0.038
3	Brick	0.12	0.84	1600	0.69

TABLE 6.9 Main Thermo-Physical and Optical Parameters of the Considered Window [4,16]

Window Elements	Thickness (m)	Specific Heat (J/(kgK))	Heat Conduction (W/(m K))	Density (kg/m³)	Optical Parameters for Solar Radiation $\tau/\rho/\alpha$	Emissivity for Thermal Radiation
Frame-wood	0.06	640	0.147	2800	–	0.9
Glass pane	0.004	840	0.78	2700	0.803/ 0.101/0096	0.84
Air gap	0.012	1005	0.033	1225	–	–

Analyzed three-layer wall has an external coating characterized by solar-radiation absorptance $\alpha_{sol} = 0.55$ and thermal radiation emissivity $\varepsilon = 0.8$ [34].

Results of the calculation indicate that, with the assumed structure of opaque envelope elements, the share of heat transfer through those parts of the envelope is very small. In fact, it is the least significant element of annual energy balance of a room. For materials and structural solutions assumed for this analysis, the share of heat transfer through opaque envelope elements is usually just a few percent of the total annual energy demand, and even in the least favorable room orientation (i.e., toward the north) it does not exceed 10%.

Even though operation of heat recuperation from a ventilation system has been assumed, the heat demand for ventilation is still one of the major component of annual heat balance. One exchange of air volume in heated/cooled room per hour is assumed, and recuperation efficiency is 80%. However, the share of heating or cooling energy needed for ventilation in the annual balance is at least more than 10% (in case of most insolated rooms) and up to 40% (northern rooms). But the dominant role in the energy balance of a room is played by windows, which become essential elements of the energy balance for each room and therefore for the whole building design as well. Simulation studies have been performed for different window sizes. Results shown in graphic form are for one size of big windows. This has been done to highlight the influence of solar radiation on the energy balance of a room. This influence proves to be very significant in modern energy-efficient buildings even in a higher latitude country. It needs to be emphasized that the evaluation of annual energy balance (annual averaged energy needs) is a significant simplification that does not reflect the true nature of dynamics of a building. Of course, the role of a window in the energy balance grows with its surface area and depends strongly on its location described by orientation and inclination, and related solar radiation impact.

TABLE 6.10 Monthly Heating/Cooling Energy Demand and Energy Losses/Gains through Envelope of South Rooms with Vertical External Walls and Small Windows

(MJ)	I	II	III	IV	V	VI	VII	VIII	IX	X	XI	XII	Year Q_h	Year Q_c
$Q_{h/c}$	−368.45	−253.42	−171.28	−19.83	101.86	146.87	185.02	183.45	83.45	−63.00	−231.34	−342.39	−1449.71	700.65
Q_{wall}	−95.85	−78.28	−68.93	−34.77	−6.48	5.49	13.33	11.12	−7.90	−39.13	−66.17	−85.89	−483.4	29.94
Q_{win}	−103.92	−28.62	32.60	99.69	152.84	160.14	178.15	185.25	133.7	61.56	−47.08	−108.64	−288.26	1003.96

Beta = 90, gamma = 0, window 1 × 1 (m²).

TABLE 6.11 Monthly Heating/Cooling Energy Demand and Energy Losses/Gains through Envelope of East Rooms with Vertical External Walls and Small Windows

(MJ)	I	II	III	IV	V	VI	VII	VIII	IX	X	XI	XII	Year Q_h	Year Q_c
$Q_{h/c}$	−406.42	−329.27	−255.72	−71.16	77.69	147.75	169.61	122.36	−1.73	−153.18	−273.45	−360.39	−1851.32	517.41
Q_{wall}	−98.81	−84.17	−75.47	−40.14	−9.52	5.75	11.49	3.34	−19.07	−48.65	−69.44	−87.29	532.56	20.58
Q_{win}	−138.92	−98.57	−45.31	53.72	131.72	160.75	164.58	131.94	59.71	−19.11	−85.92	−125.24	−513.07	702.42

Beta = 90, gamma = −90, window 1 × 1 (m²).

TABLE 6.12 Monthly Heating/Cooling Energy Demand and Energy Losses/Gains through Envelope of West Rooms with Vertical External Walls and Small Windows

(MJ)	I	II	III	IV	V	VI	VII	VIII	IX	X	XI	XII	Year Q_h	Year Q_c
$Q_{h/c}$	−395.68	−307.57	−229.94	−53.56	105.95	177.34	214.18	177.76	20.71	−141.32	−269.58	−356.97	−1754.62	695.94
Q_{wall}	−97.97	−82.49	−73.48	−38.31	−5.92	9.48	17.12	10.40	−16.16	−47.41	−69.14	−87.02	−517.9	37
Q_{win}	−129.01	−78.55	−21.51	69.50	156.38	186.61	203.52	180.28	79.24	−8.49	−82.34	−122.08	−441.98	875.53

Beta = 90, gamma = 90, window 1 × 1 (m²).

TABLE 6.13 Monthly Heating/Cooling Energy Demand and Energy Losses/Gains through Envelope of North Rooms with Vertical External Walls and Small Windows

(MJ)	I	II	III	IV	V	VI	VII	VIII	IX	X	XI	XII	Year Q_h	Year Q_c
$Q_{h/c}$	−411.01	−341.39	−287.39	−131.99	2.40	84.26	109.63	59.06	−50.16	−180.67	−282.10	−363.25	−2047.96	255.35
Q_{wall}	−99.17	−85.12	−77.93	−46.57	−19.49	−2.59	3.62	−4.97	−25.49	−51.56	−70.11	−87.51	−570.51	3.62
Q_{win}	−143.16	−109.75	−74.51	−0.67	66.40	105.61	112.48	76.95	17.70	−43.70	−93.89	−127.88	−593.56	379.14

Beta = 90, gamma = 180, window 1 × 1 (m^2).

TABLE 6.14 Monthly Heating/Cooling Energy Demand and Energy Losses/Gains through Envelope of South Rooms with Vertical External Walls and Big Windows

(MJ)	I	II	III	IV	V	VI	VII	VIII	IX	X	XI	XII	Year Q_h	Year Q_c
$Q_{h/c}$	−637.41	−276.77	8.90	360.68	642.61	701.58	795.66	821.12	558.98	188.55	−331.92	−638.09	−1884.19	4078.08
Q_{wall}	−63.90	−52.19	−45.96	−23.18	−4.32	3.66	8.89	7.42	−5.27	−26.09	−44.12	−57.26	−322.29	19.97
Q_{win}	−404.82	−78.06	189.80	468.61	691.44	716.68	793.23	826.62	606.62	300.06	−169.71	−432.97	−1085.56	4593.06

Beta = 90, gamma = 0, window 2 × 2 (m²).

TABLE 6.15 Monthly Heating/Cooling Energy Demand and Energy Losses/Gains through Envelope of East Rooms with Vertical External Walls and Big Windows

(MJ)	I	II	III	IV	V	VI	VII	VIII	IX	X	XI	XII	Year Q_h	Year Q_c
$Q_{h/c}$	−794.82	−591.30	−341.32	153.50	547.45	705.23	734.99	580.29	223.22	−175.76	−506.60	−712.74	−3122.54	2944.68
Q_{wall}	−65.87	−56.12	−50.31	−26.76	−6.35	3.84	7.66	2.23	−12.71	−32.43	−46.29	−58.19	−355.03	13.73
Q_{win}	−560.26	−388.66	−156.06	265.01	598.30	720.15	733.79	590.99	278.30	−57.91	−342.21	−506.68	−2011.78	3186.54

Beta = 90, gamma = −90, window 2 × 2 (m^2).

TABLE 6.16 Monthly Heating/Cooling Energy Demand and Energy Losses/Gains through Envelope of West Rooms with Vertical External Walls and Big Windows

(MJ)	I	II	III	IV	V	VI	VII	VIII	IX	X	XI	XII	Year Q_h	Year Q_c
$Q_{h/c}$	−750.29	−501.31	−234.31	224.84	659.35	822.38	911.36	799.36	311.99	−127.81	−490.54	−698.56	−2802.82	3729.28
Q_{wall}	−65.32	−55.00	−48.99	−25.54	−3.95	6.32	11.42	6.93	−10.77	−31.61	−46.10	−58.02	−345.3	24.67
Q_{win}	−516.29	−299.79	−50.38	335.13	707.80	834.82	906.41	805.35	365.14	−10.78	−326.35	−492.68	−1696.27	3954.65

Beta = 90, gamma = 90, window 2 × 2 (m^2).

TABLE 6.17 Monthly Heating/Cooling Energy Demand and Energy Losses/Gains through Envelope of North Rooms with Vertical External Walls and Big Windows

(MJ)	I	II	III	IV	V	VI	VII	VIII	IX	X	XI	XII	Year Q_h	Year Q_c
$Q_{h/c}$	−813.84	−641.50	−472.52	−91.99	250.97	454.83	498.40	330.92	32.88	−286.64	−542.42	−724.61	−3573.52	3729.28
Q_{wall}	−66.11	−56.75	−51.95	−31.04	−13.00	−1.73	2.41	−3.31	−16.99	−34.37	−46.74	−58.34	−380.33	2.41
Q_{win}	−579.04	−438.23	−285.62	23.81	308.47	475.32	502.45	347.16	92.24	−166.85	−377.59	−518.40	−2365.73	3954.65

Beta = 90, gamma = 180, window 2 × 2 (m²).

TABLE 6.18 Monthly Heating/Cooling Energy Demand and Energy Losses/Gains through Envelope of South Rooms at Attics with Small Windows

(MJ)	I	II	III	IV	V	VI	VII	VIII	IX	X	XI	XII	Year Q_h	Year Q_c
$Q_{h/c}$	−376.52	−257.21	−150.10	45.92	223.34	272.43	309.01	280.22	121.42	−60.95	−236.59	−348.70	−1430.07	1252.34
Q_{wall}	−99.02	−80.70	−70.30	−30.77	7.69	19.74	27.41	21.03	−6.45	−42.70	−69.36	−89.09	−488.39	75.87
Q_{win}	−108.81	−29.99	55.15	161.43	260.16	271.45	288.06	272.12	170.24	67.17	−49.14	−111.74	−299.68	1545.78

Beta = 45, gamma = 0, window 1 × 1 (m²).

TABLE 6.19 Monthly Heating/Cooling Energy Demand and Energy Losses/Gains through Envelope of East Rooms at Attics with Small Windows

(MJ)	I	II	III	IV	V	VI	VII	VIII	IX	X	XI	XII	Year Q_h	Year Q_c
$Q_{h/c}$	−411.36	−324.31	−233.71	−23.99	155.51	229.02	248.15	182.41	29.31	−141.40	−272.53	−365.14	−1772.44	844.4
Q_{wall}	−101.87	−86.19	−77.14	−38.92	−2.54	13.25	18.30	6.29	−20.50	−52.06	−72.30	−90.44	−541.96	37.84
Q_{win}	−140.80	−91.60	−21.62	99.68	202.55	234.53	236.31	189.03	92.18	−3.92	−82.14	−126.84	−466.92	1054.28

Beta = 45, gamma = −90, window 1 × 1 (m²).

TABLE 6.20 Monthly Heating/Cooling Energy Demand and Energy Losses/Gains through Envelope of West Rooms at Attics with Small Windows

(MJ)	I	II	III	IV	V	VI	VII	VIII	IX	X	XI	XII	Year Q_h	Year Q_c
$Q_{h/c}$	−400.14	−301.45	−205.97	−5.31	185.51	258.92	297.15	243.37	53.26	−128.93	−268.39	−361.53	−1671.72	1038.21
Q_{wall}	−100.96	−84.32	−74.87	−36.72	1.96	17.70	25.61	15.43	−16.86	−50.60	−71.96	−90.14	−526.43	60.7
Q_{win}	−130.50	−70.61	3.85	116.16	228.05	259.98	278.01	240.86	112.49	7.09	−78.34	−123.52	−402.97	1246.49

Beta = 45, gamma = 90, window 1 × 1 (m²).

TABLE 6.21 Monthly Heating/Cooling Energy Demand and Energy Losses/Gains through Envelope of North Rooms at Attics with Small Windows

(MJ)	I	II	III	IV	V	VI	VII	VIII	IX	X	XI	XII	Year Qh	Year Qc
$Q_{h/c}$	−419.41	−346.80	−289.78	−113.44	78.03	177.58	195.83	99.95	−58.38	−186.77	−286.11	−369.73	−2070.42	551.39
Q_{wall}	−102.53	−88.02	−81.71	−49.35	−14.43	5.37	10.28	−6.33	−33.86	−57.31	−73.40	−90.81	−597.75	15.65
Q_{win}	−148.20	−112.26	−73.12	20.66	136.96	190.97	192.01	119.21	17.86	−44.04	−94.61	−131.05	−603.28	677.67

Beta = 45, gamma = 180, window 1×1 (m^2).

TABLE 6.22 Monthly Heating/Cooling Energy Demand and Energy Losses/Gains through Envelope of South Rooms at Attics with Big Windows

(MJ)	I	II	III	IV	V	VI	VII	VIII	IX	X	XI	XII	Year Q_h	Year Q_c
$Q_{h/c}$	−680.68	−300.55	94.32	631.18	1125.50	1203.60	1292.00	1212.50	719.12	203.53	−356.41	−671.12	−2008.76	6481.75
Q_{wall}	−66.02	−53.80	−46.87	−20.51	5.13	13.16	18.28	14.02	−4.30	−28.46	−46.24	−59.39	−325.59	50.59
Q_{win}	−445.97	−100.23	276.14	736.44	1164.90	1209.20	1280.20	1211.40	765.79	317.41	−192.08	−463.86	−1202.14	6961.8

Beta = 45, gamma = 0, window 2 × 2 (m²).

TABLE 6.23 Monthly Heating/Cooling Energy Demand and Energy Losses/Gains through Envelope of East Rooms at Attics with Big Windows

(MJ)	I	II	III	IV	V	VI	VII	VIII	IX	X	XI	XII	Year Q_h	Year Q_c
$Q_{h/c}$	−824.99	−578.53	−252.10	350.98	863.42	1035.90	1056.70	834.04	362.83	−119.15	−505.35	−739.24	−3019.36	4503.87
Q_{wall}	−67.92	−57.46	−51.43	−25.94	−1.69	8.83	12.20	4.19	−13.66	−34.71	−48.20	−60.29	−361.3	25.22
Q_{win}	−588.38	−374.55	−65.72	461.67	909.62	1045.90	1050.90	842.77	418.87	0.97	−339.06	−531.08	−1898.79	4730.7

Beta = 45, gamma = −90, window 2 × 2 (m²).

TABLE 6.24 Monthly Heating/Cooling Energy Demand and Energy Losses/Gains through Envelope of West Rooms at Attics with Big Windows

(MJ)	I	II	III	IV	V	VI	VII	VIII	IX	X	XI	XII	Year Q_h	Year Q_c
$Q_{h/c}$	−778.55	−483.88	−137.20	425.67	979.08	1151.30	1246.10	1069.70	455.10	−69.17	−488.19	−724.29	−2681.28	5326.95
Q_{wall}	−67.31	−56.21	−49.91	−24.48	1.31	11.80	17.07	10.28	−11.24	−33.74	−47.97	−60.09	−350.95	40.46
Q_{win}	−542.56	−281.14	47.66	534.90	1022.30	1158.30	1235.50	1072.30	508.71	49.99	−322.12	−516.33	−1662.15	5629.66

Beta = 45, gamma = 90, window 2 × 2 (m²).

TABLE 6.25 Monthly Heating/Cooling Energy Demand and Energy Losses/Gains through Envelope of North Rooms at Attics with Big Windows

(MJ)	I	II	III	IV	V	VI	VII	VIII	IX	X	XI	XII	Year Q_h	Year Q_c
$Q_{h/c}$	−858.28	−671.58	−484.23	−8.29	562.90	836.23	853.54	514.28	23.92	−300.89	−561.50	−758.20	−3642.97	2790.87
Q_{wall}	−68.35	−58.68	−54.48	−32.90	−9.62	3.58	6.85	−4.22	−22.57	−38.20	−48.94	−60.54	−398.5	10.43
Q_{win}	−621.24	−466.38	−294.81	109.36	617.02	851.40	853.15	531.42	88.87	−177.27	−394.47	−549.79	−2503.96	3051.22

Beta = 45, gamma = 180, window 2 × 2 (m²).

In winter, during the coldest months of a heating season, in a building with small vertical windows (smaller than required by standards for ensuring proper daylighting), monthly share of heat transfer through windows is below 30% for southern windows and some 35% for the northern ones, in total energy demand. In the coldest months, the dominating element of heat balance is ventilation. Ventilation is responsible for 40–45% of the total energy needs, despite recuperation. The share of the heat losses through opaque envelope elements is around 20%.

In winter, if windows are big (with an area bigger than required by daylighting standards), then the share of the heat demand caused by heat transfer through windows grows considerably. It is the largest in case of northern windows: more than 70% of total energy needs, and in case of southern windows it is 60%. The share of opaque envelope elements is below 10%, and the ventilation is responsible for around 25% for northern rooms and 20% for southern ones.

In case of a roof window (surface plane is inclined), the share of energy transferred through such a window in total energy transfer grows by several percent. On one hand, roof windows are more exposed to solar radiation than vertical ones, and on the other they have a better "view" toward the sky. As a result, solar radiation influence is intensified during the days, and heat losses to the sky grow during the nights (maximum is observed for horizontal surfaces).

Influence of energy transferred through windows on the energy balance of a building (room) becomes particularly evident in summer. At that time, other components of the balance, like heat transfer through walls or heat losses through ventilation, are low, because the temperature difference between the interior and exterior is small. During the summer, solar irradiance is high. Monthly irradiation can be even 10 times higher than in winter months. Small inclination of the windows (i.e., roof windows) and other envelope elements considerably improves solar exposure, solar irradiation, and duration of its direct influence. During the summer, the cooling demand is primarily caused by the presence of windows.

The results of calculations show that for assumed climate conditions characterized by moderate solar irradiation and ambient temperature during a year (Table 6.9), in case of large windows for all main orientations, except for northern, more energy is gained through windows during the whole year than is lost through them. Cooling and heating demands grow with the window's surface area. Cooling demands decrease with the window slope's increase (the highest slope for vertical window). Heating demand dependence on inclination is more complicated and is explained later in this subchapter, when Figures 6.64–6.71 and Tables 6.10–6.25 are discussed. It needs to be noted that even if heat recuperation from ventilation systems is used, heating/cooling demand for ventilation remains the second major component of the energy balance.

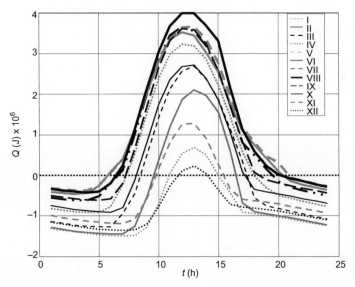

FIGURE 6.64   Daily energy demand changes in all months for rooms with windows $2 \times 2\,\text{m}^2$, vertical surfaces oriented to the south [30].

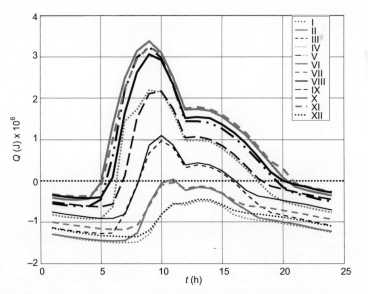

FIGURE 6.65   Daily energy demand changes in all months for rooms with windows $2 \times 2\,\text{m}^2$, vertical surfaces oriented to the east [30].

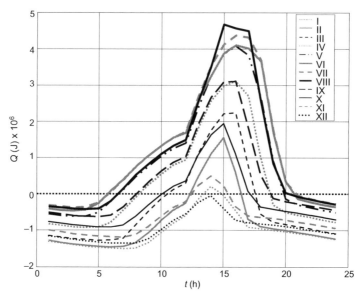

FIGURE 6.66 Daily energy demand changes in all months for rooms with windows 2×2 m^2, vertical surfaces oriented to the west [30].

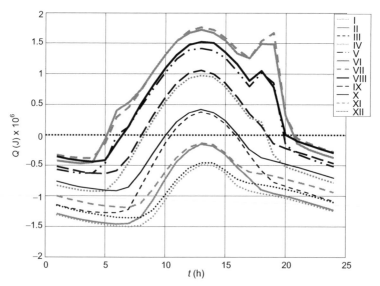

FIGURE 6.67 Daily energy demand changes in all months for rooms with windows 2×2 m^2, vertical surfaces oriented to the north [30].

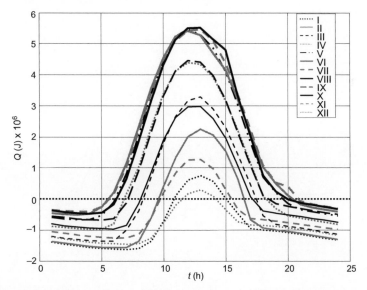

FIGURE 6.68   Daily energy demand changes in all months for rooms with windows 2×2 m^2, inclined surfaces (attics) oriented to the south [30].

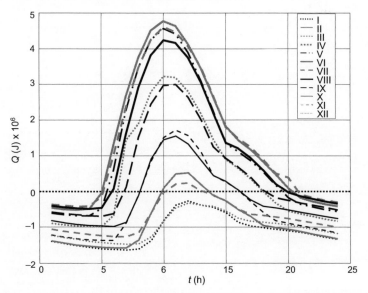

FIGURE 6.69   Daily energy demand changes in all months for rooms with windows 2×2 m^2, inclined surfaces (attics) oriented to the east [30].

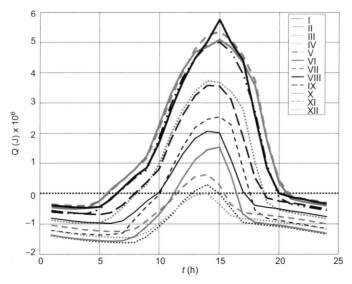

FIGURE 6.70 Daily energy demand changes in all months for rooms with windows 2×2 m^2, inclined surfaces (attics) oriented to the west [30].

Figures 6.64−6.67 (rooms with vertical walls) and Figures 6.68−6.71 (attic rooms, roof inclined at 45°) present distribution of heating/cooling demand during an average day of each month of an average year. Distributions are provided for rooms oriented toward four cardinal directions with large windows (2 m × 2 m).

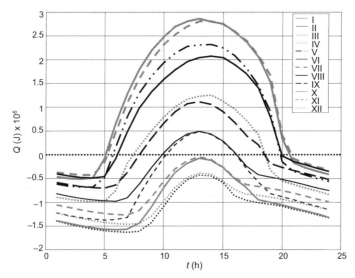

FIGURE 6.71 Daily energy demand changes in all months for rooms with windows 2×2 m^2, inclined surfaces (attics) oriented to the north [30].

Energy demand above zero stands for heat excess in the room (i.e., cooling demand), while negative values are heat losses and mean demand for heating. It needs to be noted that distributions should actually be made for energy fluxes (e.g., in the form of bars). However, heating and cooling energy per hour is presented, and because of the large amount of data (hourly sums), they are interconnected with a solid line to ensure easier interpretation.

The key feature of daily heating/cooling demand distribution in all months of the averaged year is its relatively small variability during the night and large variability during the day. This large variability during the day follows the pattern of variability of solar irradiation of a building envelope, especially its windows. It is particularly evident during the summer, when solar irradiance is high. The daily heat load distribution for analyzed rooms may be compared to solar irradiance distribution for building surfaces of corresponding orientation and inclination. Distribution of solar irradiance incident on differently situated surfaces has been presented in Section 6.3 of this chapter, in Figures 6.48–6.51 for vertical surfaces, and Figures 6.52–6.55 for inclined ones. During the day, when solar irradiance is high, maintaining constant interior air temperature requires a supply of cooling energy. Cooling demand depends on orientation of the room and especially on thermal and optical parameters of a window, its orientation, inclination, and size (dimensions). Cooling demand during summer is very evident for rooms with orientation causing increased solar radiation availability. In case of high solar irradiation, cooling demand becomes a decisive element of the energy balance. The largest cooling demand is observed in rooms in attics (with inclined envelope surfaces) with roof windows of a big size. In case of regular rooms (with vertical external walls), the highest cooling demand is observed for southern and western orientation and then somewhat smaller for eastern ones. High cooling needs are observed from May until the end of August. The peak demand can be seen in different months and different times of the day depending on the room orientations. It is characteristic that during summer from May until the end of July, the best insolated room with vertical walls is the one with western orientation. Then, from August until April, the highest irradiation is observed for southern rooms. Cooling demand is also observed in northern rooms, but it is twice lower than in western or southern ones. It should be reminded that for assumed latitude, during summer, the Sun rises in the northeast and sets in the northwest. The daily cooling demand pattern is very clearly dependent on room orientation, and peak demand can be seen at this time of day when the highest solar irradiance is observed for a given orientation. For northern rooms, characteristic peak is observed in late afternoon when the room "sees" the Sun directly and irradiance is still relatively high (beam radiation also reaches it in the morning, but, during that time, irradiance is low, as already discussed in detail in Chapter 3).

A sloped roof or other nonvertical envelope elements can "see" the Sun for a longer time and get more irradiated than vertical ones. As a result, in summer, in attics the cooling demand during the day is higher than for regular rooms on lower stories, and its distribution in a day is smoother (more even). The highest cooling demand is observed for southern and western rooms. In winter, because of the better solar radiation availability, heating demand in the attic is lower than in case of lower stories (with vertical walls), and this demand is more evenly distributed over a day. The influence of room orientation (in reference to cardinal directions) on heating demand is noticeable, but not as much as in the case of rooms with vertical walls. This means that the overheating problem will apply to all orientations, as the influence of orientation decreases when the envelope elements get more close to the horizontal surface (i.e., when inclination decreases). During the nights, heat demand of attic rooms is slightly higher, because of the "night cooling" effect (i.e., radiative heat transfer to the sky as already discussed before).

Summer cooling demand is observed from early morning until late evening in all rooms (both in the attic and lower stories). The magnitude of this demand and its distribution depend on room orientation and external partition inclination. During early spring and early autumn, when irradiance is lower than in summertime and daylight hours are fewer, cooling demand for differently located rooms starts to considerably differ. The longest duration of cooling demand during a day is observed for southern rooms. It is a bit shorter for western ones, than for eastern and, obviously, it is the shortest for northern rooms.

During winter, in February, at relatively high solar irradiation, solar gains in southern rooms (also, for quite a short time, in western ones) are higher than heat losses. Effect of heat demand reduction during a winter day intensifies with enlarging the window surface area (although this also increases nighttime heat losses). In case of south rooms with large windows (2 m × 2 m), during each month of winter, there is no demand for heating for a few hours per day. Similar effect of periodic absence of heating load during a day is observed in late autumn and winter for west rooms. In case of eastern rooms, enlarging windows only leads to certain reduction of heat load during a winter day (thanks to solar radiation reaching the room). However, larger windows cause significant increase of heating demand during nights, especially during the coldest winter months. Determining the optimal area of windows requires detailed case-by-case studies for specific rooms and user demand profiles, and may be a subject of optimization studies.

Also, studies of monthly distribution of heating/cooling demand and individual components of this demand have been carried out. Detailed graphical interpretation of obtained results may be found in the literature [12]. Tables 6.10—6.25 present monthly heating and cooling demand for rooms under consideration (for all months of the averaged year taking into account heat gains and losses through the envelope; i.e., through opaque elements and windows). As it has been already mentioned, negative values stand for heat losses, and positive ones are heat gains. It needs to be noted that monthly demand values calculated as sums of hourly heat loads resulting from heat gains or losses do not reflect true dynamics of thermal processes occurring in a building (energy flows into (positive) and out of (negative) the interior are added up together). Hourly values added up for the entire month show which thermal phenomena (losses or gains) dominate during that month of the averaged year. Thus, it is only a simplified averaged description of thermal processes in a building (room). It may be very well seen when analyzing the heat demand for ventilation. According to the monthly sums, heat demand for ventilation is present during the entire year. In reality, of course, in summer, when the ambient air temperature is higher than the temperature of the interior there are heat gains, but this is not visible in monthly sums. Monthly values provide a simplified picture of heat transfer in a building and its surroundings, and do not show variability of real parameters in time. It needs to be noted that annual sums of components of energy balance provide even more simplified information and do not permit showing actual dynamics of a building and actual variability of individual parameters. They only reflect approximated roles (shares) of individual components of energy balance in creating annual heating/cooling loads.

The highest demands for room heating in all the months in a year are of course observed for northern rooms at attics with large windows. The lowest demands are for southern rooms with small windows. Total seasonal heat loads for southern regular rooms and rooms at attics are comparable. Although, during the winter more heat is needed to ensure thermal comfort in case of rooms in attics, and in early spring or late autumn more heat is needed for regular rooms. This is also true for other orientations. During spring and autumn, inclined surfaces are better insolated (higher irradiance and for a longer time), which results with heat demand reduction. Eastern rooms require more heating than western ones during the heating season. This results from solar radiation characteristics discussed in Chapter 3.

In summer, overheating of rooms in attics (with sloped envelope planes), especially those with large windows, is evident (because of increased solar radiation availability). With the increase of the window area, heat losses increase, but heat gains increase much more. In result, cooling demand becomes a dominating element of the energy balance.

In case of a small window area, energy demand for room heating is higher than the cooling demand. The magnitude of such demand is clearly influenced by room orientation. In case of northern regular rooms, annual heating load is many times higher than the cooling load. In case of southern and western rooms, the differences are much smaller. In case of rooms in attics, the differences in energy demand for four considered orientations are reduced even further.

Increased window area increases the total heating and cooling demands. However, heat gains grow more than losses. For large windows (except for northern rooms), total heat gains in the annual balance are higher than total losses. At the same time, the heating season duration is reduced. For example, for southern rooms with vertical walls, duration of heating season is reduced from 7 months (from October until the end of April) to just 4 months (from November until the end of February). In the case of rooms in attics, this period is reduced from 6 months (from October until the end of March) to 4 months (from November until the end of February). Of course, because of the assumed constant interior temperature this means that the cooling season is extended accordingly. This again confirms the previous conclusion concerning the great impact of windows on the energy balance of a room. Thus, it shows how much attention should be paid to windows when a building is designed. This also supports the thesis that the variability of thermal balance over time should be analyzed. The smaller the time step of calculations (preferably not more than 1 h), the more accurately time variability of heating and cooling loads is reflected.

It should be noted that automatically controlled operation of heating/cooling systems in reality permits certain tolerances for desired internal microclimate conditions. This, in particular, means that it is possible to permit preset deviations ΔT_{in} from the internal temperature setpoint T_{in}, which might be equal to the internal comfort temperature (i.e., $T_{in,comfort}(t) = T_{in} \pm \Delta T_{in}(t)$). Solar energy significantly influences the thermal balance of a building. If opaque envelope elements have good insulating quality and high thermal capacity, then the influence of external surroundings through those elements is greatly reduced. However, also in this case, the influence of solar energy can be seen, even if it is rather small and directly depends on the orientation of the wall, roof plane, etc. Of course, the solar radiation influence is most evident for windows. Analysis of results presented in Tables 6.10–6.25 clearly shows that heat gains through windows are larger than heat losses for at least half a year (April through September) for all rooms under consideration, except for a case of northern room with small windows. This is a result of solar radiation flow directly into the interior of a room. Heat gains can be seen for an even longer time (8 months, March–October) in case of all considered rooms with southern windows (small and big) and western rooms with big windows.

One more interesting observation results from the presence of a window frame. Results of the calculation of heat transfer through a window (not shown in any tables or figures here, but published in a separate study [3]) show that heat is lost continuously through frames, only in southern and western rooms very small heat gains can be seen for a short time in July. This means that the increased frame area leads to notable heat losses, which is obviously not recommended in winter. Perhaps direct installation of compact glazing in opaque envelope elements could be a solution allowing the reduction of heat losses (although, obviously, this is only possible in case of fixed (not opening) windows). A number of research projects has also focused on improving the frame design and reducing heat conduction through its structure, mainly by using frames of chamber design. Of course, the best results may be achieved if there is a vacuum inside the chambers.

6.4.3 Final Conclusions on Heat Balance of Selected Rooms of a Building

So far, thermal calculations for residential buildings as well as most nonresidential ones in higher latitude countries have usually concerned heating season only. This resulted with a very one-sided approach to the energy balance of a building and as a result often with incorrect designs. In reality, a building is functioning in annual cycles and thermal comfort inside must be maintained throughout all seasons, also during summer. It turns out that for certain design solutions, materials selections, and room locations, cooling demand may become a dominating part of an energy balance if the thermal (temperature) comfort conditions are to be maintained. Curiously, this happens in higher latitude countries where until recently room cooling had not been taken into account. Currently, building envelopes are characterized by increasingly strong thermal insulation. This applies to all its components: opaque walls, windows, floors, and roofs. Forced ventilation systems are controlled in a way aimed at reducing energy consumption. Very often, heat recuperation is used. The thermal state of a building is less and less dependent on variable thermal conditions of surroundings described by ambient air temperature. However, solar radiation in the form of electromagnetic waves reaches the interior directly through glazed surfaces, which are transparent for shortwave solar radiation. Inside a building, radiation is mostly absorbed by surfaces of internal partitions and other internal objects. This increases internal energy of those materials and their temperature grows. Photothermal conversion takes place in a building and the captured heat remains inside rooms. This heat will remain in a building if it is of high thermal capacity and well insulated, heat exchange with external surroundings is strongly reduced. Because of prolonged influence of high solar irradiance during summer, excess heat is accumulated in a building, which gets overheated. Thanks to good insulating strength and leak-tightness of a building, the heat cannot "get out" naturally. It may be removed by operation of cooling/air conditioning systems, but this only increases energy consumption, instead of reducing it.

As proved by calculations presented in the previous section, overheating mainly concerns rooms in attics. For example (see results in Table 6.14), it can be estimated that a southern room (vertical walls) with an area of 80 m^2 and large windows (5 windows 2 m × 2 m) requires supply of totally 18,650 MJ of cooling energy during 8 months of a year. A room of identical size and structure made of the same materials, but located in the attic and oriented toward any of the four cardinal directions (data in Tables 6.22—6.25) would require some 30% more cooling than a southern regular room. In case of an attic room, about 23,880 MJ of cooling energy would be needed during 5—8 months of a year (in case of northern orientation—5 months, eastern and western—6 months, southern—8 months). Such a major increase of cooling demand in case of multiple apartments located in the attic in one building, or even in many buildings with such apartments, leads to considerable increase of energy demand of buildings instead of its reduction, which should be the goal.

Unfortunately, these days an increasing number of construction companies construct buildings with residential rooms within the attics in large cities, primarily because of the financial benefits. Land is expensive in urban areas, therefore, a developer wants to maximally utilize each square meter of its footprint. Figure 6.72 shows modern residential buildings with living areas in their attics.

As already mentioned, if opaque envelope elements of very good thermal insulating materials are used, their role in the energy balance decreases. Because of a trend to use large windows and glazed facades, the role of windows in energy balance (and thus also thermal comfort) increases, and in more and more cases becomes dominating (as illustrated by data Tables 6.10—6.25). Analysis of results of simulation studies shows how important solar energy is for energy gains of interiors and for thermal behavior of windows, and how important is clever placement of windows and size of their glazing for thermal energy state of a room (building). Of course, window design and choice of materials are very important and require separate analysis.

To determine the heat load and heat fluxes exchanged between the interior and exterior of a building, it is necessary to take into account variability of these energy fluxes in time, in order to ensure thermal comfort inside rooms. European Union (EU) Directive on the energy performance of buildings [35] requires performing the calculation on energy consumption in buildings for the

FIGURE 6.72   Examples of multifamily buildings with attics in Warsaw, Poland.

entire year, not only for the heating season as it used to be done for most residential buildings in higher latitude countries. Limiting the analysis to the heating season only caused that windows were considered as elements of a building envelope generating only heat losses. The methods used to include solar gains were relatively simplified. Moreover, in higher latitude countries, solar gains during winters are small (low irradiance) and were usually unnoticeable in energy balance calculations.

As proved by the results of simulation studies presented in this chapter, in case of a relatively large surface of windows, the total annual heat gains outweigh the total annual heat losses (with the exception of northern rooms) and, in specific cases, give significant increase of energy consumption. Magnitude of heat gains, of course, depends on room orientation and inclination of its envelope surfaces. Thus, the issue of supplying a sufficient amount of cooling to maintain thermal comfort during summer becomes essential.

As already mentioned, the trend of utilizing attics as living areas observed in new buildings, combined with almost exclusive use of standard roof windows, results with considerable increase of demand for cooling (air conditioning), which, in case of apartments located in attics, will be much larger than the heating load (Tables 6.22—6.25). In case of buildings with flat roofs, without attics, and with empty small ventilation spaces above the top story, the situation in top story apartments is similar or even worse. This is because during summers the horizontal or nearly horizontal surfaces (like flat or slightly sloped roofs) are most exposed to solar radiation. In summer, in such top-story apartments not provided with air conditioning, discomfort may be even more intense than in attic apartments.

Table 6.26 lists the results of calculations of the average annual heating demand E_h and cooling demand E_c for differently situated rooms selected for consideration. Those values are of indicative character only, as they have been obtained for a specific hypothetical envelope design solutions discussed in this chapter.

According to the support mechanisms for the energy efficiency policy adopted by most states, including the already mentioned EU Directive on the energy performance of buildings [35], the energy demand of buildings should be determined upon both final and primary energy consumption indexes. It is so because improvement of energy efficiency requires not only demand reduction, but also utilization of highly efficient energy conversion systems, energy generation and supply technologies, and fuel acquisition methods (especially based on fuels causing no harm to the environment) to cover existing (appropriately reduced) demand. This means that previously calculated and analyzed values of energy demand for space heating should be converted into final and primary energy consumption. Final energy consumption is determined by taking into account systems and installations used in a building in question. The more efficient systems are used, the less final energy is required. However, the final energy demand says nothing about

Solar Energy in Buildings

TABLE 6.26 Annual Heating/Cooling Energy Demand Factors of Rooms under Consideration with External Walls and Windows of Different Orientation and Inclination

$\beta = 90°$ kWh/(m²a)	$\gamma = 0°$ 2 × 2 m²	$\gamma = 0°$ 1 × 1 m²	$\gamma = 90°$ 2 × 2 m²	$\gamma = 90°$ 1 × 1 m²	$\gamma = -90°$ 2 × 2 m²	$\gamma = -90°$ 1 × 1 m²	$\gamma = 180°$ 2 × 2 m²	$\gamma = 180°$ 1 × 1 m²
$\beta = 90°$								
E_h	32.5	25.0	50.0	30.0	55.0	32.5	50.0	35.0
E_c	**70.0**	12.5	**65**	12.5	50.0	10.0	25.0	2.5
$E_h + E_c$	**102.5**	37.5	**115.0**	42.5	**105.0**	42.5	**75.0**	37.5
$\beta = 45°$								
E_h	35.0	25.0	47.5	30.0	52.5	32.5	**65.0**	35.0
E_c	**112.5**	12.5	**92.5**	12.5	**77.5**	15.0	47.5	10.0
$E_h + E_c$	**147.5**	37.5	**140.0**	42.5	**130.0**	47.5	**112.5**	45.0

Annual Heating/Cooling Energy Demand Factors above the limits required by national regulations are in bold

quality or environmental characteristic of consumed energy. For that reason also the demand for primary energy (fuel) necessary to cover this final energy needs to be determined.

Primary energy consumption depends on the type of primary fuel being utilized, especially its energy content (e.g., heating value), efficiency of the process used for fuel conversion into final energy, and on the form of that final energy: heat, cooling energy, or electricity. Primary fuel may be consumed at the consumer's site. In case of traditional fossil fuels, it may be natural gas, fuel oil, or coal. In case of renewable fuels, the possibilities include biomass and biogas, and pure renewable energy: solar energy, wind energy, hydro or hydraulic energy (rivers, tides, waves), geothermal, and energy extracted from surroundings via heat pumps. Energy demand analyses referring to energy consumed in buildings may use [36] nonrenewable energy demand factors, which assign certain priorities for energy quality of specific fuels, thus facilitating analyzing energy intensity of individual pieces of equipment, sources, and systems.

Table 6.26 lists the energy demand indicators for analyzed selected rooms. It is evident how inefficient a modern building may be due to its cooling demand, if the building design is only focused on heat consumption (heat losses). Until recently, during construction of new buildings and thermal refurbishment of existing ones, in most countries, only (adopted by valid legal regulation) maximum energy intensity factors of primary energy consumption for heating, ventilation, and water heating had been used.

For example, in Poland, according to current legal regulations [37], the maximum annual specific demand for nonrenewable primary energy for heating and ventilation is determined upon so-called building's shape factor A/V_e (new regulations adopted in 2014 have introduced new values). The maximum specific nonrenewable primary energy demand may be currently from 73 kWh/(m^2 year) in case of the shape factor $A/V_e = 1$ (compact shape) to 149.5 kWh/(m^2 a) in case of $A/V_e = 0.2$ (extended shape). Knowing these extreme values, type of utilized fuel, and efficiency of the heating system, it is possible to obtain specific heating demand factors, resulting from architectural and civil concept of a building, including its location, structure, used materials, etc. (i.e., factors analogical to those listed in Table 6.26). In order to convert the limits of primary energy demand, stipulated by legal regulations, into the maximum heat demand factors, certain simplified calculations have been performed. The following assumptions have been made: a building is supplied with heat from a local heating plant fired with natural gas (relatively clean and efficient fuel), for which index of demand for nonrenewable primary energy per energy generated and supplied to the building is 1.2 [36]. The heating system as a whole, including all its components, is characterized by a total operating efficiency of 0.8. Two extreme values of maximum annual demand for nonrenewable primary energy for heating and ventilation, upon conversion, yield two extreme values of heat demand for heating and ventilation, depending on the building shape factor. In this case, it is $100\,kWh/(m^2\,year)$ (extended shape) and $50\,kWh/(m^2\,year)$ (compact shape).

Results of the calculation presented in Table 6.26 show that the analyzed building with all its variously situated rooms enveloped by vertical external walls, is fully in accordance with the current standards concerning heating and ventilation energy demand (regardless of the shape factor). But in case of rooms in attics, in case of a small shape factor $A/V_e = 0.2$, eastern and especially northern rooms with large windows has energy demands per square meter of the heating room area higher than the limiting value, despite the assumed heat recuperation from ventilation system. Almost all rooms in a building (except for two rooms in the attic in case of an extended shape) are constructed in accordance with the current energy efficiency standards. It may be stated that building as a whole (average factor for all rooms with different orientations and location) fulfills the standards on energy-efficiency. However, the energy demand for heating and ventilation of the entire building does not reflect the actual thermal situation in individual rooms, especially in case of certain parts of its attic. Moreover, such a value only reflects part of the energy demand related to heating, while in fact there would be also cooling demand, so far not covered by legally binding standards. Only in case of small windows, and especially northern rooms, cooling demand is low. In case of rooms with a big window area, it may be larger than the heat demand, often considerably larger, and this applies to all rooms except for northern ones located in attics. Rooms with theoretically low energy demand because of the low heat load, in comprehensive analyses also taking into account the cooling demand, may prove to be quite inefficient. Purely theoretical adding up of heating and cooling demands shows that all rooms with big windows, even the northern ones with inclined envelopes, have their total heating-and-cooling demands much above the permissible limit (maximums). Moreover, cooling generation (except for free cooling) is usually much more energy-intensive with respect to primary energy than heat generation (most chilling/air conditioning equipment is powered by electricity).

Obtained results and their discussion allow us to unambiguously state that the energy analysis of a building should be carried out comprehensively. It should take into account the location of individual rooms because of the major differences in energy demands for different locations. For example, for an apartment user, it is not only important how energy-intensive will be for this apartment as a whole in a year, but also how the heating and cooling demand for individual rooms will change over time. Knowing various thermal conditions inside specific rooms makes it possible to assign them specific functions (bedrooms, living rooms, kitchen, stores, etc.).

Office buildings are occupied by humans during daytime (i.e., at the time when solar radiation is present). There may be major differences in heating and cooling demands between differently situated rooms. Moreover, during certain times of the year, especially during spring and autumn, it may happen that part of an office building needs cooling, while another part (northern) needs heating at the same time. It is especially true in a fully glazed building. Usually, such situations are not described in the civil documentation of a building, and therefore they require adapting HVAC systems for unforeseen conditions. One solution may be using a heat pump loop for one or several stories. It allows using some heat pumps as heating devices and others as chillers (by reversing heat sources and sinks, and operating cycle). It is much better to know such situations in advance, before an Heat Ventilation Air Conditioning (HVAC) system is designed, than to be forced to upgrade an existing one.

A building needs a comprehensive approach to its energy balance, taking into account dynamics of all phenomena. Energy demand for both heating and cooling should be determined with their characteristic distribution in time and with regard to the situation (function) of individual rooms in a building.

As already mentioned, currently binding EU regulations for determining energy characteristic of a building and issuing energy certificates [35] require determining the primary and final energy consumption. Necessity to supply cooling power to a building should be based on analysis of thought through energy-efficient solutions. Standard HVAC systems powered with electricity generated by fossil fuel combustion are less efficient than heat generation. As a result, most room chilling (air conditioning) technologies are also more energy intensive and have larger environmental footprints. For example, currently in Poland the index of nonrenewable primary energy consumption for generating electricity and supplying it to a building is 3 (according to Ref. [36]).

One more fact needs to be highlighted. When performing a thermal analysis of a building, taking into account solar radiation in estimated and averaged way (e.g., as monthly averages) may result with certain over-dimensioning of a heating system. In case of heat supplied from a district heating network, it does not cause major consequences. However, if unconventional low-temperature heating system is planned (e.g., with heat pumps), then not taking into account solar gains at an accurate level may have financial consequences. Over-dimensioned heating system causes unnecessary costs.

Well thought through architectural concept and design of a building envelope, its shape, and location (placement), may influence insolation conditions. Envelope elements should be designed with regard to the position of the Sun and its apparent movement in the sky, as well as local conditions that can cause shading or enhanced reflectivity. Depending on the object of the analysis and thoroughness of the study, it may be needed to know hourly and/or daily and/or monthly solar radiation sums, which may indicate danger of over-heating. This knowledge should be ultimately utilized to create a proper architectural concept of a building and to plan shading by both elements of the building architecture (e.g., overhangs) and its surroundings (vegetation). Most such solutions are applicable to vertical windows, not to the ones installed in roofs. In the latter case, heat exchange with surroundings may be controlled by proper utilization of shades, blinds, etc., which, of course, may also be used for vertical windows. Also, energy performance of a window may be influenced by its proper design, including filling the inter-pane gap and thought through choice of materials for glazing and its coating, frame, and seals.

As the results of calculations have shown, and confirmed by practice, cooling demand during summer is very prominent, especially for certain room locations. Cooling demand shows great importance of windows in a design of an energy-efficient building, especially during summer, which usually has been neglected in higher latitude countries. Admittedly, certain window shading elements (overhangs, etc.) were already used in traditional architecture also in Poland. They considerably restricted access of solar radiation during summer, while not obstructing it during winter. Moreover, traditional attics did not have living character and usually were used as storage spaces for the household. Light access for attics was provided through vertical bay windows. This does not mean that roof windows should not be used anymore. Thanks to technical progress, it is possible to apply certain materials and window designs, which ensure attaining good microclimate conditions within living spaces (rooms) with roof windows. Thus, the selection of window design and technical solutions is becoming increasingly important and a complicated issue for modern buildings.

REFERENCES

[1] Klemm P, editor. Building physics. Fizyka Budowli, vol. 2. Warszawa: Arkady; 2005 [in Polish].

[2] Grabarczyk S. Building physics. Energy efficient buildings supported by informatics *(Fizyka Budowli. Komputerowe wspomaganie budownictwa energooszczednego)*. Warsaw: Oficyna Wydawnicza Politechniki Warszawskiej; 2005 [in Polish].

[3] Chwieduk D. Modeling and analysis of solar energy conversion and use in buildings *(Modelowanie i analiza pozyskiwania oraz konwersji termicznej energii promieniowania słonecznego w budynku)*. Warsaw: Prace IPPT; 11/2006 [in Polish].

[4] Gordon J. Solar energy the state of the art. UK: ISES Position Papers; 2001.

[5] EN ISO 6946. Building components and building elements − thermal resistance and thermal transmittance. Calculation methods. CEN; 2007.

[6] EN ISO 13370. Thermal performance of buildings. Heat transfer via the ground. CEN; 1998.

[7] EN ISO 13789. Thermal performance of buildings − transmission and ventilation heat transfer coefficients. Calculation methods. CEN; 2007.

[8] Awbi HB. Calculation of convective heat transfer coefficients of room surfaces for natural convection. Energ Build 1998;28(2):219−27.

[9] Churchill SW, Chu HHS. Correlating equations for laminar and turbulent free convection from a vertical plate. International Journal of Heat and Mass Transfer November 1975;18(11):1323−9. Elsevier, http://www.sciencedirect.com/science/journal/00179310/18/11.

[10] Holman JP. Heat transfer. McGraw-Hill Higher Education; 2002.

[11] Curcija D, Goss WP. New correlations for convective heat transfer coefficient on indoor fenestration surfaces − compilation of more recent work. In: ASHRAE/DOE/BTECC conference, thermal performance of the exterior envelopes of buildings VI, Clearwater, FL; 1995.

[12]   Chwieduk D. Solar energy in buildings (Energetyka słoneczna budynku). Warsaw: Arkady; 2011 [in Polish].

[13]   EN ISO 15099. Thermal performance of windows, doors and shading devices – detailed calculations. CEN; 2003.

[14]   Duffie JA, Beckman WA. Solar engineering of thermal processes. New York: John Wiley & Sons, Inc.; 1991.

[15]   (Atlas klimatyczny Polski. Cześć tabelaryczna. Zeszyty.) Climatic atlas of Poland. Tables. Scripts. Warsaw: PIHM; 1970 [in Polish].

[16]   http://wipos.p.lodz.pl/baza/03-02.html; 2013, [in Polish].

[17]   http://energain.co.uk/Energain/en_GB/products/thermal_mass_panel.html; 2013.

[18]   Chwieduk D. Dynamics of external wall structures with a PCM in high latitude countries. Energy 2013;59:301–13.

[19]   ASHRAE. Standard method for determining and expressing the heat transfer and total optical properties of fenestration products. BSR/ASHRAE Standard 142P (Public Review Draft). Atlanta, Georgia: American Society of Heating, Refrigerating and Air-Conditioning Engineers (ASHRAE); 1996.

[20]   ASHRAE. Fundamentals, handbook. Chapter Fenestration. SI ed. Atlanta: American Society of Heating, Refrigerating and Air Conditioning Engineers; 1997.

[21]   ASTM. Standard procedures for determining the steady state thermal transmittance of fenestration systems. ASTM Standard E 1423–91. In: 1994 Annual book of ASTM standards 04.07. American Society of Testing and Materials; 1991. 1160–1165..

[22]   ASHRAE. Handbook of fundamentals; 1967. Chapter: Solar heat gain factors.

[23]   Arasteh D. An analysis of edge heat transfer in residential windows. In: Proceedings of ASHRAE/DOE/BTECC conference, thermal performance of the exterior envelopes of buildings IV, Orlando, FL; 1989. pp. 376–87.

[24]   EN ISO 10077–1. Thermal performance of windows, doors and shutters – calculation of thermal transmittance – part 1: simplified method. CEN; 2000.

[25]   Twidell J, Weir T. Renewable energy resources. E&FN SPON. London: University Press Cambridge; 1996.

[26]   Dascalaki E, Santamouris M, Balaras CA, Asimakopoulos DN. Natural convection heat transfer coefficients from vertical and horizontal surfaces for building applications. Energ Build 1994;20:243–9.

[27]   Hollands KGT, Unny TE, Raithby GD, Koniczek L. Free convection heat transfer across inclined air layers. J Heat Transfer 1976;98:189–93.

[28]   Shewen E, Hollands KGT, Raithby GD. Heat transfer by natural convection across a vertical air cavity of large aspect ratio. J Heat Transfer 1996;118:993–5.

[29]   Wright JL. Correlation for quantifying convective heat transfer between window glazings. Communication prepared for ASHRAE SPC-142 (Standard method for determining and expressing the heat transfer and total optical properties of fenestration products); 1991.

[30]   Wright JL. A correlation to quantify convective heat transfer between vertical window glazings. ASHRAE Trans 1996;106(Pt. 2).

[31]   El Sherbiny SM, Raithby GD, Hollands KGT. Heat transfer by natural convection across vertical and inclined air layers. Journal of Heat Transfer 1982;104:96–102.

[32]   Arnold JN, Bonaparte PN, Catton I, Edwards DK. Experimental investigation of natural convection in a finite rectangular region inclined at various angles from 0 to 180°. In: Proceedings of the 1974 Heat Transfer and Fluid Mechanics Institute, Corvalles, OR. Stanford, CA: Stanford University Press; 1974.

[33]   Chwieduk D. Some aspects of modeling the energy balance of a room in regard to the impact of solar energy. Sol. Energy 2008;82:870–84.

[34]   Martin CL, Goswami DY. Solar energy pocket reference. ISES; 2005.

[35]   The directive 2010/31/EU of the European Parliament and of the Council of 19 May, 2010 on the energy performance of buildings.

[36]   *Rozporzadzenie Ministra Infrastruktury z dnia 6 listopada 2008 r. w sprawie metodologii obliczania charakterystyki energetycznej budynku i lokalu mieszkalnego lub części budynku stanowiacej samodzielna całość techniczno-użytkowa oraz sposobu sporzadzania i wzorów świadectw ich charakterystyki energetycznej.* (Dz. U. Nr 201, poz. 1240 oraz z 2013 r. poz. 45). Ordinance of the Minister of Infrastructure of 6th November 2008 concerning methodology of determination the energy characteristics of buildings and apartments and method of elaboration of energy characteristics certificates (in Polish).

[37]   *Rozporzadzeniu Ministra Infrastruktury z dnia 12 kwietnia 2002 r. w sprawie warunków technicznych, jakim powinny odpowiadać budynki i ich usytuowanie* (Dz. U. Nr 75, poz. 690, z późn. zm.). Ordinance of the Minister of Infrastructure of 12th April 2002, concerning technical conditions for buildings and their location, as later amended (in Polish).

Chapter 7

Active Solar Systems in Buildings

7.1 MAIN APPLICATIONS OF ACTIVE SOLAR SYSTEMS

7.1.1 Basic Concepts of Solar Heating Systems

Building-related solar energy issues are not limited to the architectural and civil engineering concept, they also apply to building energy systems. More and more often elements of solar systems are incorporated into the envelope as its integral elements. Building integrated solar thermal (BIST) and building integrated photovoltaics (BIPV) systems are increasingly popular in modern low-energy buildings. A building is no longer just an architectural and civil structure, now it is also considered to be an energy system that gains energy from its surroundings and converts it into a useful form on the spot, for its own users. However, constructing buildings as integrated and complex architectural and energy systems is still a challenge for the future.

Solar energy technologies are already considered to be mature and have been available on the energy market for more than 30 years [1,2]. Efficiency of solar systems and related devices is continuously rising. Unfortunately, usually they are designed and installed without direct relation to any specific building. Typically, a building is built first, and only later is the concept of implementing unconventional energy systems, including solar energy technologies, developed for it.

The most popular solar technologies applied in buildings are solar active heating systems based on photothermal energy conversion. These include the following.

- Solar active (liquid) heating systems for domestic hot water (DHW), with flat-plate solar (liquid) thermal collectors or evacuated solar thermal collectors, coupled with auxiliary heaters (devices or systems), working for individual heating purposes of the user. Such systems are used in:
  - single-family houses, multifamily apartment buildings, public buildings, industrial facilities, etc., in heating systems operating for the whole year;
  - residential buildings and leisure facilities used seasonally, from spring until early autumn (e.g., cottage houses, summer resorts, and summer sport centers).

Solar Energy in Buildings. http://dx.doi.org/10.1016/B978-0-12-410514-0.00007-4

- Solar active (liquid) heating systems equipped with unglazed collectors, so-called swimming pool absorbers, for low-temperature heating applications, working seasonally for individual heating purposes of the user as:
  - outdoor swimming pools;
  - seasonal ground heat storage systems.
- Solar active (liquid) heating systems for DHW and space heating, so-called combi-systems equipped with flat-plate or evacuated solar collectors, coupled with auxiliary heaters (devices or systems), working for the whole year for individual heating purposes of the user in the residential sector, and public buildings and others in the tertiary sector [3].
- Solar active (liquid) heating systems for DHW, space heating and cooling, so-called combi-plus-systems, coupled with sorption cooling devices (chillers) or systems [4]. Depending on the temperature required by the cooling device or system, they are equipped with flat-plate, evacuated, compound parabolic concentrator (CPC) or other type of concentrating collectors, working for the whole year for individual heating purposes of the user, in public buildings and others in the tertiary sector, and multi-family residential buildings.
- Centralized solar district heating systems of various scales, for campuses, villages, small towns, and cities, forming a centralized heating plant with conventional peak heat source (plant) connected to a district heating network equipped usually with flat-plate solar collectors with an area from a few 100 to several 1000 square meters [3].
- Semiactive or active air solar heating/cooling systems for room heating or cooling, equipped with air solar thermal collectors forming solar walls; usually incorporated into a building structure; working for the whole year for individual heating/cooling purposes of the user, in small and large residential buildings, public buildings, and others in the tertiary (nonresidential) sector.

In case of solar air heating/cooling systems equipped with solar walls, good insolation may be sufficient to establish natural heat flow in envelope elements with air ducts. In case of low insolation, the flow of collected heat can be enforced by mechanical equipment (fans).

Most solar thermal systems used currently are stand-alone systems that provide energy for an individual local heating system of a single-family house, large residential building, a public building, or a housing estate. They cover heating needs locally. The energy producer is usually also the energy consumer. Energy produced locally is used locally.

7.1.2 Development of Solar Heating Systems

Thanks to considerable progress in solar technologies and materials used, solar energy heating systems cover various heating demands with increasing efficiency and reliability [1—5]. Technologies, until recently considered inefficient in countries with poor insolation conditions, are now becoming increasingly popular. This happens not only thanks to solar technologies development, but also due to the progress in the building construction sector, which caused considerable reduction of heat demand because of the decrease of the heat losses from buildings. This reduced demand may now be covered by increasingly popular low temperature heating systems (especially floor and wall heating systems) supplied by solar and other renewable energy technologies.

As already mentioned, most of the active solar systems are working for individual heating purposes of the user. They supply heat to local heating system (network) of one building or a small group of buildings. The essential working principle of such solar heating systems, regardless of their size (heat load), is focus on and control of multiple energy processes: energy generation, conversion, and storage, followed by controlled (planned) distribution of collected energy to consumers.

Technical sophistication of used solutions, type of a system, and control methods must be adapted to the consumer type and heating requirements. When the temperature of the heating medium and the heat load are high, they become more complicated, and, therefore, more expensive systems are required. When small heat loads and low temperatures of heating systems are required, and heat load distribution is predictable and relatively simple, then cheap technical solutions of the heating system may be used. The complexity of a system rises along with the heating requirements and demanded reliability in every condition.

Currently, the most popular solar systems are those for domestic hot water heating. In countries where high solar irradiance is observed and total annual irradiation is high (around $1400\,kWh/m^2$), like in the Mediterranean area, thermo siphon (natural gravity) systems are most popular. They are not equipped with any mechanical devices to force the circulation of the heat carrier. The flow occurs naturally thanks to thermal diffusion. Circulation of the working fluid is driven by the density difference between the medium heated by photothermal conversion, taking place in the collector, and the flow of cold liquid from the bottom part of the system. It is very important to install the storage tank at the appropriate height above the collector. To ensure that, the height of thermosiphon circulation to which liquid can rise thanks to the buoyancy caused by the density difference related to the temperature difference, needs to be calculated. The calculation needs to take into account the hydraulic resistance caused by friction against the pipes, which acts against the buoyancy. If a tank is located incorrectly, reverse circulation may take place because of large negative buoyancy outweighing hydrostatic pressure of heated water. An example of a typical thermosiphon system is shown in Figure 7.1.

In countries with average annual irradiation below $1200\,kWh/m^2$, thermosiphon systems are rarely used, because in winter, when irradiance is low, the natural convection conditions hardly allow for diffusive circulation to

FIGURE 7.1 Thermosiphon systems in Cyprus.

develop. It is due to the higher viscosity of nonfreezing liquid that must be used instead of water and low solar irradiance. In such countries, active solar systems (i.e., with forced circulation of the working fluid) are used instead. Currently, they are primarily used for domestic hot water heating, although the popularity of combisystems for both DHW and space heating is increasing continuously.

At an annual solar irradiation of 900−1200 kWh/m^2, single-family houses' heating systems for DHW covers some 50−65% of the DHW demand and supply around 400−500 kWh of heat per every square meter of the solar collectors' aperture area per year [3]. In case of small scale combi-systems for both DHW and space heating, usually solar energy can cover the total heat load of a building (DHW + heating) in 20−50% [1]. It needs to be emphasized that useful solar energy supplied by a solar heating system does not depend exclusively on insolation conditions. Other influencing factors are type and quality (thermal characteristics) of the solar collectors, system configuration, and the control systems, including principles of cooperation between the solar system and conventional heat source or auxiliary heating system. Another very important factor, perhaps the most significant for the effective use of solar energy, is the daily heat consumption pattern. Different behavior patterns of inhabitants and resulting different heat consumption distribution (consumption of different amounts of heat at different parts of a day) affect thermal efficiency of a solar heating system and require appropriate configuration, component dimensioning, and automatic control settings. In case of supplying heat to cover space heating demand, it is important not only to minimize the demand, but also to choose the efficient low temperature space heating system (e.g., floor heating, wall heating, etc.). In the case of public buildings, the differences between individual heat demand patterns, depending on the type of users, may be even larger, both in terms of duration and magnitude of the heat load. This is because of diverse building utilization regimes, regarding seasons (e.g., schools normally are not used in summer, while some sporting facilities might be used only in that season), days of a week (schools or offices normally are not used during weekends), and time of day (schools and offices—heat demand only during the daytime, and hospitals—continuously).

Solar collectors may be installed directly on a building, both in case of a public building and a residential one. Figure 7.2 shows a solar collector installation on a hospital roof. Solar collector systems used for heating

FIGURE 7.2   Solar collectors installed on a hospital roof in Konstancin, Poland.

residential or public buildings are usually of low- or medium-temperature type (heating fluid temperature in the heating system may be around 30 °C, but also 50—70 °C or even 70—90 °C).

Residential areas, city districts, complexes of public buildings, shopping malls, or entertainment facilities may be supplied with heat from centralized solar district heating systems. Those systems, except for solar energy, may also utilize peak heat sources based on renewables (e.g. biomass boilers), or fossil fuels (e.g. gas boilers), or based on low temperature heat sources (air, ground, water) utilized via heat pumps. In such systems, the total area of solar collectors may be from several 100 to more than 10,000 square meters. Large solar systems coupled with peak heat sources are usually of medium temperature type (heating fluid medium temperature may be around 50—90°). Figure 7.3 shows such a solar collector heating system in a centralized solar heating plant in Graz, Austria.

High-output solar systems may also be used in industrial applications, including process heat generation. It may be expected that solar energy technologies enabling the generation of process heat at higher temperature of 100—400 °C or even more will be gradually developed [5]. This development will be accompanied by an increased interest in solar cooling systems that require considerably higher temperatures than traditional heating technologies.

Combi-plus-systems (i.e., systems for DHW + space heating + cooling) must generate working fluid at temperatures of at least 80 °C to drive a cooling cycle. The exact requirement depends on the utilized heat-driven sorption cooling technology. In consequence, the type of solar collector (flat-plate, evacuated tube, concentrating) must be selected according to the utilized sorption process (different absorption and adsorption cooling technologies) [4—8].

FIGURE 7.3   Solar collectors in a centralized solar heating plant in Graz, Austria.

Solar heating systems have been developed very dynamically all over the world in recent years. Nevertheless, most of the solutions used are working for individual heating purposes of the user, supplying heat to a local network, and covering their own needs of an energy producer, who is also a consumer of this energy at the same time [9]. They are still installed mainly at single-family houses. However, medium-scale systems with solar collector areas of several 100 square meters are also becoming increasingly popular on larger residential and public buildings. There are some big scale centralized solar heating plants though.

In Europe, the most popular collector type is a flat-plate solar collector with selective absorber coating. However, in a global scale, evacuated tube collectors are dominating, mainly because of applications in Asia. In Australia, the United States, and some European countries, swimming pool absorbers are also popular.

In Central Europe and other higher latitude countries, solar thermal systems are coupled with auxiliary heat sources, which may be either renewable or conventional. System type, its complexity, and control system must be adapted to the type of consumer and heating requirements. Higher temperature requirements and heat loads require more advanced technical solutions, which are also more expensive. Therefore, for low heat loads and low temperature requirements, relatively simple and cheap solutions should be used. Complexity of a system increases not only along with thermal requirements, but also if its reliability under any conditions is demanded.

It needs to be emphasized that market penetration of such new technologies is influenced by support mechanisms enacted at specific countries [5,9]. Of course, awareness of "solar energy technologies" of the construction market decision-makers, construction and installation companies, and developers is crucial.

7.2 TYPES, FUNCTIONS AND OPERATION OF ACTIVE SOLAR HEATING SYSTEMS

7.2.1 Main Types of Active Solar Systems

A number of different active solar thermal technologies have been developed. Differences are mostly related to the heat consumption conditions determined by the type of heat consumer [10–27]. An active solar heating system is a system that converts solar energy into useful heat in an active way, utilizing solar collectors. A task of an active solar system is to collect solar energy, convert it into heat, store the heat, and then to supply it to a consumer. Operation of active systems is possible thanks to mechanical devices, which enforce the circulation of a working fluid. Those devices may be circulation pumps (in liquid heating systems) and fans (in air heating systems).

Active solar heating systems may be classified according to various different criteria, like essential characteristics, function, method of solar energy utilization, design, method of cooperation with other devices, and systems or manufacturing technology. The classification of solar active heating systems is usually based on the following features and properties.

- Type of working medium, which may be:
  - liquid (i.e., water or an antifreeze mixture, depending on the climate);
  - fluid used in both liquid and gaseous phase (e.g., refrigerant, thermal oil);
  - air.
- Function to be performed by the system:
  - DHW heating (water heating);
  - space (room) heating;
  - cooling/air conditioning.

  A solar heating system may be single- or multifunctional. In case of a single-function system, it only covers one of the possible heating functions (just mentioned). Multifunction systems may fulfill more than one function, in the following different ways:
  - common system for all functions (usually for space heating and DHW), which means that the main part of the system, including the heat storage tank is common for all functions;
  - separated system, with separate subsystems for individual functions of a system: DHW and space heating installation have separate storage tanks, auxiliary heaters, and different operating parameters;
  - combined system, where considerable part of the solar system and the main storage tank (e.g., buffer storage) are common for all functions, but there are also certain parts dedicated for specific functions (e.g., additional storage tanks or auxiliary heaters). Modern solar heating systems are of this type.
- Method of utilizing solar energy:
  - direct, where energy collected by solar collectors is supplied directly to consumers without intermediate circuits. In practice, this principle may only be used in open (flow-type) water heating systems in agriculture and in the simplest thermosiphon DHW systems;
  - indirect, where more than one circuit exists between the solar collector and the user. This always involves a closed solar collector loop and another open loop if a system supplies only DHW; or a closed loop if a system supplies only space heating. Other closed loops with their own equipment for enforcing working fluid flow (circulating pumps or fans), which are interconnected through heat exchangers. Loops may be equipped with storage tanks, auxiliary heaters, or other heating devices (including heat pumps) with their instrumentation and control systems;

- Number of heat sources used within the heating system:
  - single-source systems, which have no heat sources except for solar collectors. Examples of such systems include thermosiphon installations used for DHW heating used in very good insolation conditions. In higher latitude countries, such systems may be used seasonally (during summer), when high irradiance enables covering the demand for DHW and possibly swimming pool water totally;
  - dual-source systems (bivalent or hybrid), used primarily for DHW, although space heating applications are also possible. Except for solar energy, they also use another auxiliary heat source, usually conventional, especially in the form of various electric heaters (e.g., electric rod heater in a storage tank);
  - multisource systems, which are solar systems with multiple auxiliary (peak) heat sources, mostly of the combi type. Typical solutions involve cooperation between the solar system and a heat pump (e.g., ground heat pump) and an auxiliary conventional heater for peak loads, typically an electric heater or a gas/oil boiler.
- System manufacturing method, including:
  - factory-made solar heating systems, which are single products offered under one trade name, sold as complete ready-to-install kits with fixed configuration. This approach is used for DHW systems in the form of:
    - combined collector-storage systems;
    - thermosiphon systems;
    - systems with forced circulation- solar active heating systems;
  - custom-built solar heating systems, are either uniquely built or assembled from standard assortment of components. Subtypes include:
    - large custom-built systems for DHW and/or space (room) heating, designed by HVAC engineers and other experts for a specific application;
    - small custom-built systems for DHW generation and/or space (room) heating, offered by a supplier company, marketed by it and described in a so-called assortment file. They are assembled from standardized components and configured in a way described by the supplier's technical documentation.

As already mentioned, solar systems may supply heat at different temperature levels. Required temperature depends on the type and characteristic of a heat consumer. Technical solution and type of equipment are selected accordingly. Currently, solar heating systems used in buildings are primarily low-temperature systems with heating medium with temperature within 20−60 °C range. The lowest temperature is required for the simplest applications, like agricultural water heating (e.g., for plant watering) and for outdoor swimming pools. There are other examples of low-temperature solar heating systems.

A solar thermal system coupled with a heat pump in series, so-called series solar assisted heat pump (SAHP) system [26] has low temperature requirements. Working medium circulated in the collector loop with a temperature around 20 °C is used as a heat source for a heat pump. When a heat pump is coupled with a solar system in series, it means that after being preheated in a solar collector, the working fluid transfers the heat to a heat pump evaporator, directly or through a heat exchanger installed in the solar storage tank. A heat pump may utilize two renewable energy sources; then it is called double source or bivalent heat pump. When two heat sources are connected in series then the working fluid circulating in a ground heat exchanger is preheated there, then it is heated further in a collector loop of a solar system, and eventually enters the heat pump evaporator. In a parallel connection, a heat pump uses a ground and solar storage tank as alternative heat sources, depending on which source temperature is higher in a given moment. Seasonal underground thermal energy storage (UTES) systems may also be applied. Solar thermal collectors are used to load the ground during summer; then, in winter, the heat stored in the ground is extracted to provide heat to the heat pump evaporator.

Currently, used space (room) heating systems involve floor or wall heating, where the heating medium does not (and may not) have excessively high temperature, above 40 °C. In DHW systems, water temperature up to maximum 55 °C is required. Only in solar cooling and air conditioning systems, where solar collectors are cooperating with sorption cooling devices, temperatures of 80 °C or even much higher may be needed. Of course, in every solar thermal system, momentary temperature of a working fluid at a collector outlet may be higher than the value required to meet the thermal requirements of a consumer. This is beneficial for the heat stored in a storage tank and it does not mean that this heat is used directly (and immediately) at the same moment when it is provided by solar collectors. Usually, heat used to drive chilling devices or heat pumps is supplied from the storage tank.

In case of application of solar space heating systems in existing buildings, replacement of existing heating system with a low-temperature one (e.g., floor heating) is often impossible. In such a situation, the existing system equipped with traditional radiators designed for temperatures of 90/70 °C may be adapted for new conditions (i.e., for cooperation with a new low-temperature heat source). To achieve this, the number of radiators (their heat exchange surface) needs to be increased to enable the supply of the required heat of lower temperature (of the working fluid).

As it was noted before, depending on the function of the system, its design, and configuration may be more or less complex. Depending on the user-specific heat load pattern and required output, certain devices and auxiliaries (except for solar collectors) may be required to ensure operation of the entire system and covering heat load. Depending on the complexity of a system, its elements may include some (or all) of the following:

- solar collectors;
- supply and return pipelines;
- circulation pumps in forced circulation loops;
- heat exchangers separating closed loops in a system;
- storage tanks used to accumulate heat supplied by solar collectors or other sources;
- safety devices preventing excessive increase of pressure or temperature (e.g., expansion vessels, safety valves, vents);
- instrumentation and control systems for individual loops and devices (including temperature sensors, thermostats, and control valves, etc.);
- auxiliary conventional heating devices or systems, which may support the solar system continuously or provide a backup. These may include electric heaters or boilers running on gas, oil, or biomass;
- auxiliary equipment enabling utilization of low-temperature heat from a solar source or another renewable heat source through a heat pump;
- sorption type chillers used in solar cooling and air-conditioning systems.

Essential indoor components of a combi-type solar heating system (i.e., excluding collectors) cooperating with a ground heat pump are shown in Figure 7.4.

FIGURE 7.4   Solar combi heating system (indoor part, i.e., excluding solar collectors) cooperation with a ground heat pump.

7.2.2 Basic Configurations of Active Solar Systems

As already mentioned, the design of a solar heating system may be more or less complicated depending on its function. The most common domestic hot water systems are technically mature. Figure 7.5 shows a typical active solar liquid heating system used for DHW heating and cooperating with an auxiliary conventional heater. This system is used to cover the demand of the owner. The collector loop is closed. In a climate characterized by low ambient temperature, low solar irradiance, and frequent overcast, it is recommended to use low-flow collectors. In such climatic conditions for all-year applications, an antifreeze mixture (glycol—water solution) is used. The collector loop and heat storage tank are separated by a heat exchanger. The heat is stored in water, and storage is accomplished by using the thermal capacity of water. It is a short-term thermal storage system.

Typically for solar DHW systems, it is recommended to use storage tanks with a volume of $50-100$ L/1 m^2 of solar collector aperture area, depending on insolation conditions, collector type, and heat consumer requirements, including demand pattern. Insufficient storage capacity in reference to the installed collector area (i.e., their ability to collect solar energy), may cause problems in various buildings depending on their sizes, demand, and heating purposes. It mainly results from a failure not to take into account the real character of a building's use and, in consequence, not expected heat consumption distribution. Insufficient storage capacity may cause the temperature of the working medium in the collector and water in the storage tank to reach the maximum permissible level of temperature.

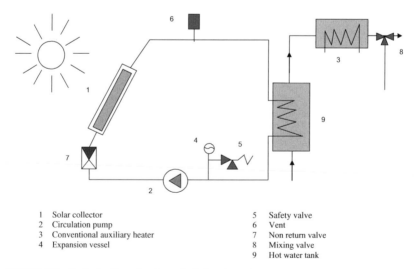

1	Solar collector	5	Safety valve
2	Circulation pump	6	Vent
3	Conventional auxiliary heater	7	Non return valve
4	Expansion vessel	8	Mixing valve
		9	Hot water tank

FIGURE 7.5 Typical active solar domestic hot water system.

The danger of an increase of water temperature in a storage tank above the upper limit is particularly visible in summer, when solar irradiation is high and heat consumption is low. To prevent this, in case of thermal systems with flat-plate collectors, reverse flow may be used during the nights. The working medium transports heat accumulated in the storage tank back to the collectors. As the ambient temperature is much below the temperature of the supplied liquid, the collector acts as a typical heat exchanger and releases heat to the surroundings. This results with a drop of temperature in the storage tank. In case of combi-systems in buildings, which are unoccupied for some time (e.g., school is closed, family is on holidays, etc.), excess heat may be used by a floor heating system. Another solution is planning additional summertime heat consumers (e.g., heating of swimming pool water, or storing heat in an underground thermal energy storage system). In addition, if there is the need to provide building cooling, then solar cooling can be applied. However, it requires more complex solutions and advanced technology.

In case of larger systems, direct cooling of the fluid circulated in the collector loop can be provided. It may be achieved by feeding the fluid into a roof-mounted chiller. The chiller releases excess heat directly to the surroundings, and already cooled down liquid returns to the storage tank. A similar effect may be achieved by coupling the collector loop to a cold water circuit via an indoor heat exchanger, but this requires considerable consumption of cold water. Working fluid may be also cooled down directly in the storage tank by a controlled supply of cold water. But, in every case, those methods lead to irreversible loss of collected heat, which might be otherwise used for some practical purpose.

Figure 7.6 shows a solar combi DHW + space heating system. Working fluid circulates in a closed collector loop between the collectors and the solar buffer tank. The solar buffer tank is integrated with a complex heating system through the heat exchanger in a main storage tank. The other heat source, such as a boiler, is used and also supplies heat to the main storage. Depending on the insolation conditions and heat consumption, the other source works as a base load, peaking, or auxiliary heater.

If water is used as collector working fluid, there is a hazard of water freezing in the collector loop or storage tank. This may lead to a pressure buildup in a system. Formed ice might block water flow in the loop. Therefore, not only an excessively high (near boiling point) temperature of the working and storage medium, but also an excessively low temperature level (close to freezing point) may cause system damage. Solar heating systems are designed to be protected from excessively high temperature and unplanned pressure. This protection is ensured by an expansion vessel, safety valve, and vents. Necessity of providing protection against overpressure applies not only to the collector loop, but also to the storage tank and its heat exchanger. According to relevant standards [17], these components should be designed for a pressure 1.5 times higher than the normal working pressure (specified by the supplier of

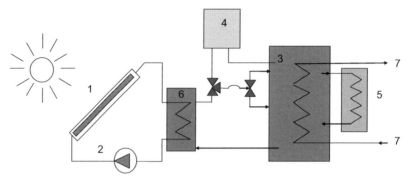

1	Solar collector	5	Space heating system
2	Circulation pump	6	Main storage tank
3	Buffer storage tank with heat exchanger	7	Hot water system
4	Other heat source, e.g., the boiler		

FIGURE 7.6 Solar combi heating system: DHW + space heating with another heat source.

collectors and entire system). Certain measures must also be taken to prevent excessively hot water from injuring system users. This may involve solutions for dumping certain amounts of DHW in case of overheating. Also, mixing valves may be used to mix hot DHW with a certain amount of cold water before a point of consumption.

In case of solar systems that are not meant to be operated during the summer (e.g., in school buildings), collectors may be covered by special reflective covers for this period. Another solution is dumping collected heat to the surroundings (e.g., through a roof-mounted cooling system). However, the best solution is to find a heat user (e.g., a swimming pool, outdoor showers, or other summer water entertainment facilities, like aqua summer centers).

It should be noted that in some solar heating system technologies, so-called drainback or draindown systems are used to prevent freezing. If there is a danger of freezing (because of low ambient air temperature), then working fluid is dumped from solar collectors and flows into a drainback chamber. It may be used again when the ambient temperature goes back to a safe level. In case of a draindown design, the liquid is entirely removed from a system through a vacuum relief valve and cannot be used again, it is permanently lost. This solution is used in lower latitude countries, where freezing conditions are very unlikely.

Thermal efficiency of a solar system depends on a number of individual factors related to the design and construction of solar collectors and the entire solar system, installation method of individual components and their interconnections, local insolation conditions, mode of operation of the solar system, and a method of its cooperation with auxiliary or peak heating device/ system. Essential issue, particularly important in higher latitude countries, is the location of the solar collectors. The collectors should be inclined and oriented in a way of maximizing the exposure to solar radiation during required periods. Recommendations for solar receiver orientation and inclination are given in Chapter 3. Specific recommendation will be different for systems intended to be used during different seasons. It needs to be noted that, in some technologies of evacuated tube collectors, inclination of tube elements is irrelevant, because the design of the absorber plates located within tubes enables their rotation to the required position.

In case of buildings in towns or cities, the investor is often forced to install solar collectors on a roof slope that is not oriented directly toward the south or even on the building envelope (e.g., because of structural considerations of a roof or local insolation conditions such as shading). It is important to choose a location that will not be shaded. The problem of shading by neighboring buildings in case of urban areas may considerably limit opportunities for using active solar systems and heavily influence their location. It is recommended to use collectors integrated with roof slopes. This facilitates collector installation on new buildings; in existing structures, such collectors may be installed during roof renovation. BIST collectors are recommended because integration into the building structure reduces heat losses from a building. Roof insulation also becomes a thermal insulation for the collectors. Another benefit is improved visual and esthetic quality of a building, as additional roof-mounted collector support structures are often unacceptable to architects and users alike. In case of sufficient insolation conditions, determined by inclination, orientation, and absence of shading, even the large part of a roof surface may be used to install solar collectors, or even the whole roof can be built as a solar collector, in case of large custom-built solar heating systems.

Many existing buildings have flat horizontal roofs. In such a case, solar collectors are installed on supporting structures, as shown in Figure 7.7. Using support structures on roofs or terraces is not favorable, not only for esthetic reasons, but also due to structural considerations. This solution requires performing additional analyses of roof structural strength because of increased mechanical load. Wind force may generate major local mechanical stress. Snow falling on a roof can form snowdrifts between solar collector rows enlarging the total weight of the system and total load on the roof increases. Moreover, heat losses from solar collectors standing alone are much bigger than from ones integrated into the building's structure.

High heat losses from solar collectors installed on supporting structures are due to the fact that they are exposed to the ambient air from all sides. They are continuously swept by wind and this increases losses because of forced convection. Moreover, already mentioned snowdrifts between rows of solar collectors can restrict solar radiation access. Another disadvantage is the possibility of the mutual shading of collector rows, if such shading is not carefully analyzed in the planning and design of the system. The cost of this type of system is high because of the additional cost of the supporting structure

FIGURE 7.7 Solar collectors installed on supporting structures.

itself. Because of thermal and structural reasons, it is recommended to carry out thermal refurbishment of a building by creating an additional story over a flat roof in the form of an attic covered by a sloped roof with roof-integrated collectors. Unfortunately, although this solution is energy-efficient, usually it is also expensive.

Another specific factor that influences the efficiency of a system is the design and location of the storage tank and its auxiliaries. The tank design itself influences the way it operates. The effective operation of most heat storage tanks is based on the heat stratification effect, where the coldest water is located at the bottom and the temperature rises along with tank height, reaching the maximum at its very top. A hot water outlet is located at the top part of the tank. Heat stratification is beneficial, because it allows the user to choose the outlet nozzle position to cover the specific temperature heating needs. In case of heating loads with different temperature requirements, it is possible to install several outlet nozzles. For example, the outlet for a low-temperature floor (wall) heating system (30–40 °C) could be located not far above the middle of the tank. The outlet for the DHW system (40–55 °C) is located closer to the top. In case of using the high-temperature room heating circuit (60–80 °C) with radiators, this outlet should be located at the top. If the high-temperature heating loop (with radiators) is not used, as is increasingly frequent, then the hot water outlet for the DHW system is at the top. Some tank types are equipped with integrated heat exchangers for DHW heating and utilize the thermosiphon effect within the tank. Full efficiency of DHW heating is achieved if the temperature at the tank bottom does not fall below a certain value and the temperature difference between the top and bottom layers is also maintained at the required level.

Two heat exchangers are often placed in a heat storage tank. The lower one is supplied with liquid from the solar collector loop, and the higher one with water from another heat source supplying water at the higher temperature. In some systems, two storage tanks may also be used: one as a buffer tank for water heated by solar energy, and another with heat exchangers supplying heat from the buffer storage and an auxiliary source.

Two storage tanks are quite frequently used in solar combi-systems with a low-temperature room heating system. There is another configuration shown in Figure 7.8. One tank is used as the main heat storage, charged from solar collectors, and another low-temperature heat source (e.g., ground via a heat pump). This tank accumulates low-temperature heat that is used to cover the demand of the low-temperature room (floor) heating system. If the temperature gets high enough, the heat may also be used for DHW applications. Another tank is used as an additional DHW heater if the temperature in the first tank

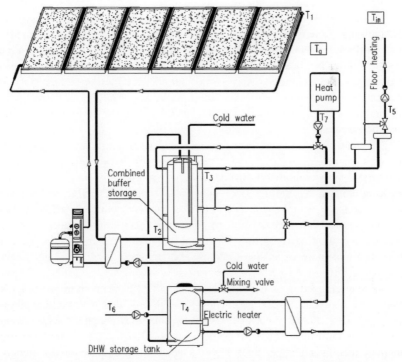

FIGURE 7.8 Solar assisted heat pump system in a single-family house in Warsaw, Poland [26].

and amount of heat stored there is insufficient. A schematic diagram of such a system is shown in Figure 7.8.

Another solution, now rarely used, is installation of a stirrer in the bottom part of the tank in order to ensure that water temperature is always the same in the entire tank. Of course, in this case, for the same energy stored in a tank, the temperature of the water supplied to a consumer is lower than in case of heat stratification, when the heating water can be taken out of the tank from the top. The solution with the continuous mixing is recommended for seasonal systems used in summertime, as it involves a higher mass flow in a collector.

Storage tanks of active solar thermal systems are installed indoors, while in thermosiphon systems used in lower latitude countries, heat storage tanks are placed outdoors (at the appropriate height above the collector). In case of a danger of freezing in the tank, a small conventional heater may be installed inside, but its operation distorts (or even prevents) the natural thermal diffusion, which is a driver of the natural convection process in this system.

It is important to ensure reduction of heat losses from a storage tank. This requires providing it with appropriate thermal insulation and using the optimal shape. Also, the thought-through design of connections is needed, as an incorrect design may lead to heat losses because of the natural convection within the pipes. To avoid this phenomenon, pipes should turn straight downward close to a tank nozzle.

In countries where ambient air temperatures drop below the freezing point during winter, a nonfreezing mixture (e.g., glycol-water solution) is used as a heat working fluid in the collector loop of all-year systems. In liquid heating systems, water is a heat storage medium. Except for water, storage tanks may contain phase-change materials (PCMs) that enhance the thermal capacity of the storage system. In case of using multiple PCMs of different phase change temperatures at different places (height of the tank), the stratification effect may be enhanced.

Depending on its function, the system may be more or less complicated. Systems for heating water in low-temperature applications (e.g., certain holiday resort applications) with low demand have a relatively simple design and working modes. Swimming pool heating systems and DHW systems are somewhat more complicated and require automatic control systems. Room heating systems are even more complex and also their control is more complicated. Increased heating requirements lead to more technologically advanced and more expensive systems, although cost increase is not linear here.

Solar space heating systems (Figures 7.6 and 7.8) are much more complex than DHW systems (Figure 7.5). They must be provided with auxiliary systems, including auxiliary peak heat sources. They may cooperate with boilers running on solid (e.g., biomass) or fluid (gas, oil) fuels. Very often they are coupled with heat pumps. Such systems need to be equipped with modern automatic control systems for all their elements, integrated with interior

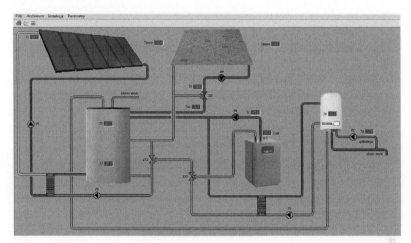

FIGURE 7.9 Print screen from software visualizing the operation of a solar heating system coupled with a ground heat pump in winter in a single-family house, in Warsaw, Poland.

microclimate sensors and controllers for heated rooms. Figure 7.9 shows a print screen from computer software used to visualize the operation of a solar heating system coupled with a ground heat pump.

Solar heating systems, especially those of high demand (e.g., applied for residential multifamily buildings), may be coupled with seasonal UTES systems [26]. Figure 7.10 shows a concept of a solar heating system with a seasonal underground thermal energy storage and a heat pump.

Automatic control of a combi heating system usually operates according to certain preprogrammed priorities for utilization of individual available heat sources. Usually, solar collectors are used as the primary source of energy for preheating and supplying heat at the lowest temperature level. Then another heat source, conventional or renewable, is used to provide more heat to the heating medium, up to the required temperature level. If a solar system is coupled with a heat pump and it is expected that in extreme weather conditions the heating output of both sources combined could be insufficient, electric heaters or conventional boilers are used as additional peak load heat sources.

In modern buildings, it is recommended to use in forced ventilation systems the heat recovery from exhausted ventilation air. Management and operation of recuperative heat exchangers should be combined with space (room) heating system (radiators, floor, and/or wall heating) supplied by renewable or conventional peak source. Different configurations of utilized heat sources require adequate settings of the automatic control system. As a result, in low-energy buildings, the demand for room heating energy is low, typically around $20-30 \text{ kWh/m}^2$ (of heated area) per year.

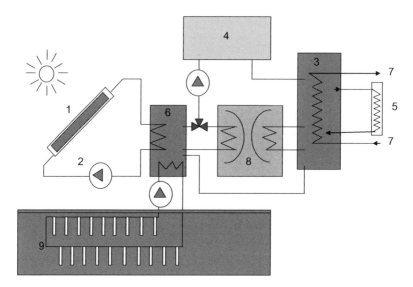

1 Solar collector
2 Circulation pump
3 Storage tank with heat exchanger
4 Peak conventional heater, e.g., boiler

5 Space heating system
6 Solar buffer storage coupled
 ground storage heat exchangers
7 Hot tap water system
8 Heat pump
9 Ground heat exchanger

FIGURE 7.10 Conceptual schematic of a solar heating system with a seasonal ground heat storage and a heat pump.

Solutions used in centralized solar heating systems differ somewhat from those utilized in small and medium-scale applications (i.e., in individual residential or public buildings). In case of centralized systems, solar collector arrays (fields) are developed. In case of housing estates or towns, the total collector aperture area may range from several 100 to more than 10,000 square meters. In such a case, the collector area per capita (at $0.5-1.0$ m^2 per person, depending on climate conditions and weather) is lower than in case of individual single-family houses (where it usually exceeds 1 m^2 per person).

7.3 SOLAR COLLECTORS

7.3.1 Main Types of Solar Collectors

A solar collector, which acts as a solar energy receiver, is the main component of any active solar thermal system. The main task of the solar collector is to collect solar energy, then convert it into the useful heat that is transferred by the working medium (e.g., water or antifreeze mixture), or air from collectors to the storage tank. Development of collector technologies has been aimed at maximizing solar radiation gains and reducing heat losses (i.e., increasing overall efficiency of converting solar radiation into heat and utilizing it). This required development of increasingly advanced and efficient solutions [11−15,21−36].

Main types of solar collectors used in buildings are listed according to their complexity and advancement of technology used:

- low-temperature unglazed collectors, so-called swimming pool absorbers in the form of black absorbing plastic panels or strips;
- flat-plate liquid solar thermal collectors;
- evacuated collectors, especially of the tube type, including evacuated collectors with compound parabolic concentrators.

Some other collectors are also used for a building's application, but they are much less popular. These include:

- storage collectors, which combine functions of a solar thermal collector and a heat storage tank;
- bifacial parabolic collectors, optionally with a transparent insulation cover;
- concentrating collectors, etc.

Active solar thermal systems in buildings may also use flat-plate or other forms of solar air collectors connected to the ventilation system or to a pebble bed heat storage system. Air heating collectors in the form of screen walls, solar walls, and simple or complex envelope systems are more frequently encountered in passive systems, as described in Chapter 5.

Flat-plate liquid solar collectors used to be the most popular type in the world. Recently, the number of installed evacuated collectors has exceeded that of the flat-plate type. This is primarily thanks to market introduction of Chinese evacuated solar collector technologies. In the simplest applications, the least technically advanced collectors without any selective coatings and transparent covers are still used; most frequently in the form of swimming pool absorbers. However, in most applications, it is essential to achieve higher temperature of a working fluid than it is required for swimming pool water heating. Then, flat-plate collectors with highly selective coatings, highly transparent solar glass (or Teflon foil) covers, or evacuated collectors are used. It is essential to choose a collector type suitable for certain applications and anticipated heat loads.

7.3.2 Flat-Plate Collector

The essential element of a flat-plate solar collector is the absorber. The absorber absorbs solar radiation and converts it into heat that is transferred through its surface into liquid or air, which flows under the entire absorber plate or in tubes A cross-section of a sample of flat-plate liquid solar collector is shown in Figure 7.11.

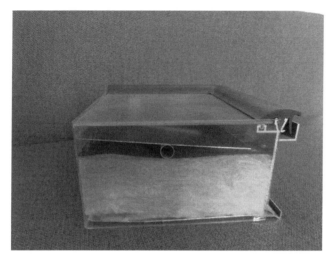

FIGURE 7.11   A cross-section of a sample of flat-plate liquid solar collector.

The main important features of a flat-plate liquid solar collector are:

- high solar radiation transmittance of the cover;
- high selectivity of the absorber plate coating;
- high thermal conductivity of the absorber plate;
- good thermal contact between the pipes (coils) and the plate;
- water-proof casing (collector box);
- good thermal insulation below the absorber plate and at the sides.

Flat-plate solar collectors have the largest absorption area of all collector types, and, therefore, also the biggest heat transfer area, which means that they are characterized by large heat losses. Depending on insolation conditions and photo-thermal conversion efficiency, temperature of the working fluid flowing through the collector piping usually may vary from several to 90 °C. Flat-plate collectors have a relatively simple structure. Their price is comparatively low because of low production cost achieved recently. Most flat-plate collectors are factory made assemblies manufactured on automated production lines. There are also custom-made designs, primarily in case of large surface collectors installed as roof-integrated elements. These are finally incorporated in the roof structure on the site.

Flat-plate collector technology is the oldest, and therefore also most mature. Quality standards for such collectors, their manufacturing, laboratory, and field testing have been developed and used for years [17–20]. Thanks to the experience gained during the long operation of such collectors and related solar heating systems, standards are updated. For example, it turned out that misting (water vapor formation) is a common problem arising during operation (mist in a collector is shown in Figure 7.12). In order to reduce the potential

FIGURE 7.12   Solar collector with accumulated mist.

for development of such defects, standard methods for testing waterproofness of solar collectors have been upgraded recently and a new regulation for standard testing will come in force soon.

Although flat-plate collectors are considered a technically mature solution, it does not mean that efficiency improvements and cost reductions are no longer pursued. New materials are researched, especially plastics, for manufacturing both transparent covers and absorber plates. Also, studies concerning collector integration with the building envelope are being developed.

The main components of a flat-plate liquid solar thermal collector (see Figure 7.11) are:

- transparent cover;
- absorber plate;
- piping (risers);
- thermal insulation;
- casing.

A transparent cover may be made of regular, hardened, or special, so-called solar glass, or of transparent plastics, including transparent insulation materials. The main function of the cover is to protect the absorber surface from adverse influence of the environment (low temperature, wind, snow, hail), while enabling high solar radiation transmittance. The cover protects the absorber from heat losses to the environment (on the face side) and mechanical damage. Therefore, it should have sufficient mechanical and thermal strength. It should be resistant to both low and high temperatures, as it must not deteriorate or sustain damage in case of being irradiated for a long time by solar radiation of high irradiance (like during multiple long sunny days in case of no heat use). The maximum operating cover temperature depends on the material it is made of. For standard glass, it is 160 °C, for solar glass 200 °C, for a double polycarbonate sheet 140 °C, for plastics in the form of Tedlar foil 120 °C, for polyester foils 170 °C, and for Teflon foils 200 °C.

The number of covers is theoretically determined by the climate conditions and function played by a collector. Multiple covers, of course, reduce heat losses, but also attenuate solar radiation incident on the absorber surface. In higher latitude countries during winter, solar irradiance is low anyway and installation of a double cover would reduce it even further. For that reason, practically all the flat-plate collectors, regardless of latitude, have single covers. High transmittance for solar radiation and low for thermal radiation are the most significant features of the cover. Typically, construction glass has high transmittance, which (for the normal angle of incidence) is equal to 81% for beam radiation and 74% for diffuse radiation. Solar glass has even better values of 87% and 80%, respectively. For polycarbonate sheets, the values are 77% and 83%, and for Tedlar foil 90% and 90%, respectively [21].

It is essential to ensure high-quality finishing of external and internal cover surfaces. In the case of the external surface, reflectivity should be reduced, while for the internal side it should be enhanced. Solar radiation upon passing through the cover may be reflected from the absorber and fall back on the internal cover surface. It is beneficial if this surface is highly reflective and returns the radiation back toward the absorber plate. The internal cover surface may be rough; often it has a structure of tiny pyramids.

Transparent insulation may be used as a material for making covers. It ensures high solar radiation transmittance, while greatly reducing heat losses. In solar collectors with a transparent insulation cover, a kind of effective heat trap can be seen between the cover and the absorber. It allows reaching the high temperature of the absorber plate, and therefore also of the working medium in the collector loop. This allows using water as a working fluid in all-year systems, even in high-latitude countries. However, collectors with transparent insulation covers also have a flaw: during hot sunny days, when no heat is consumed, excessive heating of a cover causes a danger of its potential partial meltdown. Honeycomb transparent insulations are made of poly-carbonates for which the maximum permissible temperature is 140 °C.

Influence of wind, strength of solar collector structure, and other load as-pects limit dimensions of collector aperture. A collector with a glass cover usually has an absorbing surface area of up to 3 m^2 of aperture (because of the glass weight). Collectors with acrylic covers may reach 6 m^2 of aperture area (there are supports between the cover and absorber). The distance between the cover and absorber is typically 15–25 mm (reducing convective heat transfer in the air gap). The interest in large-size collectors with large absorbing sur-faces increases, because they may be integrated into the building envelopes in a relatively simple and visually interesting manner. For that reason, many research studies focused on new technologies of large surface solar collectors that are developed.

The absorber is the most important part of a collector. In flat-plate liquid solar thermal collectors, the absorber plates are mainly made of aluminum, but also of copper, steel, or plastics. High heat conductivity might make copper a preferred solution, but, unfortunately, this material is quite expensive. The main function of the absorber is absorbing solar energy and transferring (through conduction) the collected heat to the heat transporting medium that flows through pipes in an absorber. In order to increase the ability of absorbing solar radiation and reduce heat losses because of thermal radiation, an absorber is coated with black selective coating (with high absorptance for shortwave solar radiation and low emissivity for longwave thermal radiation). It is important to ensure that the collected heat is efficiently transferred from the absorber plate to the working medium. Plastics, which tend to be cheaper, are characterized by lower heat conductivity than metals. The most popular absorber technologies are:

- panel absorber (see Figure 7.13). It is the absorber plate with straight vertical pipes bonded or pressed to metal plate (sheet, slightly finned or not) and horizontal headers; pipes are usually made of copper and the plate of aluminum;
- strip absorber (see Figure 7.14). It is formed by roll-bonding of very thin copper tubes in aluminum plates. Thin metal (copper) tube is placed between two aluminum sheets, forming a vertical pipeline for the working fluid;
- serpentine absorber (serpentine riser) with serpentine coils attached to a flat metal plate;
- vertical harp-type absorber, with vertical parallel pipes and horizontal headers, attached to a flat metal sheet;
- horizontal harp-type absorber, with horizontal parallel pipes with vertical headers, attached to a flat metal sheet. It is less popular than the vertical variation and should not be used in thermosiphon systems;
- absorber with a single surface flow channel consisting of two metal sheets thermally welded at the edges (or bonded), sometimes referred to as a sandwich structure.

In multiduct absorbers, flow resistance must be equal in every duct. Absorber dimensions are very important parameters that influence flawless operation of the system. Too short an absorber would not ensure the required flow of the working fluid, while an excessively long design would reduce the collector's efficiency. Typically, a single absorber plate has a width of 1.0–1.2 m and height of 1.8–2.2 m. The absorber design, especially its connections with collector casing, should permit thermal expansion.

Thermal insulation protects the absorber against heat losses to the environment from the bottom and sides. In case of building integrated solar

FIGURE 7.13   A sample of a panel absorber (back side).

systems, this role is played by the building envelope, usually insulation of a roof slope. Typical insulating materials of low heat conductivity are used (e.g., polyurethane foam or mineral wool). It is important to use a low-density material to prevent excessive increase of collector weight. At the same time, insulation should be characterized by other features and parameters typical for a good insulating material: water proof, including low water absorbability, resistance to weather conditions, and sufficient mechanical strength. Thickness

FIGURE 7.14   A sample of a strip absorber.

of insulation is typically 6—10 cm. Its heat transfer coefficient should not exceed 0.3 W/(m^2 K).

The casing of a typical collector is made of aluminum profiles. The task of the casing is stiffening the collector structure and combining it within a single module (device) of all the components: absorber, cover, and thermal insulation. The collector casing protects the device from the bottom and sides from direct influence of environmental conditions. In case of a collector integrated with a roof design, installation is much simpler, and its casing becomes an integral element of a roof structure. Integrated systems are mostly used in countries with poorer insolation conditions, where relatively large collector surface areas are needed, thus increasing the investment cost. However, integration of a collector with the roof structure allows reducing the total cost of both the roof and the collector system, than in case of both those elements being built independently. Integrated systems are also preferred because of the simplified installation process. In case of a single roof-integrated collector, it is installed analogically to a roof window. In case of large surface systems, the collectors may form a part or even the whole slope of a roof made according to the rules of roofing.

One of the important physical parameters of a collector is its transmission-absorption factor ($\tau\alpha$), which describes its optical efficiency. It holistically expresses the influence of the optical phenomena occurring within the cover-absorber assembly on the energy absorption within the absorber.

Another important parameter, which describes thermal performance of a collector, is its heat loss coefficient U_L. This coefficient holistically describes heat losses because of radiation, convection, and conduction. In order to maximize collector efficiency, heat losses must be minimized, which means that the thermal resistance of elements separating the essential part of a collector (absorber) from the environment should be possibly high. As for the absorbing surface itself, heat transfer processes within its structure should be intensified. It is important that the thermal resistance across the plate (between the pipes) should be as low as possible. The same principle applies to the resistance between the plate and each pipe. It is also important to ensure the already mentioned selective character of the absorber coating for solar and thermal radiation.

7.3.3 Evacuated Collector

Evacuated collectors are primarily built as tubular collectors. However, flat-plate evacuated collectors are also made and they usually are based on very large negative pressure, so-called imperfect vacuum. This type of collector is intended for large sized applications for large roof areas. Special equipment for evacuating air from the space between the cover and absorber is used at the installation site to obtain the vacuum inside.

The vast majority of evacuated collectors are of the tubular type. A single collector module consists of a series of mutually parallel tubular elements. A single module has at least six tubes, typically more than 10, sometimes even 20 or 30. There are two essential types of evacuated tubular collectors:

- single glass envelope with a metal fin in the vacuum;
- double glass envelope with the vacuum between the glass tubes, known as all glass Dewar type or glass thermos flask.

In the first case, a metal (usually copper) absorber plate of high selectivity is placed in a glass vacuum tube. A tubular channel for working fluid (it may be water) is bonded to the absorber plate. Material used as the absorber coating is characterized by very high solar absorptance of some 0.93 and very low thermal radiation emissivity, typically around 0.03. In collectors of this type, the vacuum conditions are ensured by seals between the glass tube and the metal fin with the fluid channel. The main disadvantage of this design is a threat of losing vacuum because of leaks at the glass-metal seal (because of different thermal expansion characteristics of both materials).

The concept of a single evacuated tube with a metal fin inside may be implemented in several design solutions. Most typical are solutions:

- with a metal U-tube (channel) attached to the absorber plate;
- with a single straight metal tube running along the absorber plate, with special bellows at the ends to compensate for differences between the thermal expansion of the hot metal tube and the cold glass tube of the envelope;
- with a single straight metal tube running along the absorber plate, which acts as a heat pipe. It is a very similar solution to the one described above, although, in this case, bellows may be omitted as the thermal expansion difference is small because of a small temperature difference between contacting materials.

Connections (seals) between glass tubes and metal absorber may cause leaks and problems with maintaining vacuum (or negative pressure) for a longer period. For that reason, after several years of operation, such collectors might require another air evacuation procedure to restore their efficiency.

In collectors with a single glass tube and metal absorber, the tubes are made of sodium-calcium glass. Glass temperature remains close to the ambient air temperature so no thermal shock may occur except for the area around the metal absorber element penetrations (sodium glass is susceptible to a thermal shock). That is why the careful choice of materials for seals is so important. Loss of vacuum may also result from degasification of selective coating layers on the metal fin. For that reason, so-called gas getters need to be installed in each evacuated tube.

The main cost of the evacuated tubular collector is the cost of the evacuated tube itself. In order to reduce this cost, development focuses on improving the optical and thermal efficiency of a single tube. Such improvement would enable covering the same heat load with fewer tubes. Improvement of optical performance is achieved by coating the bottom surface of the internal tube with a reflective layer. Thanks to this solution solar radiation not directly absorbed in the absorber is reflected by the bottom reflective layer and falls on the absorber plate from the bottom. Therefore, ultimately more energy is collected by the absorber.

The diameter of a glass tube is normally within the range of 7.5–30 cm (more often in the lower part of this interval). The typical length of a tube is 1.8 m, although lengths of up to 2.4 m are available on the market.

Because of the problems related to maintaining vacuum in single evacuated tubes, double-glass tubes known as Dewar tubes have been introduced. Their design follows the same principle as a thermal flask (thermos): two concentric glass tubes are arranged coaxially (tube in a tube). On one end the tubes are curved and closed, and on the other they are thermally welded. Before the weld is sealed, air is evacuated from the space between the tubes to achieve vacuum. The diameter of the outer wall may be from 3.0 to 5.7 cm and lengths may reach 2 m. In this design, there are no metal-glass connections. The double-glass tube with vacuum between the tubes is the main part of a Dewar flask evacuated solar collector (a sample of a collector of this type is shown in Figure 7.15).

This type of solar collector does not need the vacuum space to be penetrated when heat is gained (i.e., when heat is gained thanks to photothermal conversion of solar radiation and it is extracted to the user). This ensures maintaining vacuum in the intratube space. The heat carrier is circulating

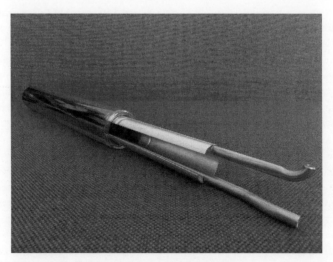

FIGURE 7.15   A sample of Dewar flask evacuated solar collector.

(directly or indirectly) in the inner space of the inner glass tube. Heat is extracted using one of the two following concepts.

- Wet tube. In this case, the working fluid is in direct contact with the inner glass tube and the heat is transferred directly. Certain disadvantage of this solution is that it cannot be drained of working medium. If one of the tubes is damaged, all liquid is lost, as no drains are present. This solution is commonly used in evacuated collectors manufactured in China [35].
- Dry tube. In this case, the working fluid does not contact the glass tube directly. Instead, it circulates through a metal U-tube or a straight pressurized tubular channel. In either case, the metal tube is connected to a metal fin. The metal fin usually has a form of cylinder (a tube) or nearly full cylinder (a part of a tube) placed within the glass tube. The heat is transferred from the inner glass tube to the fin and from the fin to the working fluid in the piping (e.g., U-tube) (see Figure 7.15). This requires ensuring good contact between the fin and the inner glass tube. The single tubular channel through which the working fluid is circulating may be a heat pipe.

Double-glass evacuated tubes may be made of soda-lime glass-like single tubes or of hard glass. It is important to ensure that the outer tube has possibly high solar radiation transmittance and, at the same time, possibly high resistance to mechanical damage (e.g., hail). The inner tube usually forms a part of the absorber. Its outer surface (i.e., on vacuum gap side) is coated with a highly selective layer (e.g., aluminum-nitrite). Absorbed solar radiation is converted into heat and conducted by the tube into the metal fin and from there to the working fluid in a pipe (U-tube).

Evacuated tubes are installed with spacing (i.e., the neighboring tubes do not touch each other). For that reason, an evacuated tube collector of the same dimensions as a flat-plate one has a smaller aperture area. In order to improve efficiency of evacuated tube collectors of either type, additional reflective surfaces are installed below the tubes in the bottom part of a collector. Without these elements, solar radiation would not be used efficiently because of gaps between the tubes. Reflective surfaces may have a form of flat diffusive reflectors or CPCs. Such surfaces allow concentrating reflected solar radiation on evacuated tubes instead of losing it. Also, internal CPC or circular reflectors (in case of fin absorbers) may be installed inside evacuated tubes.

The evacuated tubular collector with a heat pipe has a structure similar to a standard evacuated tubular design. The difference is that the working fluid is not typical antifreeze liquid, but a refrigerant, and additional heat gains because of phase change (condensation of refrigerant vapor) are utilized along with direct solar radiation gains. Heat pipes may be utilized in typical flat pressurized collectors, too.

The essential feature of solar collectors with heat pipes is their operation based on phase change. The refrigerant circulates in a close loop in piping of a solar collector. Piping installed on the absorber plate acts as an evaporator, while the piping connected to the main header pipe with heating fluid of the heating system (outside the absorber) acts as a condenser. Absorbed solar radiation increases the temperature of the refrigerant contained within the heat pipe. Refrigerants evaporate at relatively low temperature and then the vapor passes through a heat exchanger-condenser within the upper section of the collector. There the refrigerant transfers heat into the working fluid circulating in the loop between the collector and the heat storage tank and condenses in the process. Liquid refrigerant returns to the absorber-mounted heat pipe and the cycle is repeated. Collectors with heat pipes may operate efficiently even at low irradiance and low ambient air temperature and, therefore, they are recommended for applications in higher latitude countries.

Systems with heat pipe collectors are called indirect flow systems (one fluid circulated within the collector, another within the loop between the collector condenser and the storage tank). They are opposed to direct flow systems where the medium passing through the collector is supplied to the storage tank.

Evacuated tubular collectors have high efficiency at unfavorable ambient conditions (low ambient air temperature, low irradiance) thanks to considerable reduction of heat losses. Thanks to this feature evacuated collectors are recommended for application in solar space (room) heating systems, especially for medium-temperature solutions. However, there are some disadvantages of evacuated tubular collectors that can be seen in some specific conditions.

During winter, in case of heavy snowfall, there is a danger of snow lying on the collector surface for a longer time (even a few months), which makes the collector's operation impossible. In evacuated collectors, heat losses are significantly reduced. The temperature of the external surface of the evacuated tube is very close to the ambient temperature. As a result, during winter, at low ambient temperature, snow falling on the evacuated solar collectors remains there until it melts on its own. This phenomenon is rarely occurring on flat-plate collectors. In their case, when only conditions allowing for operation (sufficient irradiance) arise, the absorber temperature rises and so does the external cover temperature (because of heat losses) causing the snow to melt.

There is another problem, which may be noticed in case of operation of evacuated collectors. When solar irradiation is high and there is no use of heat converted from solar energy, then the temperature of the working fluid of the solar system could get very high. If this fluid is an antifreeze mixture not designed for high temperature ($>150\,°C$) operation, it may decompose (into water and glycol, as separate components). In this case, the working fluid loses its physical and thermal properties, especially as antifreeze mixture.

7.3.4 Other Collector Types

Low-temperature unglazed collectors are practically standalone absorbers without any transparent covers, casing or stiffening frames, and thermal insulation. They are made as flexible black pipes, panels (sheets), or in a form of strip collectors, usually of plastic. Samples of the panel and strip types are presented in Figure 7.16.

In a wide sheet (plate) of panel absorber there are many fluid passages, very close to each other, connected at the top and bottom by header pipes (absorber example in the left side in Figure 7.16). Designs may feature different pipe diameters and pitch, and also different structural materials (polypropylene, ethylene-propylene, thermoplastic elastomer). In strip un-glazed collectors, there is a very flexible extruded strip with fluid passages (absorber in the middle and in the right side in Figure 7.16). In most cases, such collectors are laid on roofs or directly on the ground, creating a kind of black flooring. Strip absorbers are cut to the required size at the site of installation, and then the header piping is connected to them. The total area can vary from 10 to 100 m². Other possibilities include tiles (of ethylene-propylene) or double-layer mats (of polypropylene or polyvinyl chloride) with the flow through the entire gap between the plates or through the embedded tubes. The special type of unglazed absorber made of hard plastic material may form sections of pathways, pavements, or even, with proper design, surface of pitches or courts for sports, etc. (absorber in the middle in Figure 7.16).

Unglazed collectors are used to heat a relatively large water mass with a required relatively small increase of water temperature. They are used for heating swimming pool water or water supplying underground thermal storage. They may also be used in agriculture.

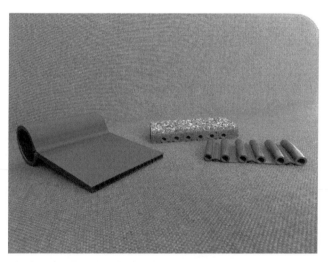

FIGURE 7.16   Samples of panel and strip types of unglazed collectors—solar absorbers.

Material used for these solar absorbers must be characterized by good solar absorptance and resistance to corrosion and weather conditions, including solar radiation, temperature variations (in range from $-20\,°C$ to $+70\,°C$), and humidity. Pool absorbers must be able to withstand anticipated mechanical loads if they are to be used as pavement of sport facilities, walkways, etc.

Pool absorbers are the simplest solar collectors: the cheapest and easiest to install, and also easy to maintain. As already mentioned, they are used in systems with seasonal ground energy storage. Energy collected by absorbers is transferred by a working medium to the ground (water can be used as a working fluid, but when a system is used all year, it must be an antifreeze mixture, and then usually regular glazed solar collectors are used). The working medium is circulating in the solar absorber—ground heat exchanger loop. In winter, energy accumulated in the ground is extracted and used for heating via a heat pump.

Unglazed collectors have nonselective coatings (their absorptance is equal to emissivity in the whole radiation spectrum), no front transparent cover and no bottom and side insulation. Therefore, their efficiency decreases relatively quickly with decreasing solar irradiation and ambient air temperature. However, for many simple applications with low heating demand, low required temperature, and short operating time mainly limited to summertime, unglazed collectors are a very good and effective solution.

Another collector type is a thermal storage collector. As the name implies, such a device combines functions of a solar collector and heat storage. The size of a storage collector is larger than of a typical flat-plate collector, especially in case of thickness. A storage tank is integrated with a solar collector and must be fitted inside its structure. This tank typically is in a form of a cylinder of a relatively large diameter (at least 30 cm) with black coating, which, at the same time, acts as an absorber. The casing of a storage collector is a box with a trapezoidal (or sometimes rectangular) cross-section. An appropriately bent aluminum (or other reflective material) panel is installed inside to act as a reflector. The black surface of the storage cylindrical tank absorbs solar energy. The reflector plate (usually curved, can be in parabolic shape) is reflecting toward the tank this part of solar radiation, which did not fall on the tank directly. Transparent glass or a plastic cover is also used as in the case of a typical solar collector. A storage collector is always bulky, with large thickness, and much heavier than a typical liquid heating flat-plate collector. This solution is usually installed on horizontal roofs able to carry sufficient load. Their transport and installation is cumbersome. Storage collectors are currently being researched and installed in Mediterranean countries, mainly in Greece and Cyprus. A good solution could be integrating them with the building structure. This would improve especially the esthetics of the collector and reduce heat losses from the casing.

Another collector type used mainly in lower latitude countries is a concentrating collector. It concentrates a beam of solar radiation directly on the energy receiver or a medium that carries heat to such a receiver. Depending on used technology (concentration method, material, structure, etc.), various concentration levels may be achieved. Concentrating collectors may be classified according to the beam concentration method or reflector type (with parabolic or parabolic dish shape) and lenticular. The main types of concentrating collectors are [38]:

- line-focus collector, which focuses solar radiation within one plane, creating a linear focus;
- parabolic-trough solar collector, which focuses solar radiation with a reflecting cylinder of parabolic cross-section;
- point-focus collector, which focuses solar radiation in a single point;
- parabolic-dish collector with a point focus and parabolic dish reflector;
- nonimaging collector, which concentrates radiation at a relatively small receiver without focusing it (i.e., without creating the Sun's projection on the receiver);
- CPC collector, which is a nonimaging type;
- faceted collector, which utilizes multiple flat reflectors to concentrate solar radiation on a small surface or along a specific strip;
- Fresnel collector, which uses Fresnel lens for focusing solar radiation on the receiver.

Detailed descriptions of physical fundamentals of concentrating collectors may be found in publications [21,22,39,41−43].

It should be noted that, according to the standard defining solar energy vocabulary [39], flat-plate collectors and evacuated tube collectors are also considered concentrating collectors if they are equipped with reflectors.

In order to effectively utilize concentrating collectors, it is recommended to collect and concentrate a direct radiation beam with a high-energy flux density. Therefore, incident solar radiation should have high irradiance and high share of beam component in global radiation. Except for ensuring good natural insolation conditions, the concentration effect may be enhanced by equipping concentrating collectors with the Sun tracking systems. Because of the required insolation conditions (i.e., high beam irradiance) at the current development stage, concentrating collectors are primarily used in low-latitude countries. However, there are research studies of the concentrating collectors to also be used more widely in higher latitude countries. The already mentioned CPC collector has relatively high efficiency and can operate in an effective way during the whole year.

A CPC collector is a nonimaging design that uses parabolic reflector segments to concentrate radiation. Its key advantage lies in the fact that parabolic segments reflect all radiation incidents on the aperture in a wide spectrum of the angles of incidence. Limits of the angle of incidence determine the acceptance angle of a collector. It should be added that the CPC name applies to many nonimaging collectors, even if their geometry is not parabolic.

In higher latitude countries, certain combinations of concentrating and flat-plate collectors are used. Those are so-called inverted flat-plate collectors, inclined inverse flat-plate collectors, and bifacial collectors. Those types are equipped with stationary concentrators.

The possibility of using flat-plate collectors at high temperatures thanks to reversing absorber plate configuration was researched by Kienzlen et al. in the late 1980s [37]. His team designed the first inverted flat-plate collector. In this solution incident, radiation falls on the concentrator, which redirects it toward the bottom of the absorber plate. This allows reducing heat losses by insulating the top part of the absorber. Convective movements within the air layer below the absorber are insignificant. However, optical efficiency of this design is lower than in the case of flat-plate collectors, because of reflector diffusion losses. Further improvement came in the early 1990s, when Geotzberger et al. proposed a bifacial collector concept [40]. This collector type was installed for the first time in the first so-called 100% solar building in Freiburg, Germany, which was also the first energy self-sufficient building. Further modifications of such collectors incorporating CPC technology were proposed in the mid-1990s by Eames and Norton [41]. Currently, there is big interest in technologies of high concentration solar collectors [42], as well as low concentration ratio solar collectors [43]. The research studies on low concentration ratio solar technologies are developing mainly in European countries, where their application potential is the largest, also in high-latitude countries.

Bifacial collector, as its name suggests, has a solar absorber that is exposed to solar radiation from both sides, directly from the top, and also from the bottom side of the absorber. Beneath the transparent insulation cover only half of the collector front is filled with double-sided absorber plates with pipes for working fluid. Below the absorber parabolic (sometimes cylindrical) a highly reflective surface is installed. Solar radiation that penetrates into the collector through its transparent cover is partially directly absorbed by the absorber (on its outer side). The other part is absorbed on the inner part of the absorber, after being reflected from the reflective surface of the internal parabolic part of the collector. Because solar energy is absorbed from both sides, the thermal efficiency of this collector type is enhanced.

Determining usefulness of collectors for practical applications requires knowledge of their thermal characteristic in various conditions of solar energy collection and energy supply to consumers.

7.3.5 Thermal Characteristics of Solar Collectors

Analysis of solar collector performance involves many parameters and values that describe its properties. One of such values is the so-called transmission-absorption factor of a collector ($\tau\alpha$). It holistically describes the influence of the optical phenomena in the system: cover absorber. It represents a property

of a cover-absorber combination and shows the rate of energy absorbed by the
absorber in the total solar energy incident on the collector (cover) face. Typical
solar collector characteristics provide the estimated value of this factor. Heat
loss coefficient U_L (also designation k_{ef} is sometimes used) is another
simplified parameter used for practical purposes. This coefficient describes the
total heat losses because of radiation, convection, and conduction from the
whole solar collector. Knowing the solar irradiance G_s incident on a tilted
collector surface, and using the averaged values of $(\tau\alpha)$ factor and U_L coefficient (e.g., from thermal characteristics), which are characteristic parameters
of a specific collector describing its ability to collect solar energy, the useful
energy gained from a collector Q_u can be determined using the Hottel-Whillier
equation:

$$\dot{Q}_u = A_c F_R \left[(\tau\alpha) G_s - U_L \left(T_{f,i} - T_a \right) \right] \tag{7.1}$$

where:

G_s, solar irradiance incident on a tilted collector surface (W/m^2);
Q_u, useful energy rate (W);
$(\tau\alpha)$, transmission-absorption factor;
U_L, collector heat losses coefficient (W/(m^2 K))
A_c, collector surface (aperture) area (m^2);
F_R, collector heat removal factor;
$T_{f,i}$, fluid inlet temperature (K);
T_a, ambient air temperature (K).

The effectiveness of a collector depends on its design, construction, materials used, and operating conditions. Effectiveness is expressed by collector
efficiency: at a specific time (momentary) and averaged over a long time (e.g.,
annual). Efficiency η at a specific time expresses the ratio between the useful
energy rate Q_u gained from a solar collector and the solar irradiance G_s
incident on its aperture surface A_c at this specific time:

$$\eta(t) = \frac{\dot{Q}_u(t)}{A_c G_s(t)} \tag{7.2}$$

Using Eqns (7.1) and (7.2), efficiency at a specific time may be expressed as:

$$\eta = \frac{Q_u}{G_s} = F_R \left[(\tau\alpha) - U_L \left(\frac{T_{f,i} - T_a}{I_s} \right) \right] = F_R(\tau\alpha) - F_R U_L \left(\frac{T_{f,i} - T_a}{G_s} \right) \tag{7.3}$$

Obtained solar collector efficiency Eqn (7.3) takes into account:

- optical properties, described by the factor $(\tau\alpha)$;
- design features, described by factors F_R and U_L;
- operation conditions, described by the fluid inlet temperature $T_{f,i}$;
- ambient conditions, described by irradiance G_s and ambient air temperature T_a.

In Eqn (7.3), the right side has two clearly distinguishable parts. The first of them describes how well the collector is absorbing solar energy. The second shows how the heat gained is lost. Increase of the first value and reduction of the second lead to improved efficiency. Heat loss coefficient U_L is considerably higher for flat-plate collectors than for evacuated tubular collectors (Table 7.1). For this reason, it is so important for flat-plate collectors to operate in the temperature range close to the ambient air temperature. When analyzing Eqn (7.3), it can be also seen that the most beneficial for thermal efficiency of solar collector, especially the flat plate one, would be the ambient air temperature higher than the working fluid temperature at the solar collector inlet.

The form of Eqn (7.3) for solar collector efficiency in a specific time is the most frequently used in research studies, including simulation analyses. The thermal characteristics of a solar collector presents the efficiency η as a function of a so-called reduced temperature, defined as:

$$t_f^* = \frac{\Delta T_{f,a}}{G_s} = \left(\frac{T_f - T_a}{G_s}\right) \tag{7.4}$$

The reduced temperature can be expressed as a ratio of the difference ΔT_f between the fluid inlet temperature T_i (in the case of Eqn (7.3)) and ambient temperature T_a to the solar irradiance G_s incident at the considered time on a tilted surface (i.e., $\eta = f(T_i - T_a)/G_s$, as presented in Figure 7.17).

The point where the efficiency curve intersects the axis of the ordinates determines value $F_R(\tau\alpha)$, which is approximately equal to the optical efficiency of the collector (more precisely the estimated optical efficiency is multiplied by heat removal factor). The higher the intersection point is located on the axis of ordinates, the better the optical performance of a collector is. In

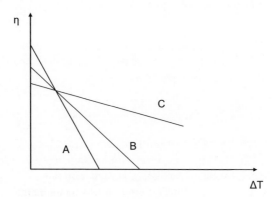

FIGURE 7.17   Exemplary characteristics of flat-plate solar collectors (A) unglazed collector; (B) flat-plate collector with single cover; (C) evacuated collector.

case of the simplest unglazed collectors, the optical performance is high, as transmittance is equal to 1 (there are no losses because of reflection or absorption in the cover, because there is no cover), and the transmission-absorptance factor is equal to the absorptance of the unglazed absorber alone. The slope of a line tangent to the curve at the intersection with the horizontal axis is equal to the $F_R U_L$ product and is a measure of heat losses resulting from the collector design and the construction materials used. Smaller slope means smaller losses.

Figure 7.17 shows that in case of the simplest collector form (A) (i.e., unglazed solar collector) the (absorber) efficiency curve drops sharply, although this design has the highest optical efficiency. For a collector based on the most advanced technologies (i.e., evacuated collector) (C) the curve is less steep. This allows reaching high fluid temperature, while preserving high efficiency even at unfavorable insolation conditions, and low ambient air temperatures. However, optical efficiency of such a design is lowest. It is primarily because of the fact that the distance that the solar radiation needs to pass within the transparent cover is longer than in a flat-plate collector (as the cover has a form of a transparent ring, and depending where (i.e., at which place of the evacuated tube), solar radiation is incident, the path through this transparent ring of the tube (at the same angle of incidence) may differ).

It should be noted that the Hottel-Whiller-Bliss equation may be expressed in different forms (i.e., for various temperature differences). As a result, the thermal efficiency (in time) formula takes different forms. Solar collector manufacturers define the reduced temperature as a ratio of the difference between the average working fluid temperature T_f in a collector loop and the ambient air temperature T_a to solar irradiance. Then, the efficiency of solar collector is expressed as:

$$\eta = \frac{Q_u}{G_s} = F'\left[(\tau\alpha) - U_L\left(\frac{\overline{T}_f - T_a}{G_s}\right)\right] = F'(\tau\alpha) - F'U_L\left(\frac{\overline{T}_f - T_a}{G_s}\right) \quad (7.5)$$

Equation (7.5) describes efficiency in a function of parameter F', which is the effectiveness of a collector as a heat exchanger. Equation (7.5) may also be expressed in a simplified way using reduced temperature concept (see Eqn (7.3)):

$$\eta = F'(\tau\alpha) - F'U_L t_f^* \quad (7.6)$$

Solar collector manufacturers use Eqn (7.5) or (7.6), because the value of F' is higher than the value Fr. As result, the point where the efficiency curve intersects the axis of ordinates, which represents estimate of optical efficiency ($F'(\tau\alpha)$) is now higher (than for Fr($\tau\alpha$)), and this looks better to a customer.

Also important is the fact that, in reality, heat loss at factor U_L is not a constant value. Instead, it depends on the temperatures of the absorber plate (working fluid) and ambient air. It may be assumed that this factor is a linear function of the temperature difference $(T_f - T_a)$, which may be expressed as:

$$U_L = a + b \left(T_f - T_a \right) \tag{7.8}$$

In this case, the linear efficiency equation transforms into the following nonlinear parabolic function:

$$\eta = \frac{Q_u}{G_s} = F'(\tau\alpha) - a_1 F' \left(\frac{T_i - T_a}{G_s} \right) - b_1 F' \left(\frac{(T_i - T_a)^2}{G_s} \right)$$

$$= a_{01} - a_{11} \left(\frac{T_i - T_a}{G_s} \right) - b_{11} \left(\frac{(T_i - T_a)^2}{G_s} \right) \tag{7.9}$$

where:

$a_{01} = F'(\tau\alpha)$;
$a_{11} = a_1 F'$;
$b_{11} = b_1 F'$.

Coefficient a_{01} represents the estimated optical efficiency, a_{11} represents the estimated linear heat losses, and b_{11} is the temperature dependence of the heat loss. Coefficients a_{01}, a_{11}, and b_{11}, of the nonlinear thermal characteristics of a solar collector, Eqn (7.9), are determined during solar collector testing, by adjusting second degree polynomials of the temperature difference to the measured efficiency in time using the least squares method.

Intersection between the efficiency curve (created either by linear or nonlinear approximation) and the axis of abscissae (Figures 7.17 and 7.18) determines the maximum achievable working fluid temperature (average temperature T_f or inlet temperature T_i), at which the collector efficiency drops to zero. At zero efficiency, there is no fluid flow through a collector, heat losses are equal to solar gains, and the maximum collector temperature, so-called stagnation temperature, may be determined. Value $\Delta T_{f,a}/G_s$ for $\eta = 0$ may be read from a graphic interpretation of the collector thermal characteristics. Stagnation temperature T_{max} may also be calculated using the efficiency equation. Thus, this temperature is expressed as:

$$T_{max} = G_s \left(\frac{\Delta T_{f,a}}{G_s} \right)_{\eta=0} + T_a \tag{7.10}$$

Stagnation temperature is a function of solar irradiance and ambient air temperature. According to the international standard ISO 9806-2 in European countries, stagnation temperature for flat-plate collectors is around 150–180 °C, while for evacuated tube collectors it is higher than 350 °C.

Another important parameter that needs to be determined is when the thermal characteristic is created, is the collector time constant. The time

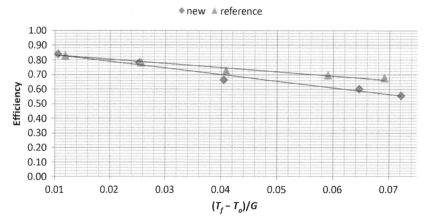

FIGURE 7.18 Flat-plate solar collector efficiency characteristics determined during tests under solar simulator at an indoor laboratory, Institute of Heat Engineering, Warsaw University of Technology.

constant needs to be found to identify a time after a change of certain selected parameter (e.g., irradiance G) after which the influence of the collector thermal capacity on its balance becomes negligible.

Long-term thermal efficiency of a solar collector is determined by the equation analogical to Eqn (7.2), as a ratio of useful energy gained from a solar collector in a certain period of time (day, month, year), to solar irradiation of the front surface of the solar collector in the time under consideration. Apart from thermal characteristics, also flow characteristics of solar collectors are determined. Useful energy gained from the solar collector is then expressed as a function of the mass flow of the working fluid and its temperature rise in the collector. Table 7.1 presents some exemplary parameters and coefficients of the nonlinear thermal characteristics of some selected liquid solar thermal collector types.

The presented discussion applies to the general issues of solar collector design and construction. However, there are also other important considerations that may affect collector effectiveness. One of them is the location of the collectors. The influence of the location (tilt, orientation) is related to the solar radiation availability, described by the solar irradiance G_s incident on the given surface. The highest irradiance in an all-year cycle is achieved for surfaces oriented toward the south. The recommended tilt depends on the latitude and is particularly important for higher latitude countries. Unfortunately, the collector location often results from the orientation and slope of an existing roof. It is therefore important to consider the collector installation and location during the design stage of a new building.

TABLE 7.1 Exemplary Parameters of Basic Types of Liquid Heating Collectors

| Collector Type | Optical Efficiency, (%) | Indicators of Heat Loss | | The Maximum Pressure (bar) | The Maximum Stagnation Temperature (°C) |
		a_1 (W/m² K)	a_2 (W/m² K²)		
Flat, selective	82	3.37	0.017	6	211
Vacuum tube	79	1.71	0.009	6	300
Vacuum tube with heat pipe	72	1.28	0.009	6	150

7.4 APPLICATION OF PHOTOVOLTAICS IN BUILDINGS

7.4.1 Physical Fundamentals of the Internal Photovoltaic Effect

Photoelectric phenomena involve the change of electrical conductivity, creation of electromotive force, or electron emission when a body is irradiated by electromagnetic waves. Those phenomena take place because of deviation from equilibrium between the electrons and the crystal structure.

The internal photoelectric effect is a change of energy distribution of electrons in solids and liquids because of the absorption of light. Because of the influence of light (electromagnetic) radiation, concentration of current carriers within the medium changes. In result, photoconductive phenomena and voltage are created [38,44−47]. Absorption of a single photon of light excites one electron. In the external photoelectric effect, electromagnetic radiation causes electron emission from a body into vacuum or another medium.

When a photon is absorbed, the energy and momentum conservation laws must be obeyed. The momentum of a photon is usually very low in comparison to the momentum of an electron. Therefore, during electron transition (lifting) caused by just one photon, electron momentum remains practically unchanged, it is the so-called straight transition. There are also oblique transitions, where both the momentum and energy of the electron are changed. Those transitions result from interactions of electrons and holes with thermal vibrations of the crystal structure. The width of the optical energy gaps between bands in which electron transitions take place may be determined experimentally by adjusting the energy of light photons and determining the minimum photon energy ($h\nu_{min}$) for which the internal photoelectric effect occurs. Determining the minimum photon energy, so-called activation energy, is important for finding the band gaps, especially the one between the valence band and the conduction band in semiconductors and dielectrics. The band gap energy at the temperature of 25 °C for selected semiconductors is: for crystalline silicon 1.12 eV, for amorphous silicon 1.75 eV, cadmium telluride 1.44 eV, and gallium arsenide 1.43 eV.

When a photoelectric effect occurs within a semiconductor, if the solar radiation photon has sufficiently high energy, it lifts an electron from the valence band to the conduction band. Electrons lifted from the valence band become free electrons inside the semiconductor. If the energy is too low, the lifted electron is unable to overcome the band gap and returns to the valence band. But if the photon energy is higher than the band gap energy, the electron is lifted and heat is generated. During the operation of a PV system, heat is always generated because the photon energy is hardly ever exactly equal to the band gap energy. This heat may be considered "waste energy" and is usually dissipated into the surroundings. Heat generation causes the temperature of the PV cell to grow, adversely affecting its efficiency.

The internal photoelectric effect is used in many semiconductor devices, including PV cells used for converting solar radiation into electricity [21,23,38,44−47]. In PV devices, according to [44] "photovoltaic phenomenon of creating electromotive force at a junction between two different semiconductors or between a semiconductor and metal occurring upon absorption of electromagnetic radiation quanta quantum within optical frequencies." As the irradiance grows, so does the electromotive force.

PV devices use mostly photoelectric effect at a p−n junction (i.e., at the interface between p-type and n-type semiconductors). The solar radiation incident on a semiconductor may create an electron−hole pair. In the case of pure monocrystalline silicon, there are four electrons on a valence shell. The elements (except for hydrogen and helium) have a tendency to fill their valence shells with eight electrons. In result, the crystal structure is formed through covalent bonds (with common electrons for neighboring atoms within the crystal). If the influence of light is terminated, holes and electrons recombine, and the atom returns to an unexcited state and releases previously absorbed energy. In order to avoid recombination, the electrons and holes should leave the semiconductor flowing to the external circuit.

A p−n junction is created of one or two differently doped semiconductors. Doping involves introducing another material, donor, or acceptor. In case of doping with acceptors (e.g., B, Al, Ga, In, and Tl; three valence electrons), the number of holes is higher than that of electrons and a p-type semiconductor is created. The most typical acceptor dopant for silicon is boron. Doping involves adding a single boron atom per 10,000,000 atoms of silicon. In case of doping with donors (e.g., Sb, P, As, and Bi; five valence electrons), one free electron remains and an n-type semiconductor is created. The most popular donor dopant for silicon is phosphorus; single atom is doped per 1000 atoms of silicon.

Because of solar radiation, electrons from a filled valence band are lifted to the conduction band, thus increasing the amount of electrons within this band. Holes appear in place of lifted electrons. They also take part in the conduction process. Free charge carriers—electrons and holes—move toward areas where their concentration is lower. In the area of the p—n junction, electrons and holes separate. The junction electric field directs the electrons toward the n-type semiconductor, and holes toward the p-type semiconductor. The carrier flow through the junction, which concentrates electrons in the n-type semiconductor and holes in the p-type semiconductor, ceases when the n area is charged negatively and the p area is charged positively. The created electric field prevents further charge concentration. The current starts to flow across the junction to balance the diffusive flow of carriers, it is so-called reverse current. When both currents equalize, dynamic equilibrium occurs. The potential difference present at this condition is known as the photoelectromotive force.

If both layers of the p- and n-type semiconductor are made of the same material (e.g., the one crystal of silicon), then the cell is called monocrystalline with p—n homojunction. In case of both layers of p- and n-type semiconductor being made of multicrystals of the same material (e.g., silicon), then a cell is called polycrystalline with p—n homojunction. If the p- and n-type layers are made of different materials, then a cell is called polycrystalline with p—n heterojunction. If a cell material does not have the crystalline structure, then the cell is called amorphous.

In order to describe the energy processes within semiconductors with p—n junctions, including PV cells, current—voltage characteristics of p—n junctions are used. They are usually expressed graphically in the form of curves characteristic for the given reverse current. If there is no light, current—voltage curve passes through the origin, $I_L = 0$ (reverse current). Intersections of the characteristic curve with the current axis I determine the values of the short-circuit current for $R_0 = 0$ (external resistance) and intersections with the voltage axis V stand for the open circuit and denote open-circuit voltage V_{oc} of the system. A characteristic feature of the internal photoelectric effect is strong dependence between the ratio of the short-circuit current to the incident light flux (spectral sensitivity) and spectral composition of the light. This means that the spectral curve has one clear maximum.

A thin sheet of semiconductor material into which p—n junction has been formed is usually protected by a transparent cover. There are two conductor contacts, which are connected to electrodes of the p and n areas of the junction. Because of the solar irradiation, the photoelectromotive force is formed across the p—n junction. It is (within certain limits) proportional to irradiance. If the electrodes of a PV cell are connected through an external load with certain resistance, a photoelectric current starts to flow through the circuit. The value of the current depends on the resistance of the external load.

Current-voltage characteristics of a typical PV cell (e.g., silicon cell) are made for solar radiation corresponding to the air mass of 1.5. Part of the curve corresponding to the negative irradiance values is a characteristic of an unlighted cell, which behaves like a semiconductor diode.

Cell efficiency is defined as a ratio of the electrical power density of a cell to the solar irradiance incident on the cell surface. Efficiency depends on the load (i.e., current and voltage in the external circuit). It needs to be noted that the cell output power drops if the temperature rises. Therefore, it is important to maintain possibly low cell (module) temperature during operation (as already mentioned). The output power grows with the increase of solar irradiance; this occurs by the increase of current (voltage increase is negligible).

PV cells are combined into modules and modules into panels (Figure 7.19 shows micro modules). In case of cells connected in parallel, the voltage value remains unchanged, while current values are added up. In case of series connection voltage values add up, while the current is not changed. A single cell usually has an output of $1-2$ W. The output of a module (with a surface area of $0.3-3.0$ m^2) is within the range of $30-300$ W. The electric current created by PV cells is direct current.

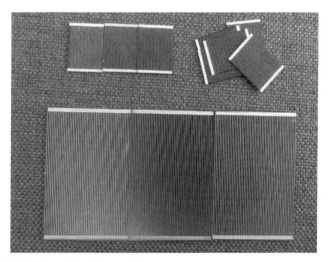

FIGURE 7.19 Photovoltaic micro modules.

7.4.2 Technologies of Solar PV Systems

Solar energy is absorbed by a cell made of semiconducting material, primarily within the visible spectrum. Irradiation of a cell connected to an external load generates a potential difference and electric current flows through a circuit. In such conditions, a cell works as a generator. Electricity is generated in a clean, silent, and reliable way. Silicon is the most popular material used for PV cells

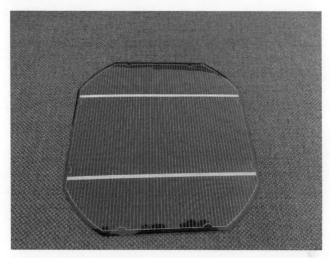

FIGURE 7.20   Monocrystalline PV cell.

production. It may be crystalline (monocrystalline or polycrystalline) or amorphous type. A monocrystalline cell is shown in Figure 7.20, some (broken) pieces of polycrystalline cells in Figure 7.21 and amorphous type in Figure 7.22. Cells may also be made of more complex semiconductors, like copper indium diselenide (CiS). This technology is still rarer than silicon-based PV cells.

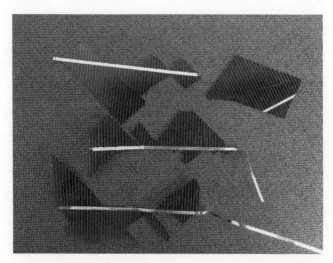

FIGURE 7.21   Pieces of polycrystalline PV cells.

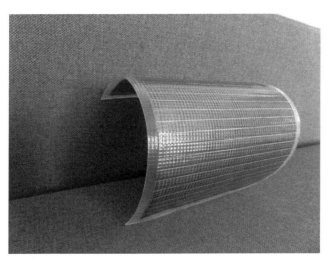

FIGURE 7.22   Amorphous silicon in red color.

Monocrystalline cells have the highest energy conversion efficiencies among all silicon cells, but their production is also most expensive. In laboratory conditions, efficiencies of individual cells approach 28%, but mass-produced cells available on the market have efficiency around 18%. Modules made of amorphous silicon are manufactured in different shapes, sizes, and colors, and their efficiency is lower than 10%. Thin-film cell production, except for already mentioned CiS also involves cadmium telluride (CdTe). European Union puts a lot of effort into research of innovative materials, both those mentioned above and also organic cells (Europe is currently a leader in development of this very advanced technology).

Currently, basic silicon technologies: monocrystalline, polycrystalline, and "ribbon", form 90% of the PV market. The remaining 10% are thin-film technologies using amorphous silicon, CdTe, and CiS.

Thin-film cells are produced by depositing extremely thin semiconductor layers on a substrate made of glass, steel, flexible steel, or plastic foil. The cost of the substrate is low. The cost of semiconductor material is also low in comparison to a typical silicon cell because of much lower material usage. Energy consumption for the production of thin film cells is also lower than in case of monocrystalline cells, because the production process does not require such high temperatures. Labor cost is lower and mass production becomes easier, as the thin film cells are produced as very large and integrated (series-connected) modules. In case of typical silicon technologies, it is necessary to connect cells individually and interconnect them with cables, which is more difficult and time consuming. However, the essential advantage of typical silicon-based PV modules is their higher efficiency of energy conversion from solar energy into electricity.

PV systems may be classified according to their operating regime and cooperation with the power grid as follows: autonomous (island) isolated from the grid, home-built and professional, distributed connected to the grid, and centralized connected to the grid.

Autonomous island systems only use electricity generated by connected PV cells. Such a system must be equipped with a battery with solar charge controllers. The battery is disconnected once fully charged or if there is a threat of discharging it too deeply. Small autonomous, standalone devices and systems are used to power telecommunication equipment, road sign lighting, information signs, parking meters, and other consumers located away from the power grid. An example of such a device is shown in Figure 7.23.

Systems utilized in buildings may be either operating in island mode (isolated) or connected to the grid. Island systems supply their own demand of the consumer and must be equipped with batteries to ensure covering electricity demand when the PV system is not operating, and with an automatic control for system operation and battery charging controller. If the energy consumers require an AC power supply, inverters need to be incorporated in the system. In case of systems connected to the grid, generated electricity may be fed directly into the traditional electrical installation of the building or into the public grid, also via inverters and converters ensuring the proper voltage level of supplied energy. Whenever panels are not generating the required electricity, the supply function is taken over by the grid.

FIGURE 7.23   A standalone PV system.

Small- and medium-scale systems in form of so-called BIPV should be increasingly popular in buildings. They can be used either for autonomous systems or connected to the grid. BIPV systems are integrated with the building envelope, primarily with its main facade and roof on the southern side. They can be treated as a kind of a building envelope ("construction") element, and electricity generator (device that simultaneously generates electricity). This dual nature of building elements, with coherency between architecture and energy production, will be standard in solar self-energy sufficient buildings. PV technologies should be recognized as interdisciplinary technologies, being a combination of architecture, civil engineering, electrical, and heat engineering.

7.4.3 Application of PV Systems in Buildings

As already mentioned, PV panels that form an element of a building envelope may also play other functions besides power generation. They are often used as shading components. Already presented Figure 3.16 and Figure 7.24 show examples of such applications. PV modules may be incorporated into transparent glazing of a building over a patio, winter garden, or an overhang. They also may be installed directly in windows (e.g., in the gap between glass panes) (Figure 7.24).

If the windows are of a box structure (old traditional type of double glazed windows with possibility to separate two glass panes when needed; e.g., for cleaning), additionally, air gap between the outer and inner glazing may be used as an area for ventilation air preheating. The solar radiation incident on a glazing with integrated PV panels is partially converted into electricity, while its part is transmitted into the room's interior. At the same time, absorption of solar radiation in glazing and PV conversion in PV cells lead to increased temperature of both cells and glazing. As a result, also the air in the gap is heated. Heated air with a lower density rises and may be utilized in ventilation systems as preheated air. In such a case, it is directed toward a ventilation-heating center. Constant inflow of air of lower temperature than the PV module temperature (from the bottom) ensures cooling PV panels, which improves their efficiency (which drops if the cell temperature grows). If there is no need to supply preheated air to the ventilating center, then it may be naturally transferred out of the building by opening ventilators in the upper part of the windows or glazed facade.

PV cells in form of transparent panels integrated with a building envelope are often used in low-energy buildings. Besides energy gains in the form of generated electricity, they also ensure light access and provide partial shading by reducing solar radiation access. They may also be incorporated into opaque elements of the envelope. They may constitute elements of roof slopes. Also, a hybrid solution in the form of PV/T thermal collectors may be

FIGURE 7.24   Example of PV acting as a shading element.

utilized [49,50]. In this case, the external roof surface is made of PV panels. The same surface is also an element of the solar collector. Beneath the PV panels there are air ducts to which ventilation air is supplied from the outside. The air is heated up in the ducts and naturally rises because of the reduced density, reaching a ventilation-heating center. Gained heat is used for room heating and at the same time PV panels are cooled, which prevents efficiency reduction.

PV panels may also be used as a kind of roof tiles. This kind of application is enabled by modern amorphous silicon technologies. Moreover, PV cells may come in different colors, including red, which makes them look quite similar to traditional ceramic tiles.

To be used in building structures, PV modules should be integrated with roof slopes and building facades, thus becoming incorporated into external elevations. In industrial and public buildings, roof-mounted systems are currently most widely used, but also facade-integrated panels are getting increasingly popular. To a large extent they complement typical building glazing. Especially large-surface PV modules are needed to make integration easier with facades.

Presented examples of PV panels and also solar collectors integrated with a building envelope clearly show that installation of solar systems in buildings should be taken into account already at the stage when architectural and civil engineering concepts are created. This allows integrating such elements with the envelope and internal structure of a new building in a planned and effective way during building construction. It is beneficial for both technical and esthetic reasons.

7.5 SOLAR COOLING

7.5.1 Possible Applications

Interest with solar cooling technologies has been growing recently. Periods of the highest cooling demand coincide in time with the highest solar irradiance. Solar energy is therefore a very coherent energy source for covering cooling and air conditioning needs. During the hottest period, chilling and air conditioning devices and systems operate at full output. Solar cooling technologies allow reducing energy consumption from conventional sources or even stopping such consumption.

Quite popular application of chilling systems in tropical areas is the use in medicine and food storage facilities. In case of solar technologies, this mainly involves PV solutions. More and more solar cooling systems are also used in various buildings: residential, public, commercial, and industrial. PV systems that drive compressor chillers, as well as solar collector systems that drive sorption chillers may be used as both base-load or peaking systems to support already existing devices and systems operating during the hottest periods [4−9].

It may be expected that solar cooling and air conditioning systems will find a lot of applications in buildings, both in tropical and moderate climate zones. Cooling systems are supposed to maintain required air temperature in rooms, while air conditioning systems are also responsible for keeping certain interior air humidity. Cooling systems primarily use generators of chilled water with a temperature of several degrees Celsius. Refrigerating systems are intended for maintaining low temperatures in industrial facilities (refrigeration halls, storage spaces). They use ice generators that reach temperatures below the freezing point.

In highly developed countries where high cooling demand is typical for summer, like the United States, Japan, or Australia, increased electricity demand during the afternoon is causing considerable problems in the power system [51]. In such countries, air conditioning systems are used not only at workplaces but also in residential buildings. In European countries, this kind of problem only starts to appear. Unfortunately, practically in all countries, regardless of latitude, cooling demand increase during summer months is observed.

7.5.2 Solar Cooling and Air Conditioning Technologies

A simplest way to utilize solar energy for cooling involves connecting a conventional vapor compression chiller to a PV power supply system. A more complex technology involves driving sorption chillers by heat generated at a solar collector system. Heat from collectors may also be used for desiccative and evaporative cooling (DEC) systems. Individual solar cooling technologies are described below.

A vapor compression chiller driven by direct current generated by PV modules is a relatively simple solution, but it requires using a battery. This limits the size and rating of such a system. As a result, this technology is used mainly in small scale applications, often in portable devices used to store medicines or food in tropical areas. Figure 7.25 shows a schematic of a vapor compression chiller driven by electricity generated by PV modules [52]. This electricity is used to power a direct current motor that drives the chiller compressor. The cycle of the refrigerant is a typical cycle of a vapor compression chiller.

The chiller shown in Figure 7.25 contains no battery. In fact, a battery with a solar charge controller may be placed between the PV modules and the motor. Such a solution enables charging the battery and consuming electricity by the compressor at the same time and enables operation of the compressor motor at steady power supply parameters. If a vapor compression chiller is an alternating current-driven design, then an inverter needs to be placed after the battery to convert direct current into alternating current. An alternating current chiller may also be powered directly from a power grid during periods when no solar radiation is available or when irradiance is too low to be used effectively. In this case, no battery is needed.

Except for PV technology, solar thermal collectors coupled to sorption chillers are also seen as a very promising solution. Sorption chillers are divided into absorption and adsorption technology. The general principle of operation is similar in both cases. Heat is removed from a colder heat sink and discharged to the warmer one thanks to a driving energy supplied in the form

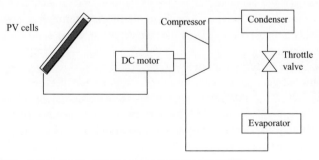

FIGURE 7.25   Schematic of a PV-driven vapor compression chiller.

of heat. In cooling technologies, a phenomenon is called absorption if the refrigerant is absorbed in a liquid called absorbent (refrigerant can be in gaseous and liquid phase), while adsorption involves refrigerant being absorbed in a solid medium called adsorbent (refrigerant can be also in gaseous and liquid phase). Refrigerant absorption or adsorption occurs at low pressure and low temperature. Reverse process (i.e., desorption of a rich solution) occurs at higher pressure and high temperature. In this way, the compression process is carried out. In a desorber (also called a generator), compressed refrigerant vapor is released. Then, just like in the case of an electrically-driven cooling device (refrigerator), the refrigerant vapor is flowing into the condenser, where it releases heat at a constant temperature. The created condensate flows through a throttling (expansion) valve, where it decompresses and cools down. Then, cooled down and decompressed refrigerant is supplied to the evaporator, where it takes heat from the low temperature heat source (cooling this source down). This causes the refrigerant to evaporate; vapor flows into absorber/adsorber, and the cycle is repeated.

Absorption chillers may operate continuously while adsorption devices operate in cycles. In absorption chillers, lithium bromide and water are the most typical working pairs, while in adsorption devices following pairs of adsorber-refrigerant are typically used: silica gel—water, activated carbon—ammonia, or methanol, and zeolites—ammonia. An example of a simple solar-driven adsorption chiller is shown in Figure 7.26.

In the adsorption chiller shown in Figure 7.26, the role of adsorber and desorber is played by solar collectors filled with adsorbent. During the day, solar collectors work as desorbers. They absorb solar radiation and heat the refrigerant that evaporates out of the adsorbent. Then a refrigerant vapor flows to the condenser. There it condenses, releases heat, and as a liquid enters the refrigerant tank. During the night, when there is no solar radiation, temperature and pressure in the collector are dropping. It may also happen during the day, if insolation conditions are changed and solar irradiance reduces considerably. If the collector pressure drops to a level below the evaporator pressure, then the throttling valve controlled by the pressure difference enables the flow of refrigerant into the evaporator. There it evaporates and collects heat from the low-temperature heat source (directly or indirectly, through a cooling medium), thus cooling it down. Then the refrigerant as vapor is absorbed by the adsorbent. When solar radiation appears again with sufficiently high irradiance, the collector turns back into the desorber.

All this means that in the considered system cooling takes place during nighttime (when the refrigerant evaporates in the evaporator). If there is no cooling demand during the night, the cooling energy must be stored for later use (e.g., during the daytime). It may be stored in ice blocks, hence the name of such equipment: ice generators.

Heat generated by solar collectors may also be used for desorption. In such a solution, the solar collector is no longer directly used as the adsorber and

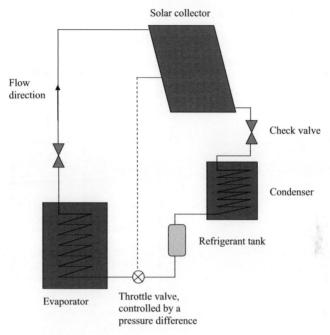

Solar collector

Flow
direction

Check valve

Condenser

Refrigerant tank

Evaporator

Throttle valve,
controlled by a
pressure difference

FIGURE 7.26 Schematic of a solar-driven adsorption chiller.

desorber. Instead, heat from solar collectors is supplied to an adsorption chiller to the adsorbate chamber through a heat exchanger. When most refrigerant is released from the adsorbate, the heat supply is closed and refrigerant vapor flows to the condenser. In the condenser it condensates—like before—and is directed to the evaporator. In order to increase the number of cycles during 1 day, a design with a cooling tower may be used. Cooling water from the cooling tower decreases the temperature in the chamber that contains the adsorbate. This enables supplying the refrigerant vapor returning from the evaporator back into the chamber and repeating the cycle in the same day. In case of small devices, the cooling tower may be replaced by normal heat exchangers, which releases the heat to the environment (ambient air, ground).

In order to ensure cooling continuity, it is possible to use two coupled chillers that work alternately. This idea is recommended for cooling/air conditioning systems. In this case, it is necessary to release the heat into the environment (e.g., via a cooling tower). Figure 7.27 shows a simplified diagram of such a system [52].

In case of an idea of solar adsorption system operation, presented in Figure 7.27, the solar collector does not also play a role of absorber or desorber, but just supplies hot water [52]. The working fluid of the collectors supplies heat to the absorption and desorption processes through a heat

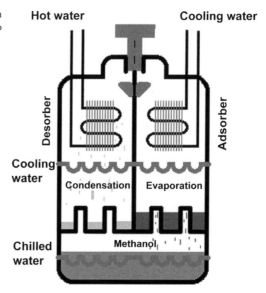

exchanger. Solar collectors may also play their normal heating function (e.g., for DHW). Two chillers are working alternately, with desorption occurring in one and adsorption in the other. Heat is supplied to one chamber and absorbed at the other one. Additionally, the desorbed refrigerant is immediately condensed and cooled down. The switch between the chillers occurs when most refrigerant in one chamber is adsorbed, while most in the other is condensed. The chamber revolves and another working cycle starts.

Operation of an absorption chiller is continuous. For this reason, such systems are often used for space (room) air conditioning. Figure 7.28 presents a schematic of an absorption chiller driven by energy generated by a solar collector system.

In the system shown in Figure 7.28, there are two connected but distinct cycles: the refrigerant cycle and absorbent cycle. In the absorber at low pressure, refrigerant vapor is absorbed by the absorber. This decreases the mass fraction of lithium bromide in the solution. Heat is removed from the absorber. Then a pump forces the flow of the enriched absorbent solution to the desorber where the pressure is higher. Heat gained from solar collectors is supplied to the desorber. Because of the increased temperature, the refrigerant is released from the absorbent. The mass fraction of lithium bromide rises in the desorber. The absorbent returns to the absorber, while the refrigerant vapor flows to the condenser. Pressure in the condenser is high and because of heat removal the refrigerant is condensed and directed through a throttling valve to the evaporator. The actual chilling effect occurs there. Refrigerant vapor at low

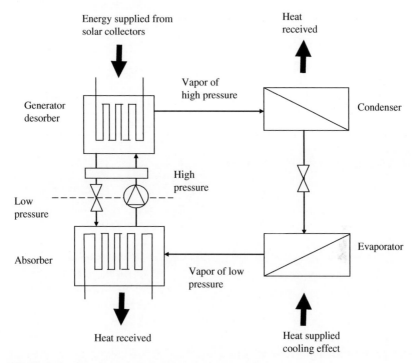

FIGURE 7.28   Schematic of a solar-driven absorption chiller.

pressure is then fed to the absorber closing the cycle. In real cycles, medium circulated in the absorber loop is always a solution, with changing refrigerant content.

The most popular working fluids used in absorption refrigerants are lithium bromide as an absorbent and water as a refrigerant (as in the example discussed above). Less popular are devices with ammonia—water solution—in this case water acts as absorbent. The advantage of this concept is the ability to reach temperatures below the freezing point in the evaporator. This is not possible in case of using water as a refrigerant. In ammonia-based solutions, a rectifier must be used to separate water vapor from ammonia vapor created in the desorber [55]. This problem does not occur in solutions based on lithium bromide and water.

In order to increase coefficient of performance (COP) of a system, a regenerative heat exchanger may be used. The exchanger is placed between the absorber and desorber in such a way that the heat removed from the poor solution is supplied to the rich one.

Modern solutions may also use combined cooling cycles. Heat recovered from the condenser is used for desorption (release) of more refrigerant. But in order to make the heat from condenser useful, the temperature in the first

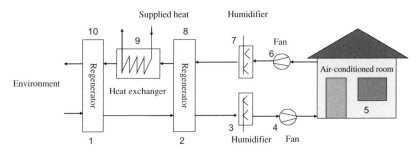

FIGURE 7.29 Schematic of a DEC system.

desorber must be higher, above 120 °C. Then the temperature in the other desorber may be around 90 °C and COP may exceed 1. This type of solution is known as a two-stage absorption chiller.

DEC systems are used for room air conditioning [54]. Operation of such systems is based on the effect of water evaporation in the air. A simplified schematic of such a system is shown in Figure 7.29. Air taken from the ambient atmosphere reaches desiccant rotor (regenerator, point 1) where the air exhausted from the room heats up the incoming flow. The inlet airflow is dried in the rotor thanks to hygroscopic adsorbents. The desiccant rotor is a rotating heat exchanger. Adsorbent saturated with humidity (from the inlet air) revolves together with the rotor. After rotating 180°, humid adsorbent is placed on the level of the exhaust air where it is regenerated. Warm air flowing from the room releases moisture from the adsorbent. The inlet air after the regenerator is directed to another regenerator (i.e., heat recovery heat exchanger; point 2) and there it is precooled. Then it proceeds to a humidifier (point 3), where it is humidified and cooled by water evaporation. Afterward, it is supplied (by a fan 4) to the room (5). The exhausted air (removed by a fan 6) is first supplied to a humidifier (7) where it is cooled down and humidified, and then is heated in the regenerator (8), and then in the heat exchanger (9), and finally releases its heat in the final regenerator (10). Then it is removed to the ambient atmosphere.

The considered system requires a supply of energy to heat the air in the exchanger. This energy may have a form of a medium-temperature heat (around 50 °C), and therefore it may be supplied from the simplest solar collector technologies. Disadvantages of this technology include high water consumption and difficulties with maintaining constant temperature of air supplied into the room. An advantage is easy maintenance of constant air humidity. It should be noted that in the areas where ambient air humidity is high, the drying flow may be unable to remove a sufficient amount of moisture from the ambient air, thus reducing the amount of water evaporated in the humidifier. As a result, the temperature of the air supplied to the rooms may fail to reach the required level. In such a situation, the temperature of regenerating air in the rotor must be increased. To solve the problem and maintain the temperature at the required level, while decreasing water consumption, installation of a vapor compression chiller may be used. In this case, heat from the collectors is only used for inlet air drying and heating. This idea permits only limited reduction of energy consumption in reference to a traditional compressor-driven cooling system. Air drying in DEC systems may be performed by solids: silica gel, lithium chloride salt, molecular network, synthetic polymer, etc., or liquids: lithium chloride, calcium chloride, and lithium bromide.

7.5.3 Development Prospects for Solar Cooling Technologies

Presented solar cooling technologies have both advantages and disadvantages. In case of coupling PVs to compressor-based chillers, it is possible to construct very compact and simple devices. Such technologies should be used in standalone (autonomous) systems of various purposes and in transport. As for the applications in residential and public buildings, an important role will be played by systems supplied with energy by solar collectors—sorption technologies and DEC. It should be noted that available technologies may be divided according to their task into air conditioning and cooling (in buildings) and refrigerating.

It should be emphasized that selecting the correct type of solar collectors for solar cooling and air conditioning technologies is essential. The highest coefficients of performance are reached by the systems where the driving heat is supplied at high temperature levels. Such a high temperature may be achieved in concentrating collectors, but to operate efficiently they need solar radiation of high irradiance and with a large share of beam radiation. In the areas where irradiance is lower (especially in case of beam component), it is better to use evacuated or flat-plate collectors. Unfortunately, these technologies, especially flat-plate collectors, provide output at much lower temperatures. Temperature ranges for various types of collectors used in cooling and air conditioning systems are:

- for flat collectors 55−85 °C;
- for evacuated collectors 75−150 °C;
- for single-axis concentrating collectors 120−250 °C.

Appropriate sorption cooling technologies may be assigned to specific collector types used to drive thermal cycles and related temperature levels [52−58]. The most important data concerning single- and two-stage absorption technology, their COPs, cooling powers, and applicable collector types are listed in Table 7.2.

Data presented in Table 7.2 shows that flat-plate collectors may be sufficient for driving adsorption systems if they operate in good insolation conditions. Such systems have already been put into operation in Europe for some

TABLE 7.2 Comparison of Main Parameters for Absorption and Adsorption Technologies

Type of Sorption Technology	Adsorption	Absorption	
		Single Effect Cycle	Double Effect Cycle
Mode of Operation	Discontinuous (Batch)	Continuous	
Range of temperature of heating fluid	55° ÷ 95 °C	70° ÷ 110 °C	110° ÷ 170 °C
Minimal cooling temperature	About −15°C	LiBr–water: 1 ÷ 2 °C Water–ammonia: −5° ÷ −10 °C	
Range of COP	0.4 ÷ 0.7	0.6 ÷ 0.8	0.8 ÷ 1.2
Type of solar collectors to be applied	Flat plate, vacuum tube, concentrated	Vacuum tube, concentrated	
Range of cooling capacity	50 ÷ 1000 kW	LiBr–water: 15 ÷ 6000 kW Water–ammonia: 10 ÷ 20 kW	

COP, coefficient of performance.

time, even on a larger scale. Figure 7.30 shows refrigerating devices with the output of 545 kW driven by flat-plate solar collectors with a surface area of 1579 m^2 Lisbon, Portugal.

As already mentioned, the main advantage of an absorption system in reference to an adsorption one lies in its ability to operate continuously. This removes a need for cooling energy storage. Absorption systems are also characterized by higher coefficients of performance. However, absorption chillers using lithium bromide and water may only be used for generating chilled water. Ice may only be produced by adsorption chillers or absorption devices using ammonia and water. So far, the latter technology only allows reaching low outputs. The advantage of absorption chillers is their compact size in comparison to adsorption equivalents. Absorption technology is also currently cheaper.

In case of large output installations, cooling towers or heat storage systems are needed for both technologies. Construction cost of these towers may be a major element in the total system cost. Disadvantage of cooling towers is their high water consumption and quite inconvenient operation, necessity to clean them, possibility of legionella contamination and vapor plume generation [57].

FIGURE 7.30 Cooling system driven by flat-plate solar collectors, commissioned in 2008, Lisbon, Portugal.

As already mentioned, air conditioning systems are supposed to maintain at a preset level not only air temperature but also its humidity. When considering air conditioning systems, especially DEC, technology needs to be considered.

Figure 7.31 presents the possibility of using different solar technologies to drive different cooling systems. Individual efficiencies are presented, depending on the specific technology utilized: sorption technologies are taken into account together with vapor compression chillers powered by electricity generated in PV cells. Model and assumptions were proposed by Kima and

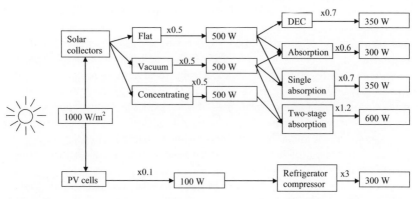

FIGURE 7.31 Energy efficiency of different solar technologies applied to drive different cooling systems. DEC, desiccative and evaporative cooling [52].

Infante Ferreira [52]. Even if their model is very simple and estimates are very rough, it still gives a clear picture of the possibilities of applying different solar cooling technologies at the present state of their development.

Figure 7.31 shows consecutive energy conversion processes from solar radiation energy to useful energy gained in the form of cooling (for different technologies). As it has been mentioned, it is of course greatly simplified. Identical insolation conditions described by irradiance of 1000 W/m^2 have been assumed for all technologies. The determined installed power of cooling systems are valid for operation of solar systems with collectors or PV panels with an area of 1 m^2. Efficiency of all collector types has been assumed as 50%, and of PV panels as 10%. Averaged COP values have been assumed for individual cooling technologies. Total efficiency of conversion of solar energy into cooling energy for specific technologies may be specified as a ratio of the final cooling system output to the solar irradiance incident on a surface of solar collectors or PV panels. Efficiency is the highest for an absorption two-stage system—it is 0.6. The lowest value has been obtained for an adsorption system and a system with PV-powered vapor compression chiller.

Analysis of Figure 7.31 allows calculating what solar collector area is needed to achieve the required cooling output. Ensuring a sufficiently large solar collector area is a considerable challenge when designing solar cooling and air conditioning systems. Simplified calculations show that, in case of two-stage absorption chillers, 1 kW of output only requires 1.67 m^2 of solar collectors. However, in order to reach high COP concentrating, collectors need to be used to enable reaching high temperatures of the heating fluid (which is not possible at lower latitudes). A larger collector area is needed for single-stage absorption chillers or DEC systems. In their case, 1 kW of output already needs some 3 m^2 of collector area. The advantage of DEC technology is its very simple design and low required temperature of the heating medium, but, as already mentioned, such devices consume considerable amounts of water, which is significant when planning larger systems. Thus, it should be expected that, in practice, rather small DEC solar systems will be used. Single-stage absorption systems do not have those disadvantages, and evacuated collectors are sufficient to drive them (or even flat-plate ones, if solar irradiation is high), although their COP may not be so high. For adsorption systems and vapor compression chillers, the required solar receiver (collector, PV module) area is the largest; 1 kW of cooling power requires over 3 m^2 of receiver surface. Adsorption chillers also have the lowest COP values and may only operate in cycles, which may lead to temperature variations of the air supplied to a room. On the other hand, they are able to operate at low supply temperatures, starting already at 60 °C, which is very attractive for applications at higher latitudes.

Recently, a number of research projects focused on low-output (several kilowatts) absorption and adsorption chillers has been carried out [52–60]. Development of such systems might help in popularizing and refining sorption-based technologies. It would enable installing solar air conditioning systems at single-family houses, thus greatly enhancing the popularity of such technologies.

In solar buildings providing space heating and hot water supply, the surface area of solar collectors need to be big, especially in high-latitude countries. Very often, solar combisystems are overdimensioned and there is a serious problem of working fluid overheating in the solar collector loop and storage. Solar cooling based on sorption cooling technologies is a very good solution in such situation and seems to be a very prospective technology for the future.

REFERENCES

[1]   Weiss W, Wittwer V. Contribution of solar thermal to the EU SET plan. Consolidated position of ESTTP. Presentation at the AGE Hearings, Brussels, V; 2007.
[2]   Weiss W, Bergmann I, Stelzer R. Solar heating worldwide. Market and contribution to the energy supply 2007. IEA, International Heating & Cooling Programme; 2009.
[3]   Dalenback JO. Take off for solar district heating in Europe. Pol Energ Słonecz 2010;1—4/2009, 1/2010:9—13. PTES—ISES.
[4]   Crema L, Bozzoli A, Cicolini G, Zanetti A. A novel retrofittable cooler/heater based on adsorption cycle for domestic application. Pol Energ Słonecz 2010;1—4/2009, 1/2010:43—53. PTES—ISES.
[5]   RETD renewable energy technology deployment. OECD/IEA Renewables for heating and cooling. Untapped potential; 2007.
[6]   Critoph R. Solar Thermal Cooling Technologies. In: Renewable Energy. Innovative Technologies and New Ideas. Warsaw: Warsaw University of Technology and Polish Solar Energy Society; 2008. pp. 55—62.
[7]   http://www.iea-shc.org/task38/index.htm.
[8]   Henning HM, Critoph R. Renewable energy for refrigeration and cooling; 2007. Presentation of the WREN seminar 2008, Brighton.
[9]   Chwieduk D. Przygotowanie mechanizmu wsparcia dla wytwarzania ciepła i chłodu z odnawialnych źródeł energii ze szczególnym uwzględnieniem wykorzystania kolektorów słonecznych. Expert study for the Polish National Energy Conservation Agency commissioned by the Ministry of Economy; 2009.
[10] TERES Report. Perspektywy energetyki odnawialnej w Unii Europejskiej i krajach Europy Wschodniej do roku 2010. ALTENER EC; 1994.
[11] Chwieduk D. Wykorzystanie energii promieniowania słonecznego do celów użytkowych, Instalacje wewnętrzne w budynkach", Część 10. Rozwiązania niekonwencjonalne. Rozdziały: 10/2—10/4, liczba stron 130. Praca zbiorowa, wydanie książkowe wymienno — kartkowe, Wyd. WEKA Sp. z o.o., Warsaw; 2000.
[12] Chwieduk D. Budownictwo niskoenergetyczne. Energie Odnawialne. Rozdział 13. ss.1065—1151. Budownictwo Ogólne. Tom 2. Fizyka Budowli. Red. P. Klemm P., ARKADY; 2005.
[13] Chwieduk D. Rozwój wybranych zastosowań energetyki słonecznej. Pol Energ Słonecz 2006;3—4/2006:4—11. Wyd. PTES—ISES, Warsaw.
[14] Chochowski A, Czekalski D. Słoneczne instalacje grzewcze; 1999. Wyd. COIB, Warsaw.
[15] Pluta Z. Słoneczne instalacje energetyczne; 2003. OWPW, Warsaw.

[16]   Chwieduk D, Critoph RE. Solar energy educational demonstration system for schools "mini solar laboratory". In: Proceedings. World renewable energy congress VIII. Denver (CO, USA): Elsevier Edition, CD; August 29–September 3, 2004.

[17]   EN 12976-2:2006. Thermal solar systems and components – factory made systems. Part 1: general requirements.

[18]   EN 12976-2:2006. Thermal solar systems and components – factory made systems. Part 2: test methods.

[19]   EN 12975-1:2006. Thermal solar systems and components – solar collectors. Part 1: general requirements.

[20]   EN 12975-2:2006. Thermal solar systems and components – solar collectors. Part 2: test methods.

[21]   Gordon J. Solar energy the state of the art. ISES position papers, UK; 2001.

[22]   Duffie JA, Beckman WA. Solar engineering of thermal processes. New York: John Wiley & Sons, Inc.; 1991.

[23]   Twidell J, Weir T. Renewable energy resources, E&FN SPON. London: University Press Cambridge; 1996.

[24]   Anderson EE. Fundamentals of solar energy conversion. Reading (MA): Addison-Vesley Publ. Co.; 1982.

[25]   Chwieduk D. Słoneczne i gruntowe systemy grzewcze. Zagadnienia symulacji funkcjonowania i wydajności cieplnej, Studia z zakresu inżynierii. Nr. 37. Warsaw: The Committee on Civil Engineering and Hydroengineering of the Polish Academy of Sciences; 1994.

[26]   Chwieduk D. Solar Assisted Heat Pumps in Comprehensive Renewable Energy. Solar Thermal Systems: Components and Application, Vol. 3. Elsevier; 2012. pp. 495–528.

[27]   Clean energy Project analysis: RETScreen® Engineering & Cases Textbook. Solar water heating project analysis, Minister of Natural Resources Canada; 2001–2004.

[28]   Esbenssen T, Gramkow L. Integrated solar technologies. Pol Energ Słonecz 2006;1–2:22–6.

[29]   Sayigh AAM, editor. Solar energy engineering. London: Academic Press; 1997.

[30]   Schulz H, Chwieduk D. Wärme aus Sonne und Erde Energiesparende Heizungssysteme mit Erdwärmespeicher, Solarabsorber und Wärmepumpe. Staufen bei Freiburg: Okobuch Verlage; 1995.

[31]   Smolec W. Fototermiczna konwersja energii słonecznej. Warsaw: PWN; 2000.

[32]   Zapałowicz Z. Instalacje słoneczne w Polsce na progu XXI wieku. In: 5th Int. conf. on unconventional, electromechanical and electrical systems; 2001. Supplement, Szczecin, 233–238.

[33]   Zawadzki M, editor. Kolektory Słoneczne, Pompy Ciepła – Na Tak, Polska Ekologia; 2003.

[34]   Tripanagnostopoulos Y. Innovative solar energy systems for aesthetic and cost effective building integration. Pol Energ Słonecz 2006;1–2/2006:9–14.

[35]   Yin Z, Xue Z, Zhang J. The evacuated absorber tube industry in China. In: Proceedings of ISES solar world congress, Taejon; 1997.

[36]   Duff WS. Experimental results from eleven evacuated collector installations. IEA solar heating and cooling programme. TVI-4; 1986.

[37]   Kienzlen V, Gordon JM, Kreider JF. The reverse flat plate solar collector: a stationary, nonevacuated, low technology, medium-temperature solar collector. ASME, J Sol Energy Eng 1988;110:23–30.

[38]   Quaschning V. Understanding renewable energy systems. EARTHSCAN, London; 2006.

[39]   EN ISO 9488:2003. Solar energy – vocabulary.

[40]   Goetzberger A, Dengler J, Rommel M, Goettsche J, Wittwer V. New transparently insulated, bifacially irradiated solar flat-plate collector. Sol Energy 1992;49:403–11.

[41] Eames PC, Norton B. Detailed parametric analysis of heat transfer in CPC solar energy collectors. Sol Energy 1993;50:321−38.

[42] Hoffschmidt B, Alexopoulous S, Goetsche J, Sauerborn M, Kaufhold O. High concentration solar collectors in comprehensive renewable energy. Solar thermal systems: components and application, vol. 3. Elsevier; 2012. p. 165−210.

[43] Kalogirou SA. Low concentration ratio solar collectors in comprehensive renewable energy. Solar thermal systems: components and application, vol. 3. Elsevier; 2012. p. 149−164.

[44] Encyklopedia Fizyki t.1. Warszawa: PWN; 1974.

[45] Halliday D, Resnick R, Walker J. Fundamentals of physics. 10th ed. Wiley; 2013.

[46] Lasnier F, Ang TG. Photovoltaic engineering handbook. Bristol: Hilger; 1990.

[47] Luque A, Heqedus S. Handbook of photovoltaic science and engineering. Hoboken: John Wiley & Sons; 2003.

[48] Zapałowicz Z, Szyszka D. Exploitation of photovoltaic installation in Ostoja in summer. Pol Energ Słonecz 2009;1−4/2009 I 1/2010:32−7. Wyd. PTES, Warszawa.

[49] Tripanagnostopoulos Y. Photovoltaic/thermal solar collectors in comprehensive renewable energy. Solar thermal systems: components and application, vol. 3. Elsevier; 2012. p. 255−300.

[50] Souliotis M, Tripanagnostopoulos Y, Kalogirou S. Thermosiphonic hybrid PV/T solar systems. Pol Energ Słonecz 2010;1−4/2009, 1/2010:28−31. PTES−ISES.

[51] http://oasishis.caiso.com/ − California Independent System Operator (ISO) Open Access Same-Time Information System (OASIS) web site.

[52] Kima DS, Infante Ferreira CA. Solar refrigeration options − a state-of-the-art review. Int J Refrig 2008;31:3−15. November 2006.

[53] Zhai XQ, Wang RZ, Wu JY, Dai YJ, Ma Q. Design and performance of a solar-powered air-conditioning system in a green building. Appl Energy 2008;85:297−311. January 2007.

[54] Kabel AE. Solar powered air conditioning system using rotary honeycomb desiccant wheel. Renewable Energy 2007;32:1842−57. April 2006.

[55] Fan Y, Luo L, Souyri B. Review of solar sorption refrigeration technologies: development and applications. Renewable Sustainable Energy Rev 2007;11:1758−75. January 2006.

[56] Gonzalez Manuel I, Rodrıguez Luis R. Solar powered adsorption refrigerator with CPC collection system: collector design and experimental test. Energy Convers Manage 2007;48:2587−94. July 2006.

[57] Helm M, Keil C, Hiebler S, Mehling H, Schweigler C. Solar heating and cooling system with absorption chiller and low temperature latent heat storage: energetic performance and operational experience. Int J Refrig 2009;32:596−606. November 2008.

[58] Henning HM. Solar cooling and air conditioning − thermodynamic analysis and overview abort technical solution. EUROSUN, Glasgow, UK; 2006.

[59] Le Pierrès N, Mazet N, Stitou D. Modelling and performances of a deep-freezing process using low-grade solar heat. Energy 2007;32:154−64. March 2005.

[60] Vargas JVC, Ordonez JC, Dilay E, Parise JAR. Modeling, simulation and optimization of a solar collector driven water heating and absorption cooling plant. Sol Energy 2009;83:1232−44. July 2008.

Chapter 8

Buildings "Aware" of Solar Energy Impact: Summary

Effective functioning of a building throughout its lifetime with regard to the energy it consumes depends on its location, situation in reference to cardinal directions, architecture, civil engineering solutions, installations, interior arrangement, and usage patterns. A building in a city center has different environmental conditions than one located in the suburbs, the countryside, or far away from any settlements. Often the situation in reference to cardinal directions is forced by the specific local environment. It is most difficult to adjust this situation in urban areas, while it is easiest in the countryside. Usually city inhabitants have limited influence on the architectural and structural solutions of a building they live in or on the design of its installations. Conversely, outside towns and cities we may make decisions concerning those elements quite freely, provided that we have the required knowledge and awareness concerning energy efficiency and the possibilities of using unconventional solutions and technologies. In towns and cities, access to heat or electricity sources is typically easy, but also specifically determined or even dictated by the presence of centralized energy sources, district heating systems, and power grids to which buildings are duly connected. Outside urban areas, usually no heating networks are present, and access to other networks such as a gas supply may also be limited. This opens more opportunities for using unconventional solutions, including solar energy, but these also require appropriate knowledge. However, not everyone is an architect, designer, civil engineer, or expert on conventional and renewable energy systems. Moreover no single person can be an expert in all of these fields at the same time. Yet it is knowledge and experience in all of these areas that ultimately enable energy-efficient features of a building. Therefore, in order to achieve a considerable reduction of energy consumption in buildings, it is essential to ensure collaboration of experts from all of these areas.

Inhabitants of an already existing or newly built building should have access to its essential energy characteristics, including variability of energy demand and consumption over time. Of course, an energy characteristics certificate provides key energy consumption values in reference to both final and primary energy, and rates the building according to its energy quality. This enables determining whether using the building will be energy-intensive and

Solar Energy in Buildings. http://dx.doi.org/10.1016/B978-0-12-410514-0.00008-6

estimating energy consumption costs. However, it would be good if an inhabitant also had access to individual elements of the energy balance of the building as a whole, and also of all its rooms, not only as annual values but also for individual months, and even for daily energy distribution hour by hour for averaged months of a year. All of the elements of an energy balance, i.e., heat gains or losses through the envelope—opaque walls, transparent elements (windows), doors, roofs, and floors—as well as through ventilation, are important here; and so is the influence of internal heat gains from various devices and human beings, and solar radiation gains taking into account shading. This kind of data would support making decisions about purchasing apartments within larger buildings, attached houses, or single-family houses within residential developments or about planning a stand-alone single-family house construction. In particular, knowledge about availability and impact of solar radiation becomes significant for modern energy-efficient buildings with high thermal capacity and very good thermal insulating quality of envelope elements, mainly walls, windows, and large glazed facades. Solar radiation reaches the interior primarily through transparent envelope elements and is converted in a photothermal process; generated heat may no longer escape the building, as this process is restricted by the envelope with high thermal resistance. Information about potential overheating of individual rooms should be accessible for future inhabitants or users of rooms within an office building, school, hospital, or any other facility.

In addition to the energy-related aspects, individual thermal comfort is also important. Thermal comfort is an individual matter. A future user should know how the interior temperature might change in time during summer, if no air conditioning systems are planned. Data concerning impact of solar radiation on energy balance of individual rooms also enables assigning them to proper functions. Also aesthetic features of architectural and structural solutions aimed at restricting solar radiation access (external shades, overhangs) are important; this also applies to external surroundings (large glazed areas, if unshaded, may lead to overheating and the necessity of using air-conditioning systems). Data concerning energy balance components, including the influence of solar gains and their variability over time, provide a basis for concepts and actual designs of modern buildings as not only low-energy structures but also as buildings "aware" of solar energy influence.

Utilization of solar energy in a building requires not only more conscious shaping of its structure, but also careful design of its internal systems. Obviously a solar building must be compliant with all energy-efficiency standards, but it also must have a good access to solar energy. In building solar systems, this access must be possibly good throughout the year. Solar radiation receivers within solar systems (both thermal and photovoltaic) should be integrated with the building envelope and form its important elements. Nowadays such integration mainly applies to roof slopes. Components of modern passive solutions are incorporated into southern façades (which may be either flat or rounded, stretching from the southeast to the southwest, as described in detail in Chapter 3), and of course they should be shaded during certain periods. Thus design of a building shape is now increasingly a matter of integrated architectural, civil, and installation engineering, and the building itself is no longer a construction structure only. Instead it becomes an energy system producing useful energy on the spot, according to the needs of its user. An example of such a building is shown in Figure 8.1.

Solar energy utilization in a building results in a coherent and multidisciplinary approach to the entire structure and its surroundings. Both the buildings and us exist within a particular natural environment, which we should utilize in a reasonable and well-thought-out manner. Solar energy utilization in buildings is friendly to both people and the environment, and it not only serves the human by providing useful energy but also by teaching a smart and friendly approach to the environment. Teams designing solar buildings need expertise in construction physics and civil engineering, and also knowledge of properties of construction materials available on the market and

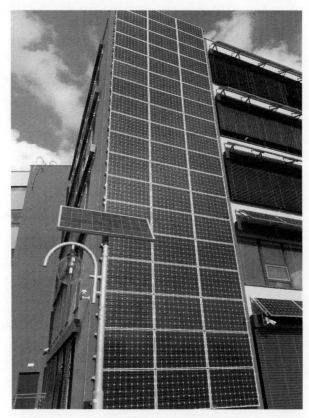

FIGURE 8.1   Passive and active solar building, Euro-Centrum, Katowice, Poland.

operating principles of modern energy systems and devices. It is essential to take into account incorporation of many process solutions and mutually coupled systems already at the architectural concept stage. Moreover, it is necessary to properly integrate the building with its surroundings and take into account existing environmental constraints. Application of solar energy solutions in a building leads to a wider look at the issues of utilizing energy resources and the environment. As a result, it enforces considering buildings in a sustainable way. This means issues to be considered are not limited to energy consumption, but also include the terrain, ground, and water-use related features and building material properties. It is essential to consider all of these elements as variables in time. The basis for such considerations is the analysis of the energy balance of a building and components of that balance as parameters variable in time due to the dynamic interaction between the building and its surroundings, including the impact of solar radiation.

Solar radiation always reaches the earth surface during the day and should be effectively converted into other forms of energy to be used for practical purposes. A feature that distinguishes solar energy technologies from other energy solutions, especially conventional ones, is the fact that utilizing solar energy at a particular place by a particular building does not affect the availability of solar radiation elsewhere—neither in the direct surroundings and neighboring buildings nor at any other locations, no matter how far away.

Index

Note: Page numbers followed by f indicate figures; t, tables; b, boxes.

A

Absorbent, 339–340, 342–343
Absorber, 313. *See also* Desorber
  absorptance, 122
  dimensions, 313
  flat-plate solar collector, 309
  metal, 316
  serpentine, 313
  transparent covers and, 311
  vertical, 123
Absorbing surface
  emissivity, 103
  influence of, 125–128
  system of transparent cover and,
    121–125
Absorption, 98–101, 339–340
  chillers, 340
  effect on transmission, 121
  flat-plate solar collectors, 310
  infrared solar radiation, 7–8
  reflection in transparent plate, 108f
  transmission-absorption factor,
    323–324
  wave, 101
Activation energy, 329–330
Active solar system, 97. *See also* Passive solar
    system
  configurations, 300–308
    domestic hot water system, 300f
    heat pump system, 305f, 308f
    print screen, 307f
    solar collectors, 304f
  PV application in buildings, 329–338
  solar collectors, 308–328
  solar cooling, 338–349
  solar heating system, 289–290
    development, 290–295
    thermosiphon system, 292f
    types, 295–299
      solar combi heating system, 299f, 302f
Aerosol scattering, 11
Air conditioning technologies, 339–345
  DEC system, 344f
  PV-driven vapor compression chiller, 339f

solar-driven adsorption chiller, 341f
Air mass, 8. *See also* Biomass
Albedo, 7
American society of heating, refrigerating,
    and air conditioning engineers
    (ASHRAE), 227
Amorphous cell, 331
Anisotropic radiation model, 43–46, 45f.
    See also Isotropic radiation model
Anisotropy index (A_i), 43
Aphelion, 5
ASHRAE. *See* American society of heating,
    refrigerating, and air conditioning
    engineers
Assortment file, 297
Atmospheric aerosol, 6–8
Atmospheric window, 14
Auxiliary heating system, 292–293
Azimuth-altitude dual axis tracker, 35

B

Beam radiation, 1, 21
  correction factor for, 40
  equivalent angle of incidence,
    119–120
  incident angle, 25–26
  loss of flux, 2–3
  profile angle, 49
  radiation corrections, 38
Bifacial collector, 323
Biomass, 18
BIPV. *See* Building integrated photovoltaics
BIST. *See* Building integrated solar thermal
Bivalent heat pump, 298
Buffer space, 137
  advantage of system with, 153
  indirect solar gain system with, 145–153
  in traditional building, 152f
Buffer system, 149, 151
Building energy consumption reduction,
    133–138
Building integrated photovoltaics (BIPV),
    289, 336
Building integrated solar thermal (BIST), 289

Printed and bound by CPI Group (UK) Ltd, Croydon, CR0 4YY

13/05/2025

01870167-0001